Control of Power E ers and Systems

Control of Power Electronic Converters and Systems

Edited by

Frede Blaabjerg

ELSEVIER

ACADEMIC PRESS

An imprint of Elsevier

Academic Press is an imprint of Elsevier
125 London Wall, London EC2Y 5AS, United Kingdom
525 B Street, Suite 1800, San Diego, CA 92101-4495, United States
50 Hampshire Street, 5th Floor, Cambridge, MA 02139, United States
The Boulevard, Langford Lane, Kidlington, Oxford OX5 1GB, United Kingdom

Library of Congress Cataloging-in-Publication Data
A catalog record for this book is available from the Library of Congress

British Library Cataloguing-in-Publication Data
A catalogue record for this book is available from the British Library

ISBN: 978-0-12-805245-7

Publisher: Mara Conner
Acquisition Editor: Sonnini R. Yura
Editorial Project Manager: Mariana L. Kuhl
Production Project Manager: Vijayaraj Purushothaman
Cover Designer: Harris, Greg (ELS)

Typeset by SPi Global, India

Contents

Contributors

Numbers in paraentheses indicate the pages on which the authors' contributions begin.

Ahmed Abdelhakim (153), University of Padova, Padua, Italy

Pablo Acuna (31), RMIT University, Melbourne, VIC, Australia

Ricardo P. Aguilera (31), University of Technology Sydney, Ultimo, NSW, Australia

Yeongsu Bak (175, 301), Ajou University, Suwon, South Korea

Frede Blaabjerg (117, 207, 269), Aalborg University, Aalborg, Denmark

Tao Ding (243), Xi'an Jiaotong University, Xi'an, China

Mojtaba Forouzesh (3), Aalborg University, Aalborg, Denmark

Ngoc Ha Pham (3), University of Technology Sydney, Ultimo, NSW, Australia

Katherine A. Kim (243), Ulsan National Institute of Science and Technology (UNIST), Ulsan, Korean

Georgios Konstantinou (31), University of New South Wales, Sydney, NSW, Australia

Kyo-Beum Lee (175, 301), Ajou University, Suwon, South Korea

Jose I. Leon (31), University of Seville, Seville, Spain

Rafael Peña-Alzola (207), University of Strathclyde, Glasgow, United Kingdom

Zian Qin (69), Delft University of Technology, Delft, The Netherlands

Ariya Sangwongwanich (153), Aalborg University, Aalborg, Denmark

Yanfeng Shen (69), Aalborg University, Aalborg, Denmark

Hye-Ung Shin (301), Ajou University, Suwon, South Korea

Yam P. Siwakoti (3), University of Technology Sydney, Ultimo, NSW, Australia

Yipeng Song (117, 269), Aalborg University, Aalborg, Denmark

Sergio Vazquez (31), University of Seville, Seville, Spain

Huai Wang (69), Aalborg University, Aalborg, Denmark

Gaolin Wang (331), School of Electrical Engineering and Automation, Harbin Institute of Technology, Harbin, China

Bo Wang (331), School of Electrical Engineering and Automation, Harbin Institute of Technology, Harbin, China

Dianguo Xu (331), School of Electrical Engineering and Automation, Harbin Institute of Technology, Harbin, China

Yongheng Yang (93, 153, 243), Aalborg University, Aalborg, Denmark

Ming Yang (331), School of Electrical Engineering and Automation, Harbin Institute of Technology, Harbin, China

Yong Yu (331), School of Electrical Engineering and Automation, Harbin Institute of Technology, Harbin, China

Guoqiang Zhang (331), School of Electrical Engineering and Automation, Harbin Institute of Technology, Harbin, China

Keliang Zhou (93, 153), University of Glasgow, Glasgow, United Kingdom

Dao Zhou (117, 269), Aalborg University, Aalborg, Denmark

Part I

Fundamentals of Power Converters

Chapter 1

Power Electronics Converters—An Overview

Yam P. Siwakoti*, Mojtaba Forouzesh† and Ngoc Ha Pham*
**University of Technology Sydney, Ultimo, NSW, Australia, †Aalborg University, Aalborg, Denmark*

1.1 INTRODUCTION

Nowadays, power electronics converters play an important role in the power processing and control systems of various applications from small scale ones like in energy harvesting and implantable devices to large scale ones like in power generation and transmission. Power electronics converters use electronic components based on semiconductor switches operated at different frequency levels from 50 to 60 Hz mains frequencies to 100 MHz radio frequency. In order to optimally support different features of diverse applications, power electronics converters should benefit from different characteristics such as nonisolated/isolated, voltage-fed/current-fed and hard-switched/soft-switched. There are various types of power conversion (AC-DC, DC-AC, DC-DC and AC-AC) applicable to different applications. For instance, most home appliances (e.g., television and personal computer) connected to the electric grid need an AC-DC converter in their power processing system. The vast fields of renewable energy grid-integration and motor drives need DC-AC converters to provide constant/variable AC voltage at the output. Many battery-powered devices utilize DC-DC converters to provide the required DC voltage for different loads. In addition, DC-DC converters are suitable for maximization of energy harvesting of renewable energy sources like photovoltaics and wind turbines. In order to change the voltage or frequency of an alternating current source, AC-AC converters are needed (e.g., light dimmer and mains frequency changer) [1–5].

1.1.1 Nonisolated/Isolated

Power electronics converters are generally consist of only semiconductor switches and energy storage elements. Nonisolated converters are often

Control of Power Electronic Converters and Systems. https://doi.org/10.1016/B978-0-12-805245-7.00001-9

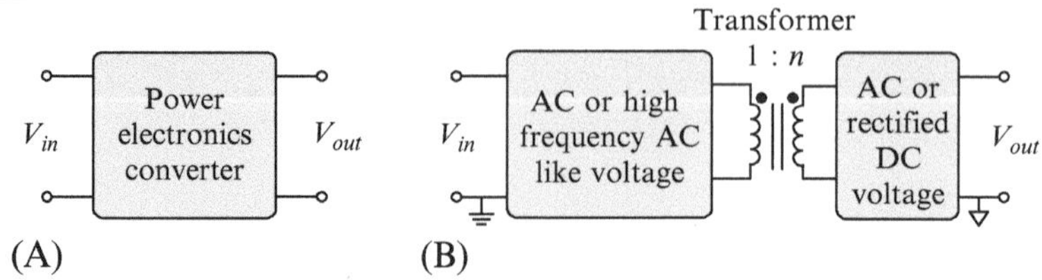

FIG. 1.1 General layouts of (A) nonisolated and (B) isolated power electronics converters.

preferred in applications that electrical isolation is not a necessity, because they are less bulky and costly, and more efficient and reliable. Fig. 1.1A shows a general layout of nonisolated power electronics converters in which the circuit do not consist of magnetic or electric isolation. On the other hand, isolated power converters often use either transformer or coupled inductor for multiple purposes such as voltage level shifting, obtaining multiple outputs, providing galvanic isolation and ground loop avoidance. Fig. 1.1B illustrates a general layout of the isolated power electronics converters in which a transformer is employed to provide electrical isolation. The AC transformer in those converters need an AC voltage or a square/quasisquare wave voltage at the primary side for proper operation and saturation avoidance. A transformer can be used in buck type converters to allow voltage step-up feature. In addition, a simple way to achieve multiple outputs in a single converter is to use multiple windings transformers or coupled inductors. Moreover, employing transformers in the circuit allows electrical isolation that provide galvanic isolation between the input and output modules of a converter. When two or more electrical devices share a common ground in a power system, the current from the return path of one device can disrupt the operation of other devices, which can be avoided by utilizing transformers [3–5].

1.1.2 Voltage-Fed/Current-Fed

Depending on their input circuitry, power electronics converters can be classified as either voltage- or current-fed converters (see Fig. 1.2). Most power electronics converters are supplied form voltage sources, however, some converters operate with current sources. In voltage-fed converters, a capacitor is connected in parallel with the source so the input voltage cannot change instantly and in current-fed converters, an inductor is connected in series with the source so the input current cannot change instantly. Due to the fact that two voltage sources with different magnitude cannot be connected in parallel, the load of voltage-fed converters is like a current source and usually a filter inductor (load inductance) is connected in series with the load. Analogously, because two current

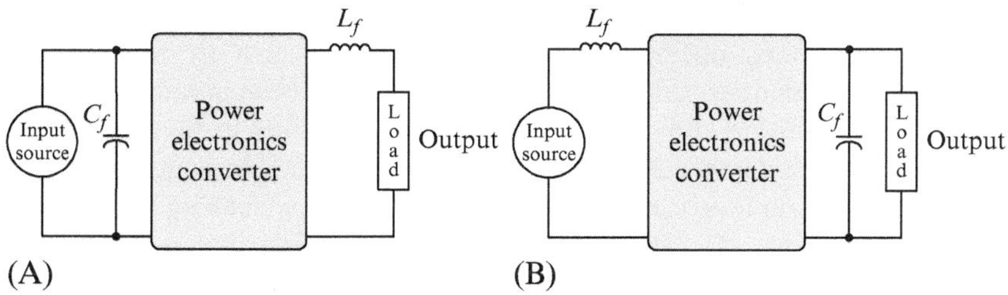

FIG. 1.2 General layout of (A) voltage-fed and (B) current-fed power electronics converters.

sources with different magnitude cannot be connected in series, the load of current-fed converters is like a voltage source and usually a filter capacitor is connected in parallel with the load.

In power converters with inverter bridge (one-phase or three-phase) the switching devices at each leg of a voltage source type are not allowed to turn ON at the same time (a condition known as shoot-through), the switching patterns must include a dead-time between the high and low side switches of each leg. On the other hand, as the switching devices of all legs of current source type should not turn OFF simultaneously (known as open-circuit), the high and low side switches must always include an overlap [4]. Unlike voltage-fed and current-fed converters, impedance source based converters are immune to both shoot-through and open-circuit [6]. Current-fed converters are very popular for low voltage renewable energy applications such as photovoltaics (PV) and fuel cells (FC) because their input inductors can provide a continuous input current, typically with low ripple. This feature reduces the negative impacts of high-ripple current on low voltage high-current sources. By contrast, the lack of an input inductor in voltage-fed converters results in considerable ripple current at the input; however, as these converters have no right half plane (RHP) zero in their control to output transfer function, they have faster dynamic response than current-fed converters with input inductors and RHP zero [4].

1.1.3 Hard-Switched/Soft-Switched

Power electronics converters generally deal with semiconductor switching devices and energy storage components. In basic power converters switched at high frequencies, the transition between switching states can be associated with voltage ringing and spike on the semiconductors. These type of power electronics converters are hard-switched and they can produce electro-magnetic interference (EMI) emission as a result of high dv/dt and di/dt at switch turn ON and turn OFF, which can spread throughout the circuit from the input to the output [7]. In addition, a large part of power loss is due to switching power loss in hard-switched converter that can significantly reduce the efficiency of the power converter. In order to overcome the aforementioned problems in hard-switched converters, different switching techniques such as zero voltage switching (ZVS) and zero current switching (ZCS) have been introduced. These switching methods are known as soft-switching techniques as they reduce the switching stress and hence do not produce significant EMI in switching transitions, also the power loss occurred due to high-frequency switching is negligible with these methods [4]. The choice of selecting soft-switching techniques depend on the specific characteristics of the power electronics circuit such as switching frequency, required package and size, type of switching device used, and system control complexity [1]. Although, soft-switched converters are advantageous to hard-switched converters, but due to the significant improvement in the development and commercialization of the new wide band gap (WPG) semiconductor switches with fast switching capability and low conduction loss (i.e., SiC and GaN), hard-switched converters are sometimes preferred in order to reduce system complexity and cost.

1.1.4 Conversion: Types of Power Converters

The main aim of power electronics converters is to perform a power conditioning to meet certain requirements of different applications. In order to achieve this goal, they should be able to reliably change one form of electrical energy to another form by supplying voltage and current of different magnitude and frequency. Fig. 1.3 illustrates four different power conversion between alternating current and direct current form of electrical energy. AC-DC converters or rectifiers convert the alternating current to direct current. DC-AC converters or inverters operate in reverse of rectifiers and convert direct current to an alternating current of desired magnitude and frequency. DC-DC converters or choppers convert the constant/variable direct current to variable/constant direct current at different voltage level. AC-AC converters convert alternating current of the mains supply to alternating current of the desired frequency and

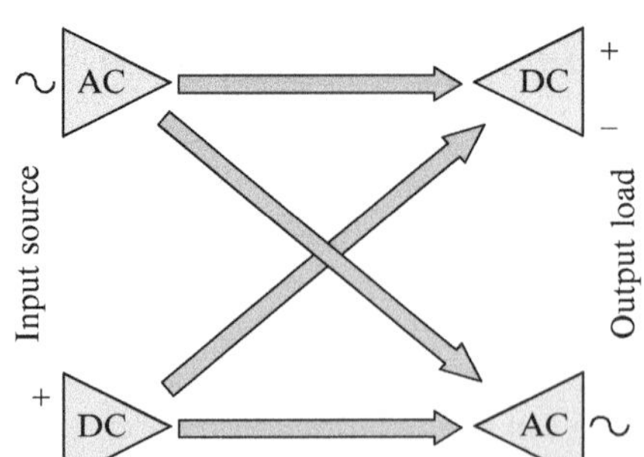

FIG. 1.3 Different types of power conversion in power electronics converters.

magnitude [1,2]. In the following of this chapter, an overview of the mentioned power converters is provided.

1.2 AC-TO-DC CONVERTERS

An AC to DC converter converts AC voltage to a stable DC output voltage. These converters are very popular in our daily life because the grid is mainly AC while all of our electronic devices run on DC voltage. Applications of AC-DC converters range from low power such as AC adapter (charger) for personal computer and smart phone, medium power such as air conditioner and washing machine with inverter technology, to high power such as rectifier for variable speed drive.

Unidirectional AC to DC converters are also called rectifiers. The simplest construction of a rectifier is made from diodes. However, it is well-known that a diode rectifier creates nonnegligible distortion in its input current and thus requires a significant effort in filter design so that the distortion does not affect the AC grid [8].

Single-phase diode rectifier is very popular in our daily life. Fig. 1.4 shows the construction and operating waveforms of the simplest diode rectifier: half-wave rectifier. It converts single-phase AC voltage to DC voltage via a single diode. The output voltage takes only the positive cycle of the input voltage and is then smoothen using an output capacitor. However, the input current does not take negative value and thus contains a DC component. As a result, the total harmonic distortion in the current is THD = 158%, much higher than IEEE standard of 5%. The output voltage ripple is also large and contains 50-Hz ripple. On the other hand, a clear benefit of this circuit is simplicity. Its usage is therefore limited to low power applications, which do not have much effect on the grid or require less filtering effort.

An improved version of the half-wave rectifier is a full-wave rectifier shown in Fig. 1.5. Full-wave rectifier employs four diodes to converts both positive and negative cycles to the DC side before stabilizing the output voltage with a capacitor. The resulting voltage ripple is therefore doubled to 100 Hz with half of the ripple of the half-wave rectifier. The input current contains no DC

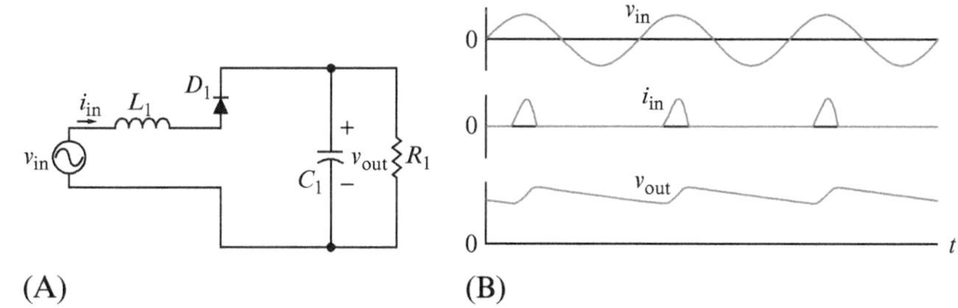

FIG. 1.4 Half-wave rectifier: (A) circuit diagram and (B) its waveforms.

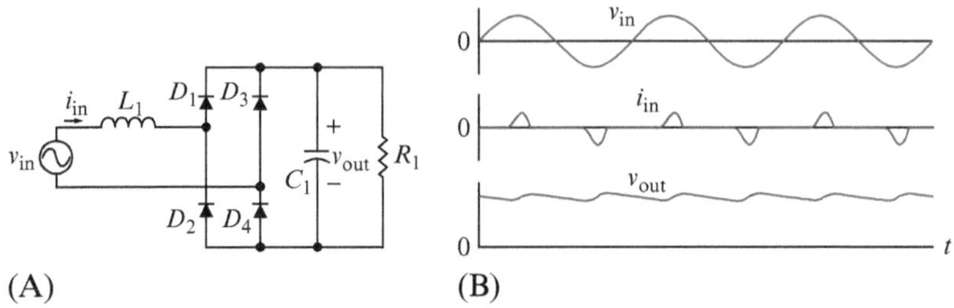

FIG. 1.5 Full-wave rectifier: (A) circuit diagram and (B) its waveforms.

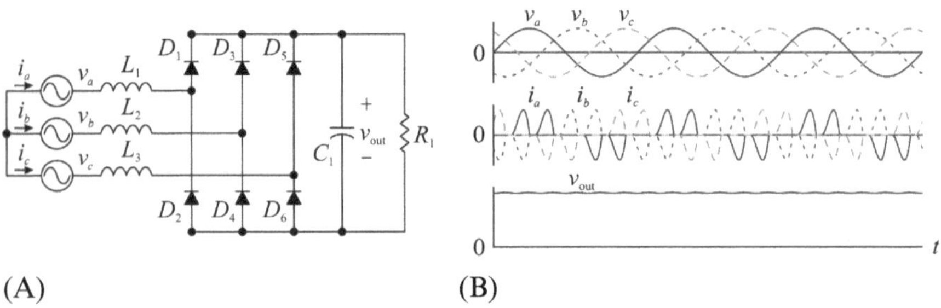

FIG. 1.6 Three-phase rectifier: (A) circuit diagram and (B) its waveforms.

component but the distortion is still high as THD = 113%. Since diode is cost effective, full-wave rectifier is therefore more popular than half-wave rectifier in low power applications.

The ripples in the single-phase diode rectifiers can be further reduced by using three-phase circuitry as shown in Fig. 1.6. It is also known as six-pulse rectifier because there is six pulses of output voltage during one cycle. The output voltage contains very low ripple while the input current has a lower distortion of THD = 92%. It should be noted that the output voltage here has the peak of the line-to-line voltage instead of line-to-neutral like the single-phase circuit.

It is possible to reduce the current distortion even further using a 12-pulse rectifier circuit as shown in Fig. 1.7. It employs multiwinding transformer with Y-Y and Y-Δ connections and appropriate turn ratios to achieve two three-phase voltage sources with the same amplitude at 30-degree phase shift. This arrangement will create a total of 12 pulses per cycle at the output. It can be seen that the output voltage is quite stable and the input current becomes more sinusoid with a relatively small THD as low as 11%. The drawback of this approach is that it requires a multiwinding transformer which is bulky and heavy.

Since all diode rectifiers cannot achieve the IEEE standard of 5% THD, additional filters should be used at the input. However, those filters are heavy and bulky, and could reduce the power factor of the circuit, which will introduce negative effects to the efficiency of transmission line. The advancement of

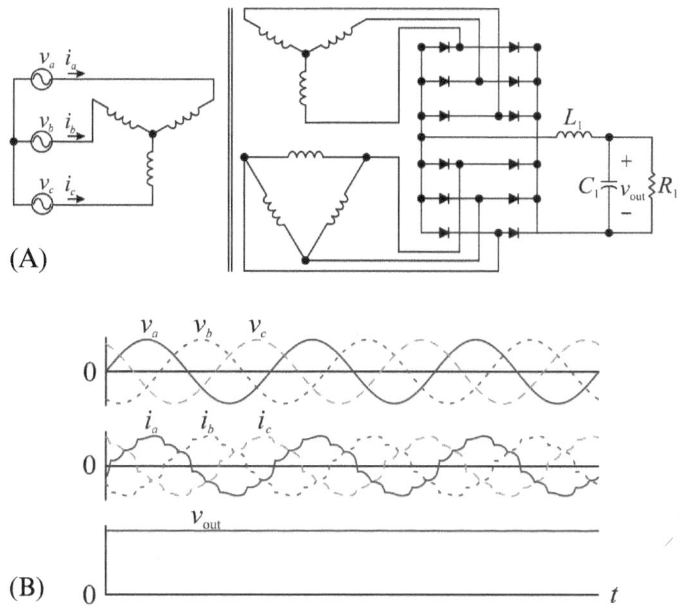

FIG. 1.7 12-pulse rectifier: (A) circuit diagram and (B) its waveforms.

power electronics technology enabled possible replacement of the bulky solutions with compact ones called power factor correction (PFC) circuitries which can ensure a sinusoidal input current with a THD below 5% as well as unity power factor.

1.2.1 Single-Phase PFC

Fig. 1.8 shows the most common approaches for single-phase PFC. They all have boosting capability and can control the output voltage to be higher than the peak of input voltage. The first one is called bridgeless topology as shown in Fig. 1.8A. It has a very high efficiency with two active switches and an AC inductor at the input. Some derivations are totem pole in Fig. 1.8B and AC switch in Fig. 1.8C which rely on similar operating principle: treating each positive and negative cycle as an input to a separated boost converter. Another popular solution in Fig. 1.8D is single-switch boost type PFC that employs a simple full-wave rectifier in series with a boost converter with a DC inductor. This construction has higher conduction loss but only needs one active switch which is simple to control.

The waveforms of single-phase PFC is shown in Fig. 1.9 where the input current is in phase with the input voltage and its distortion is within the limit THD = 5%. Notice that the output voltage contains a ripple of 100 Hz which caused by the single phase AC power fluctuation at the input.

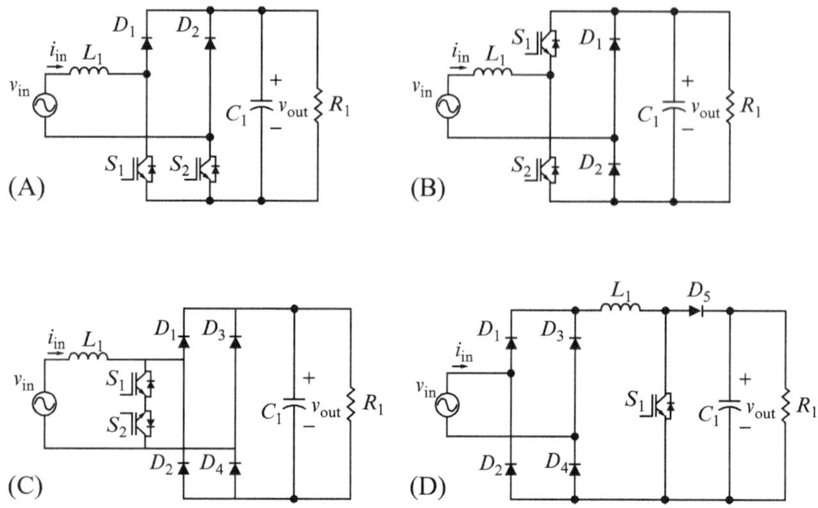

FIG. 1.8 Single-phase PFC: (A) bridgeless, (B) totem pole, (C) AC switch, and (D) single-switch boost types.

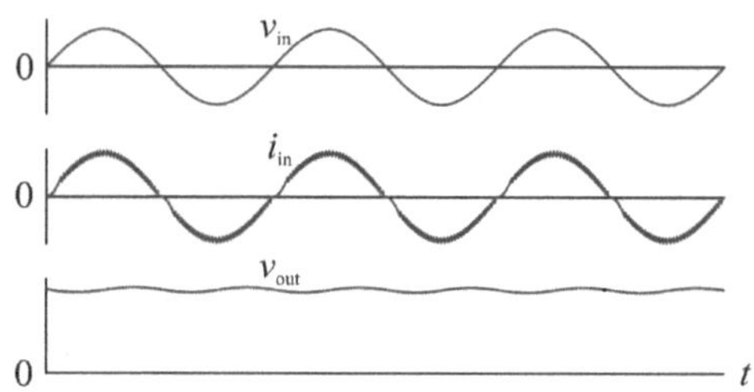

FIG. 1.9 Operating waveforms of a single phase PFC.

1.2.2 Three-Phase PFC

Three-phase PFC is important for the industry because of its capability to deal with high power. This chapter only introduces the basic constructions of three-phase PFC. Details of operation will be discussed in Chapter 13.

The simplest approach to achieve three-phase PFC is to combine three modules of single phase PFC as shown in Fig. 1.10. Each module converts single phase voltage to an intermediate voltage before converting all to a common output using isolated DC to DC converters at a second stage. There are two types of connection from the three modules to the three-phase grid: Y connection as in Fig. 1.10A and Δ connection as in Fig. 1.10B where the input voltage to each module is 1.7 times higher than the case of Y connection. Each module can apply phase shift modulation to get multilevel output and reduce the size of passive filters.

Fig. 1.11A and B show two standard approaches to three-phase PFC using bidirectional DC to AC inverter: voltage source type with AC inductor and

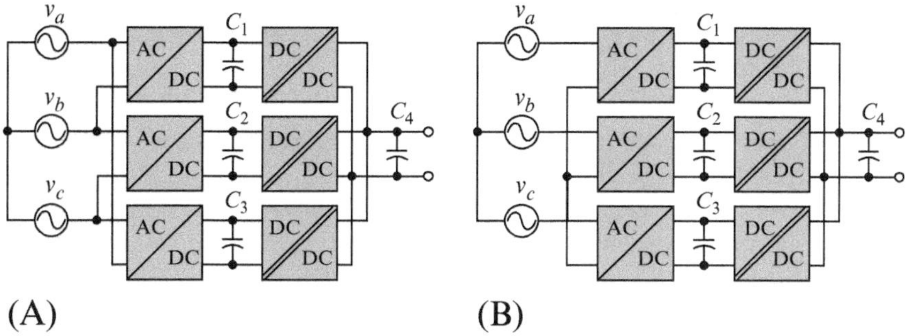

FIG. 1.10 Modular three-phase PFC: (A) Y connection and (B) Δ connection.

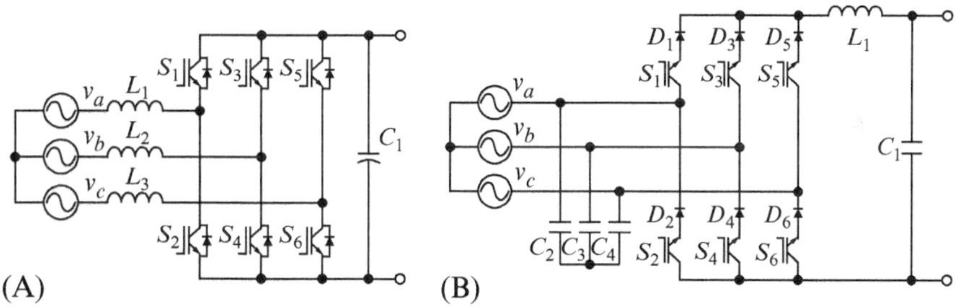

FIG. 1.11 Standard three-phase PFC (bidirectional): (A) voltage source (boost) and (B) current source (buck) types.

current source type with DC inductor. The voltage source type can boost the input voltage up to a higher value compared to the peak input. On the other hand, the current source type converter can lower the output voltage (buck type) but requires a reversible output voltage to enable reversed power flow.

Another approach to achieve PFC is to compensate distortion with harmonic compensation techniques (third harmonic injection). The harmonic injection could be from voltage sources as shown in Fig. 1.12A where voltage ripple is compensated by an active filter such as H-bridge structure in series with

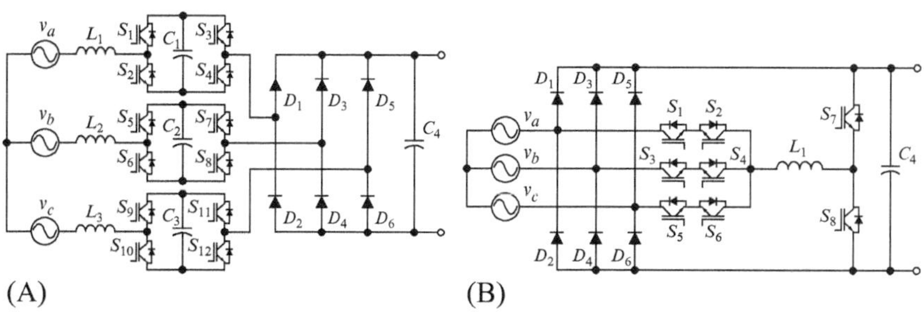

FIG. 1.12 Harmonic compensation PFC: (A) voltage injection type and (B) current injection type.

an inductor [9]. A more efficient way is current injection method as shown in Fig. 1.12B where a suitable amount of current is injected into the main lines via bidirectional switches made by two switches in series. The benefit of this approach is that all bidirectional switches have low voltage rating as well as very low conduction loss and zero voltage switching performances. Notice that the output voltages in both types are not controlled.

The current injection approach can be modified to have a controllable output with two additional boost structures in combination with bidirectional switches as shown in Fig. 1.13A or passive network as shown Fig. 1.13B. The later approach is called "Minnesota rectifier" and only needs two active switches [10]. However, it requires big and bulky line-frequency transformer.

Fig. 1.14A and B show two popular boost type 3-phase rectifiers with bidirectional Δ or Y switches. The Δ switch rectifier charges up the line currents via only two active switches and thus has relatively low conduction losses. However, it only permits two-level modulation. The Y switch topology has another name of "Vienna rectifier" [11]. The Y connection has its neutral point connected to a middle point of two capacitors in series and can perform as a three-level boost converter which can reduce the size of input AC inductors.

Fig. 1.15 shows a hybrid current injection Buck-type PFC rectifier or Swiss rectifier [12]. Compared to those in Fig. 1.14, it employs two buck converters to control the output voltage to be lower than the peak input voltage, and thus does not require the AC filter inductors.

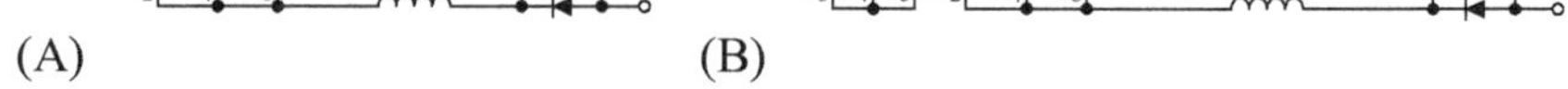

FIG. 1.13 Current injection PFC with boosting capability: (A) active type and (B) passive network.

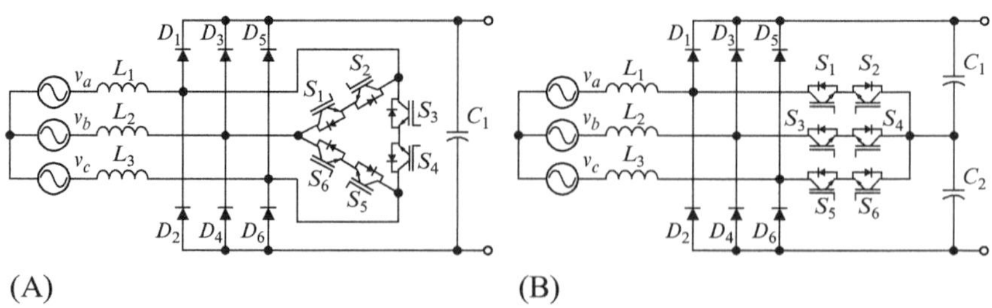

FIG. 1.14 (A) Δ switch rectifier and (B) Y switch type (Vienna rectifier).

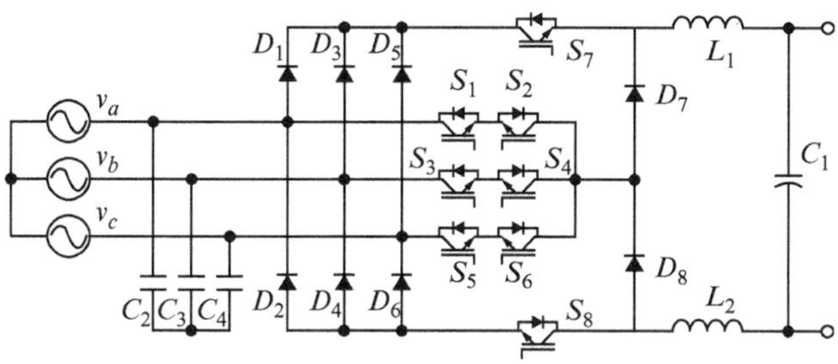

FIG. 1.15 Hybrid current injection Buck-type PFC (Swiss rectifier).

1.3 DC-TO-AC CONVERTERS

DC-AC converters or an inverter is a device produce an AC output of a definite phase, frequency and magnitude from a DC source. Based on the type of the source and load as shown in Fig. 1.16, inverters are segregated into two distinct categories: voltage source inverters (VSIs) and Current Source Inverters (CSIs). The name VSIs comes from the fact that the output voltage of the converter is independently controlled. Similarly, the output AC current waveform is controlled in CSIs. VSIs can perform only buck operations, while CSIs can perform only voltage boost inversions. Although single-phase CSIs are possible, three-phase CSIs are more practical and prevalent.

Based on the type of output waveform as shown in Fig. 1.17, inverter are classified into three categories: (1) square wave inverter, (2) modified square

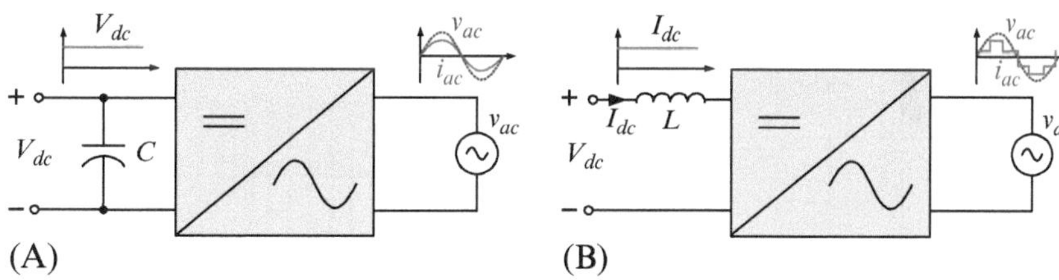

FIG. 1.16 DC-AC converter topologies: (A) VSI topology and (B) CSI topology.

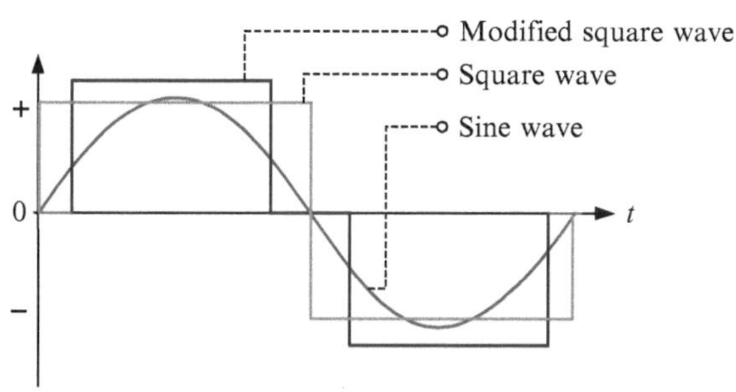

FIG. 1.17 Different types of AC signals produced by inverters.

wave or quasisine wave inverter, and (3) pure sine wave inverter. Inverter, which produces square-wave, are largely obsolete, as the waveform shape is not well suited for running most modern appliances. With the introduction of digital controller ICs, superior modified sine wave and true sine wave inverters are possible with various modulation techniques. In addition, advancement of semiconductor technology, circuit integration techniques, and availability of advanced digital controller helps to realize complex inverter topologies with higher efficiency, reliability, better power quality and advance application specific control capabilities.

1.3.1 Single Phase DC-to-AC Converter

Single-phase DC-AC inverters with power <5 kW are widely used in many applications such as AC motor drives, uninterruptible power supplies, integration of small-scale renewable sources and residential loads. The most common topology are half bridge or full bridge as shown in Fig. 1.18. They are also used to form a cell for high-power static power topologies, such as for multilevel or multicell configuration that are reviewed in Section 1.3.3.

Half-bridge inverter generates bipolar voltages ($-V_{dc}/2$ or $V_{dc}/2$), while full-bridge inverter generates monopolar voltage (0 to V_{dc} or 0 to $-V_{dc}$. In addition, unlike half-bridge topology where the output harmonics occur around the switching frequency, the harmonics generated by the full-bridge inverter are

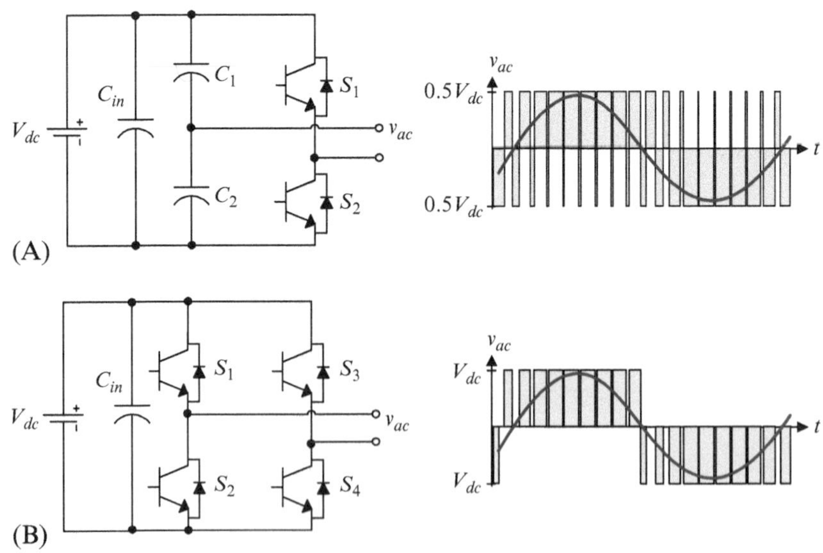

FIG. 1.18 Voltage source inverter showing its AC output voltage for the SPWM: (A) half-bridge and (B) full-bridge.

lower and they appear at twice the switching frequency. Therefore, the reduction in switching device in half-bridge is penalized by large output filter inductor and capacitor across the bridge.

1.3.2 Three-Phase DC-to-AC Converter

Three-phase DC-AC converter or 3-ph inverter are generally used for medium to higher power application (>5kW), e.g., variable speed drive, high-voltage DC (DC transmission), grid-integration of various renewable energy sources and various industrial processes. Fig. 1.19 shows two basic three-level three-phase inverters with three half-bridge inverter legs, which are connected to a three-phase load. The circuit configuration is similar to the single-phase inverter with additional phase leg. In general, the problems related to high-power applications can be overcome by the increasing the number of phases. This not only increase the power output, but also increases the degree of freedom to control the inverter. When modulated appropriately, the inverter generates a line-to-line output waveform that has three-levels with 120 degrees out of phase with each other. There are various topologies and control and modulation techniques to reduce the switching losses, electromagnetic interference (EMI), harmonics and stress on the switching devices.

1.3.3 Multilevel DC-to-AC Converter

Multilevel inverters exhibit some interesting advantages compared to two-level VSIs, especially for high-voltage power conversion, where lower switch voltage stress and lower harmonic content are needed. For grid connected applications, e.g., photovoltaic inverters and motor drives, multilevel topologies are more common due to their advantages regarding an improved output current, lower switching losses and reduced electromagnetic interferences. In multilevel topologies low voltage switches can be used instead of high-voltage switches in two-level inverters. Low voltage switches are normally smaller and cheaper and can handle higher switching frequencies. Therefore, the conduction losses can be reduced with low R_{DSon} and/or forward-voltage drop switches. Further, to achieve the same output quality in two-level topologies they need

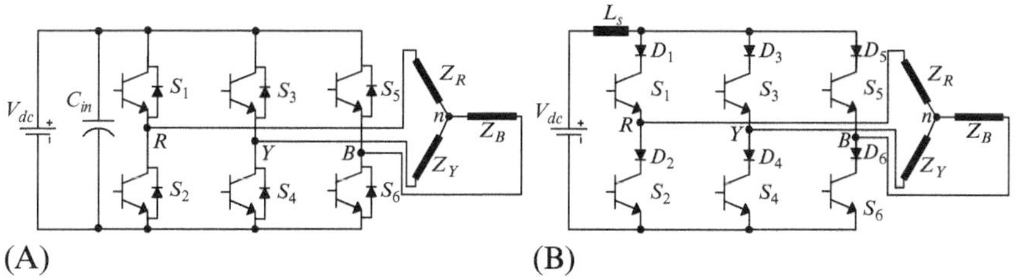

FIG. 1.19 Three-phase three-level inverter topologies: (A) VSI topology and (B) CSI topology.

to switch more often than multilevel topologies, thus the switching frequency could be reduced in multilevel topologies and by association the switching losses.

Various multilevel converter topologies have been reported in the literature since 1970s [1]. Subsequently, several multilevel converter topologies have been developed with different features [13–17]. The general classification of the multilevel inverter is shown in Fig. 1.20. With several voltage levels a better approximation to a sinusoidal waveform can be achieved which comes with a reduction in the passive filter components and therefore a lower THD as shown in Fig. 1.21. However, besides these advantages, the main drawback of multilevel inverters is their complexity regarding the structure and control technique. In addition, multilevel inverter requires a higher DC-link voltage (two times the peak AC output voltage), which needs an additional front-end boost DC-DC converter or string of series connected PV modules to lift the DC-link voltage up to 800 V for active power control to the grid. However, the multistage power conversion reduces the efficiency and reliability, whilst increasing the size and cost of the system. The additional boost stage can be eliminated by connecting

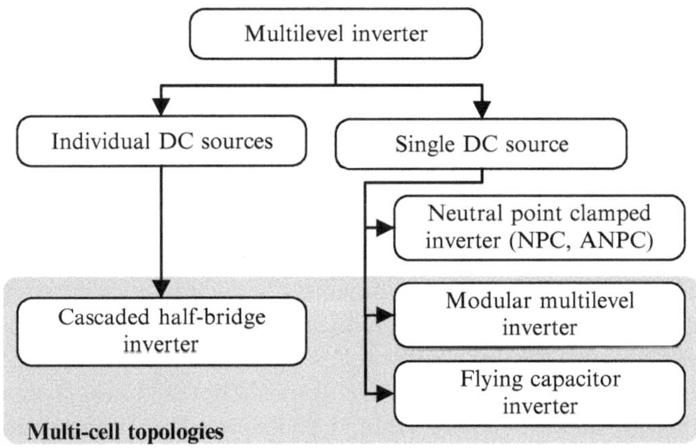

FIG. 1.20 Classification of multilevel inverter.

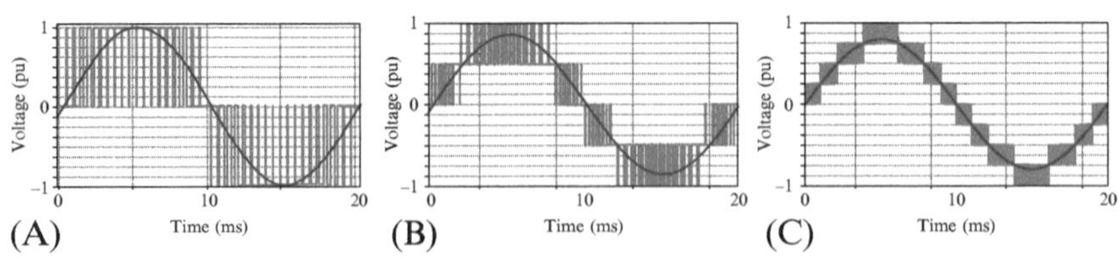

FIG. 1.21 Comparison of output voltage waveforms: (A) three-level inverter, (B) five-level inverter, and (C) nine-level inverter.

PV modules in series (string) to produce high-DC-link voltage, whereas the losses due to mismatch between the modules and shading relatively forfeits the energy gain from the system.

1.3.3.1 Neutral Point Clamped Topologies

Neutral-point-clamped (NPC) inverters are the most widely used topology of multilevel inverters in high-power applications. Fig. 1.22A and B shows diode clamped NPC topology for a three-level and five-levels respectively. It can achieve better harmonic reduction than traditional two-level VSIs and the associated control strategies help to minimize semiconductor losses. However, in diode clamped NPC the power losses are unevenly distributed among the switching devices, which limits the output power capability and could affect to the power converter reliability. This led to development of the three-level active neutral-point-clamped (3L-ANPC) topology as shown in Fig. 1.22C and D. The 3L-ANPC replaces the clamping diodes by switching devices with antiparallel diodes to provide a controllable path for the neutral current. Various higher levels (7L, 9L, …) NPC and ANPC topologies has been developed in the literature to meet the requirements of various applications (e.g., rolling mills, fans, pumps, marine appliances, mining, tractions, and most prominently grid-connected renewable energy, etc.) and available semiconductor technology.

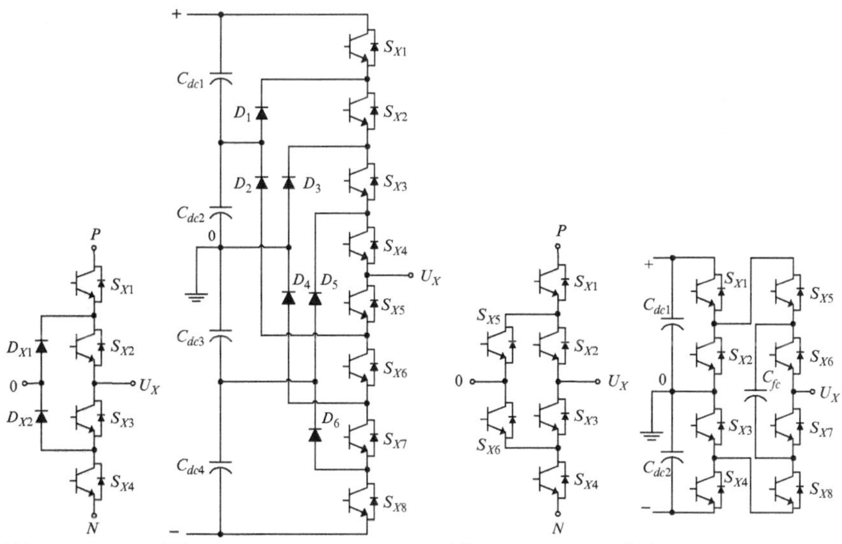

(A) (B) (C) (D)

FIG. 1.22 Phase leg of NPC inverter topologies: (A) 3L-diode clamped NPC, (B) 5L-diode clamped NPC, (C) 3L-ANPC (D) 5L-ANPC.

1.3.3.2 Multicell ML Inverter Topologies

Multilevel inverter based on multicell topologies utilizes a basic power-cell, which are connected in several configurations to produce the required number output voltage levels as shown in Fig. 1.23. Multicell topologies such as Cascaded H-Bridge (CHB) require isolated power supply for each basic cell, whereas multilevel modular converter (MMC) and Flying Capacitor share a common voltage source. The advantage with the multicell ML topologies is that they are modular which reduces the design as well as maintenance cost. However, it requires large number of active and passive components in the system design. In addition, the voltage balance in the capacitor is quite challenging, which require additional components and/or sophisticated control techniques.

1.4 DC-TO-DC CONVERTERS

A DC-DC converter changes the DC input voltage to a higher or lower DC output voltage. In addition, it should regulates the output voltage to the line and load variations. On the other hand, many applications that need to control a DC motor such as DC traction system and DC drive demand a DC-DC converter to provide variable DC voltage to control the motor speed. This type of power converter is widely known as chopper, which is due to the fact that they chop the voltage by semiconductor switches operated at high frequencies. The output voltage of choppers can be controlled by tuning the switch ON time, which is known as Pulse Width Modulation (PWM) control method. With respect to the application's demand, DC-DC converters might have electrical isolation between the input and output, which can be obtained by utilizing a transformer or coupled inductor [4,18]. In the following of this subsection, step-down, step-up and step-up-down DC-DC converters are segregated by their isolation condition.

1.4.1 Nonisolated DC-to-DC Converter

The fundamental early choppers consists of small number of components such as a switch, a diode and at least one energy storage element (an inductor and/or a capacitor). This type of power converters usually use an inductor or coupled inductor for inductive energy storage to shift the voltage up or down. Due to the inherent magnetic storage feature of these converters, they are usually heavy/bulky and their power density and efficiency are low in high-power applications. These converters can operate either with a continuous current in so-called continuous conduction mode (CCM) or including a zero current state in discontinuous conduction mode (DCM). In general, CCM operation is more prevalent owing to the load dependent voltage gain, high-current ripple, and low efficiency of DCM operation. However, due to smaller inductor implementation in DCM operation and in applications with special control requirements such as fast dynamic response, DCM operation of DC-DC converters

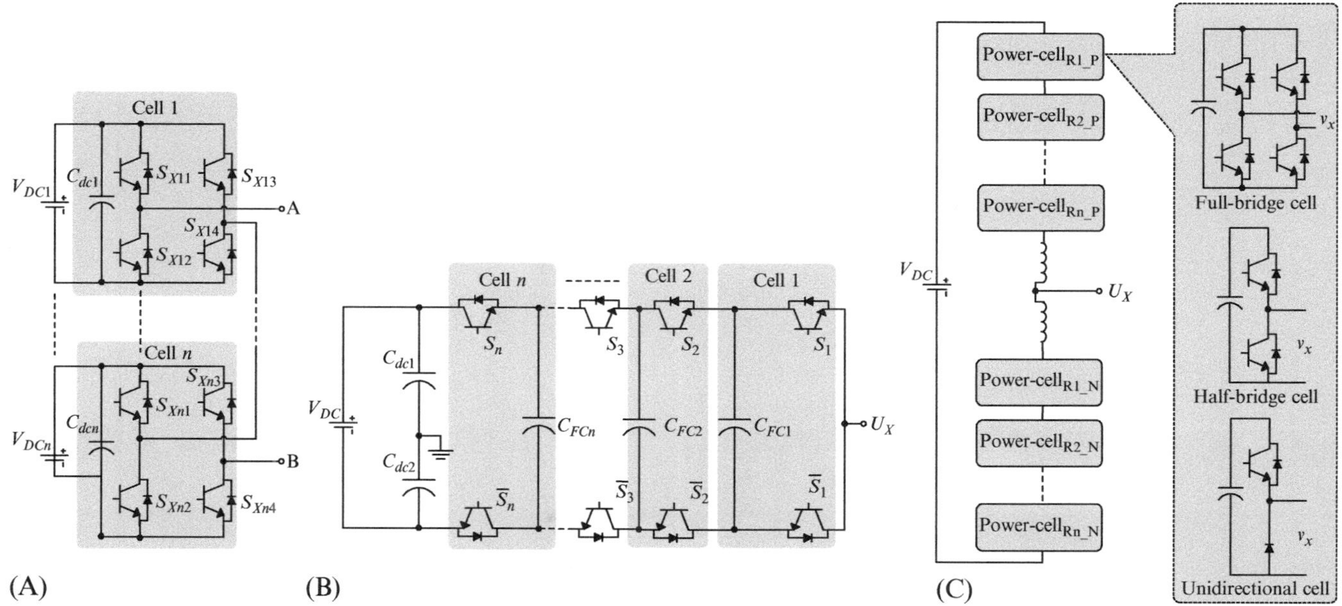

FIG. 1.23 Multicell multilevel inverter topologies (A) cascaded H-Bridge, (B) flying capacitor (FC), and (C) modular multilevel converter (MMC) illustrating different power cells.

is preferable. As shown in Fig. 1.24 nonisolated DC-DC converters can have common ground between the input and output or can have floated output as in three level boost converter. Common ground connection is more preferable in sensitive applications that demand low level of noise [4,18]. Some of the fundamental nonisolated DC-DC converters are shown in Fig. 1.25.

1.4.1.1 Buck Converter

The basic structure of step-down DC-DC converter is show in Fig. 1.25A. The average output voltage of this converter is below the input voltage. This converter has two main states in CCM operation, the ON state of the switch and the OFF state of the switch. The diode act as a freewheeling diode in OFF state of the switch, which allow load current to pass through it. The output voltage of a buck converter can be controlled by tuning the ON time of the switch. As this converter do not have a right half plane zero in its control to output voltage transfer function, it is an interesting solution for the Point of Load (POL) applications, which usually demand a very large conversion ratio [8].

1.4.1.2 Boost Converters

The configuration of the basic step-up converter is shown in Fig. 1.25A. The average output voltage of this converter is more than the input voltage. Similar

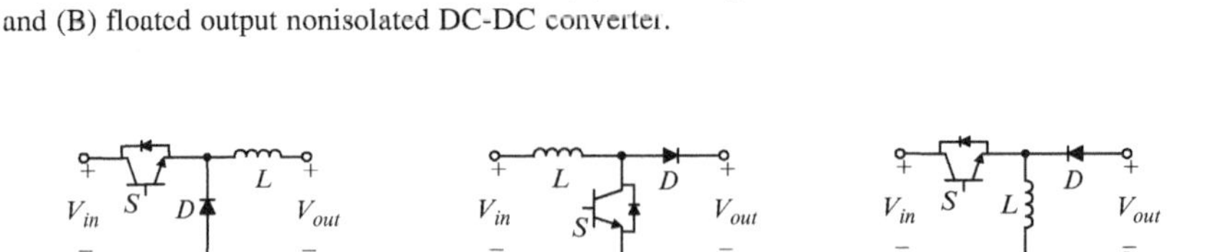

FIG. 1.24 Nonisolated DC-DC converters, (A) common grounded nonisolated DC-DC converter and (B) floated output nonisolated DC-DC converter.

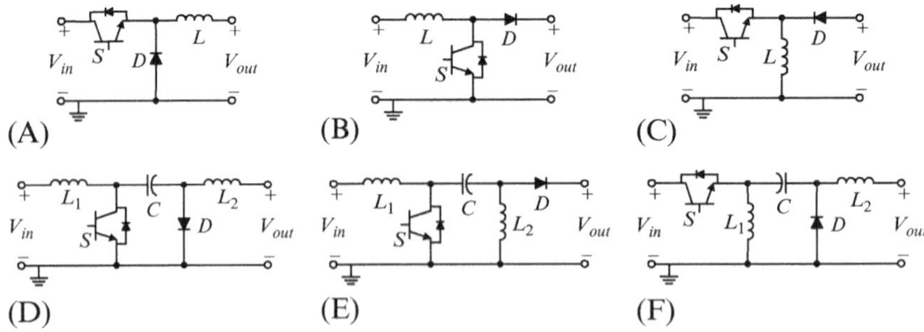

FIG. 1.25 Various fundamental nonisolated DC-DC converters. (A) Buck converter, (B) boost converter, (C) buck-boost converter, (D) Ćuk converter, (E) SEPIC converter, and (F) zeta converter.

to the buck converter this converter also has two main states, switch ON and OFF states. In the ON state of the switch the input voltage is on the inductor, when the switch turns OFF the diode conducts, and hence the output voltage is equal to the input voltage plus a ratio of the input voltage. Due to the input inductor this converter has a right half plane zero in its control to output transfer function that makes the closed loop control difficult [4].

1.4.1.3 Buck-Boost Converter

A buck-boost converter can simultaneously step-down and step-up the input voltage. Fig. 1.25C illustrates the configuration of this converter. Like the buck and boost converters the buck-boost converter uses an inductor for both step-down and step-up functions. By tuning the duty cycle below 0.5 this converter features like a buck converter and by tuning the duty cycle above 0.5 and below 1 this converter acts as a boost converter. This type of DC-DC converters are beneficial when the input voltage variation is very large in which both buck and boost operation are needed [4,18].

1.4.1.4 Ćuk Converter

This converter is a step-down-up converter with inverting output. Unlike other basic types of DC-DC converters that use as inductor for energy storage, in the Ćuk converter a capacitor is used for that purpose. Fig. 1.25D illustrates the configuration of the Ćuk converter. One of the benefits of this converter over buck-boost converter is that the switch has a grounded source, which facilitates and simplifies its drive circuit. Another advantage of the Ćuk converter is that it can be implemented in an isolated form by using an AC transformer and an additional capacitor. Moreover, by coupling both discrete inductors of the Ćuk converter on a single core, lower output ripple can be achieved and the size and cost of the converter can be reduced. In addition, using multiple secondary windings transformer can lead to multiple DC outputs in the Ćuk converter [18].

1.4.1.5 SEPIC Converter

The single ended primary inductor converter (SEPIC) is shown in Fig. 1.25E. It is essentially a step-down-up converter with noninverting output, which uses only one grounded switch. This converter transfers energy between the capacitor and inductors to change the voltage level and hence a large capacitor with high-current capability is needed. Similar to the buck-boost converter the output current of SEPIC is pulsating. Similar to the Ćuk converter both inductors of SEPIC can be coupled and form an isolated DC-DC converter [18].

1.4.1.6 Zeta Converter

Zeta converter is a fourth-order DC-DC converter with step-down and step-up ability. Fig. 1.25F shows the configuration of the Zeta converter. Similar to SEPIC it produce a positive voltage from the input voltage. As this topology is derived from a buck converter, it has a floated switch and hence demand extra components for the drive circuit. Unlike the Ćuk converter, Zeta converter has a pulsating input current. In literature, this converter sometimes known as dual SEPIC [19].

1.4.1.7 Emerging Hybrid Topologies

The permutation and combination of different voltage shifting techniques such as switched capacitor/inductor, implementing cascading and interleaving concepts, and using voltage multipliers and coupled magnetics in the circuit can lead to diverse novel DC-DC converters with different characteristics. With multistage emerging technologies, a higher power density and integration can be achieved compared to the traditional single stage converters. In fact, multidomain hybrid structures can split the voltage conversion range into multiple smaller ranges and deliver power in multiple tracks [20].

1.4.2 Isolated DC-to-DC Converter

Electrical isolation is an important feature for grid tied DC-DC converters and for some other applications that require reliable power transfer with low noise and reduced electromagnetic interference (EMI). The safety standard indicates the voltage level of electrical isolation between the input and output of a DC-DC converter, which can be achieved by means of either transformer or coupled inductor [4]. Some sensitive loads such as those used in medical, military, and avionics applications are vulnerable to faults and noise; as safety is also a major concern for these applications, electrical isolation is typically necessary. Isolated DC-DC converters can be single- or two-stage structures and can be implemented using either a coupled inductor or transformer. Fig. 1.26A shows schematics of single-stage isolated DC-DC converters and an isolated DC-DC converter with a coupled inductor. In this category, the coupled inductor will store energy in one cycle and then power the load in the other cycles; such converters usually operate at high frequency in order to reduce the size of the magnetic components. The switching concept in isolated DC-DC converters varies by topology, with forward, push-pull, half-, and full-bridge converters being examples of well-known transformer-based isolated DC-DC structures [4,18]. As shown in Fig. 1.26B, an auxiliary converter can be employed in the first stage of a two stage isolated DC-DC converter to preregulate the voltage level demanded. This auxiliary circuit, which can be a single DC-DC converter with separate modulation and control or can comprise an impedance (Z-) source network, benefits from integrated modulation and control [4,5]. Some of the fundamental isolated DC-DC converters are illustrated in Fig. 1.27.

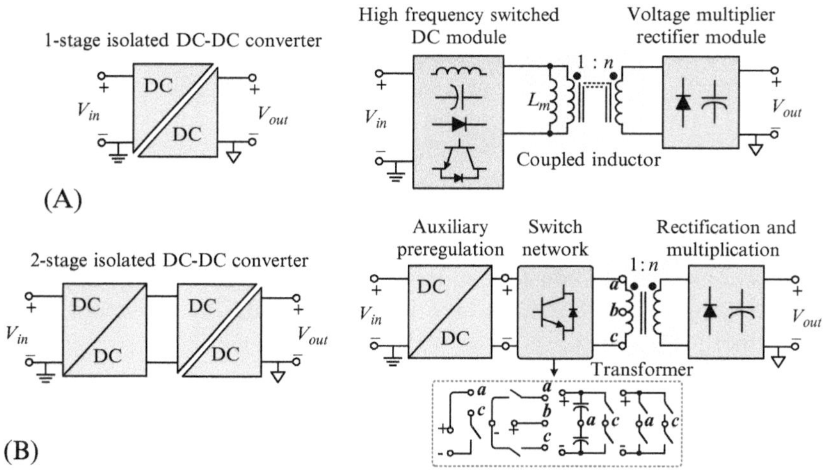

FIG. 1.26 Isolated DC-DC converters, (A) single-stage isolated DC-DC converter and (B) two-stage isolated DC-DC converter.

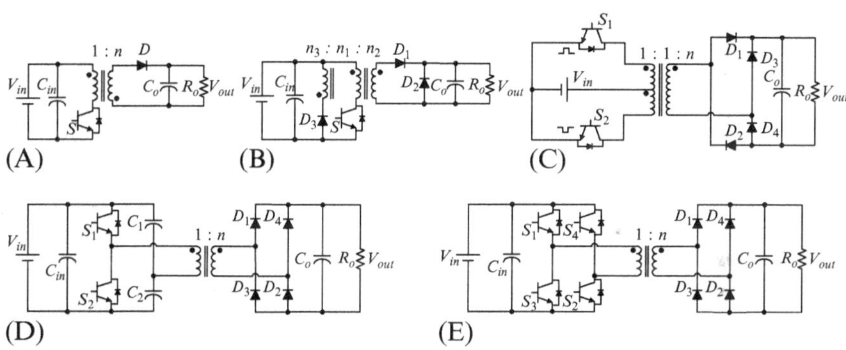

FIG. 1.27 Various fundamental isolated DC-DC converters. (A) Flyback converter, (B) forward converter, (C) push-pull converter, (D) half-bridge converter, and (E) full-bridge converter.

1.4.2.1 Flyback Converter

Flyback converters are a type of isolated buck-boost DC-DC converter that use a coupled inductor instead of an isolation AC transformer that is shown in Fig. 1.27A. The flyback converter store energy in the ON state of the switch while transferring it to the load in the OFF state of the switch. The voltage gain of the flyback converter is theoretically infinite. Due to the fact that in this converter the energy transfer is not directly it is only practicable for low power applications (usually below 100 W). Moreover, by tapping the secondary winding of the coupled inductor, this converter can provide multiple outputs [18].

1.4.2.2 Forward Converter

The forward converter is another kind of isolated DC-DC converter that can transfer energy directly without energy storing in switch turn on state. Fig. 1.27B shows the structure of the forward converter. Unlike the flyback converter that uses couple inductor, the forward converter use AC transformer with no air gap and hence it is more energy efficient and amenable to medium power applications (up to 200 W with single switch and up to 500 W with two-switch). The voltage conversion type of the forward converter is essentially as the buck converter and constrained by the transformer turn ratio, however, it can be changed to different voltage levels by tuning the turn ratio of the transformer [18].

1.4.2.3 Push-Pull Converter

Fig. 1.27C illustrates the configuration of the push-pull converter. This kind of DC-DC converter alternates the current in the transformer by the operation of its two switches at the primary side of the transformer. The advantage of the push-pull converter is that both switches are grounded and there is no need for high-side driver circuits. The term push-pull is because it has bidirectional excitation of the transformer. This converter has a steady input current flow and hence low noise at the input side [18].

1.4.2.4 Half-Bridge Converter

Half-bridge converter is an isolated DC-DC converter with two switches that uses two capacitor at the primary side of the transformer to split the input voltage. Fig. 1.27D illustrates the half-bridge DC-DC converter. Similar to the push-pull converter switches S_1 and S_2 turn ON alternately to excite the transformer. When S_1 turns ON $V_{in}/2$ is on the primary side of the transformer and when S_2 is ON $-V_{in}/2$ is on the primary of the transformer. Unlike the push-pull converter that has a switch voltage stress twice the input voltage, the switch voltage stress in the half-bridge converter is half of the input voltage. The half-bridge DC-DC converter is suitable for application up to 500 W [18].

1.4.2.5 Full-Bridge Converter

Full-bridge DC-DC converters use four switches that symmetrically excite the primary winding of the transformer. Similar to the half-bridge converter the voltage conversion of the full-bridge converter is buck type. During the first switching period switches S_1 and S_2 conduct and V_{in} applies on the primary winding, and in the following switching period switches S_3 and S_4 conduct and $-V_{in}$ applies on the primary winding of the transformer. Hence, for a given power, the primary current of the transformer of the full-bridge is half of in the half-bridge converter. The full-bridge DC-DC converter usually is used for kW level applications. Moreover, due to the complete utilization of the core, in which the magnetization current can flow in both positive and negative direction, there is no need to reset the core [18].

1.4.2.6 Emerging Hybrid Topologies

In order to enhance the voltage conversion ability, reliability, efficiency and power density of isolated DC-DC converters, a number of studies have focused on developing new types of DC-DC converters. As some examples we can name impedance (Z-) source-based isolated DC-DC converters with simultaneous buck-boost ability and seamless control [21], dual active bridge (DAB) and dual half bridge (DHB) topologies with wide soft switching ability [22] and multistage/-level DC-DC converters suitable for wide voltage and power range [4].

1.5 AC-TO-AC CONVERTERS

Many industrial applications demand an AC-AC converter to take power from an AC source and deliver it to an AC load with voltage/current of different frequency, amplitude and phase. The most conventional AC-AC converters consist of a rectifier followed by an inverter. These converters use a voltage/current DC link between the two stages and are known as two level indirect AC-AC converters. This subsection mainly discusses direct AC-AC converters that are used to change the frequency of input source (voltage/current) to a different desired level of the load. Fig. 1.28A illustrates the indirect AC-AC converter with an energy storage element (capacitive or inductive) and Fig. 1.28B illustrates the direct AC-AC converter without energy storage element. Sometimes, AC-AC converters are used to change the magnitude of the input voltage as well as the frequency. Two well-known direct frequency changer are cycloconverter and matrix converter. The cycloconverter is a naturally commutated converter that usually use controlled rectifiers and matrix converter is a forced commutated converter that usually use fully controlled bidirectional switches. These are mainly used for tuning the speed of AC drives [23].

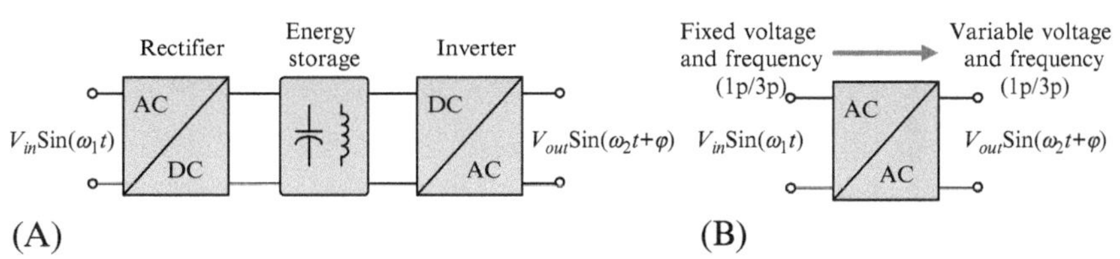

FIG. 1.28 General layouts of AC-AC converters, (A) indirect AC-AC converter and (B) direct AC-AC converter.

1.5.1 Cycloconverter

Cycloconverters are frequency changers that convert AC power of specific frequency and voltage to different frequency and voltage of AC power without any intermediate DC link. A cycloconverter is a naturally commutated converter in which the output frequency and voltage can be controlled independently and continuously. It consists of back to back connected controlled rectifiers whose output voltage and frequency can be controlled by tuning firing angles of rectifiers. With respect to the connection of rectifiers, its structure can comprise of half-wave or full-wave bridge. Fig. 1.29A illustrates a single-/three-phase to single-phase bridge cycloconverter. It consists of two back-to-back connected rectifiers, in a way that the positive converter (P type) conducts with positive currents and the negative converter (N type) conducts with negative currents. This type of cycloconverters is widely used in electric traction applications. Fig. 1.29B shows a three-phase to single-/three-phase half-wave cycloconverter. Similar to the single-phase cycloconverter, P and N converter types conduct positive and negative currents respectively. This type of cycloconverters is widely used for large motor drive. To reduce the output oscillation and have

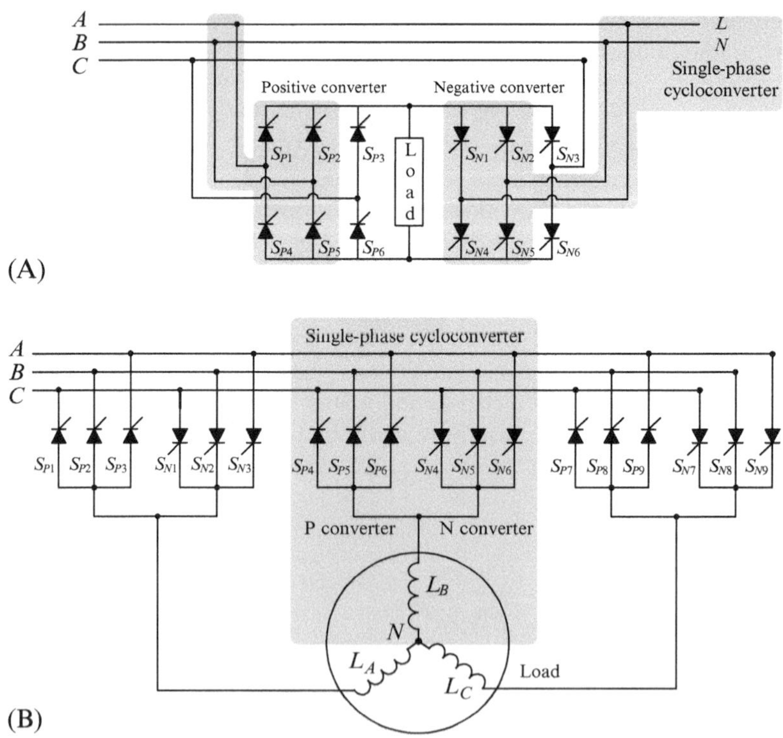

FIG. 1.29 Illustration of (A) single-phase bridge cycloconverter and (B) three-phase half-wave cycloconverter.

a smoother output voltage, usually high number of pulses are required in cycloconverters (6-pulse, 12-pulse, etc.). This is due to the lower commutation time that is required when the higher number of pulses are generated in cycloconverters [24].

1.5.2 Matrix Converter

Matrix converters are force commutated AC-AC converters without large energy storage components and DC-link voltage. Main features of the matrix converter that makes it a popular choice for AC-AC conversion are the followings: simple and compact structure, sinusoidal input and output current, operating with unity power factor with any type of load and regeneration capability. The matrix converter is a single stage DC-DC converter that uses $m \times n$ bi-directional switches to connect an m-phase source to an n-phase load. Typically, the matrix converter is voltage-fed and should not be short circuited at the input and at the load side they are usually inductive (motor drive applications) and hence should not be open circuited [25]. Fig. 1.30A illustrates a 2×2 matrix converter that converters a single-phase input voltage source to a single-phase load. The instantaneous output voltage ($V_{out}(t)$) of this converter has a maximum value identical to the maximum value of the input voltage, it has fundamental and additional high-order harmonics, and its main harmonic has a stepped-up frequency and a stepped-down amplitude [26]. Fig. 1.30B shows the structure of a 3×3 matrix converter that connects a three-phase AC voltage to a three-phase load like an AC motor. From Fig. 1.30B, it is clear that each phase of the three-phase input source can be connected to each phase of the load, and hence the combinations of the switching patterns of the switches can produce 27 switching states [25].

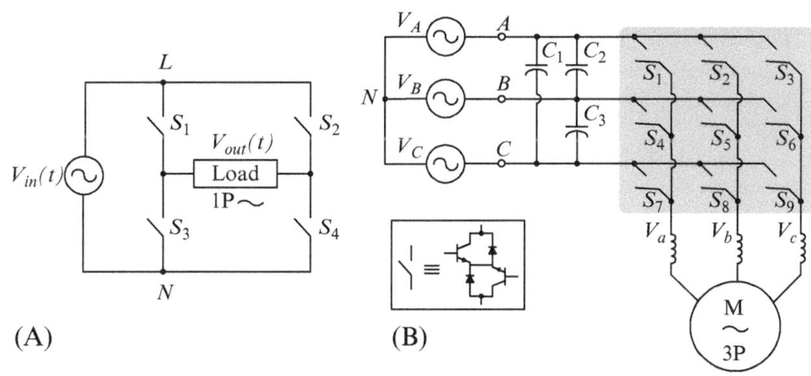

FIG. 1.30 The structure of matrix converters, (A) a 2×2 matrix converter and (B) a 3×3 matrix converter.

REFERENCES

[1] R.L. Steigerwald, Power electronic converter technology, Proc. IEEE 89 (6) (2001) 890–897.

[2] T.G. Wilson, The evolution of power electronics, IEEE Trans. Power Electron. 15 (3) (2000) 439–446.

[3] F. Blaabjerg, Z. Chen, S.B. Kjaer, Power electronics as efficient interface in dispersed power generation systems, IEEE Trans. Power Electron. 19 (5) (2004) 1184–1194.

[4] M. Forouzesh, Y.P. Siwakoti, S.A. Gorji, F. Blaabjerg, B. Lehman, Step-up DC-DC converters: a comprehensive review of voltage-boosting techniques, topologies, and applications, IEEE Trans. Power Electron. 32 (12) (2017) 9143–9178.

[5] S.B. Kjaer, J.K. Pedersen, F. Blaabjerg, A review of single-phase grid-connected inverters for photovoltaic modules, IEEE Trans. Ind. Appl. 41 (5) (2005) 1292–1306.

[6] Y.P. Siwakoti, F.Z. Peng, F. Blaabjerg, P.C. Loh, G.E. Town, Impedance-source networks for electric power conversion—part I: a topological review, IEEE Trans. Power Electron. 30 (2) (2015) 699–716.

[7] H. Chung, S.Y.R. Hui, K.K. Tse, Reduction of power converter EMI emission using soft-switching technique, IEEE Trans. Electromagn. Compat. 40 (3) (1998) 282–287.

[8] J.W. Kolar, T. Friedli, The essence of three-phase PFC rectifier systems—part I, IEEE Trans. Power Electron. 28 (1) (2013) 176–198.

[9] J.A. Wiik, F.D. Wijaya, R. Shimada, Characteristics of the magnetic energy recovery switch (MERS) as a series FACTS controller, IEEE Trans. Power Delivery 24 (2) (2009) 828–836.

[10] M. Rastogi, R. Naik, and N. Mohan, "Optimization of a novel DC-link current modulated interface with three-phase utility systems to minimize line current harmonics," in Proc. IEEE PESC'92, vol. 1, pp. 162–167.

[11] J. W. Kolar, Dreiphasen-Dreipunkt-Pulsgleichrichter, Filed Dec. 23, 1993, File No.: A2612/93, European Patent Appl.: EP 94 120 245.9-1242 entitled "Vorrichtung und Verfahren zur Umformung von Drehstrom in Gleichstrom".

[12] T.B. Soeiro, T. Friedli, J.W. Kolar, in: Swiss rectifier—a novel three-phase buck-type PFC topology for Electric Vehicle battery charging, Proc. Twenty-Seventh Annual IEEE Applied Power Electronics Conference and Exposition (APEC), Orlando, FL, 2012, , pp. 2617–2624.

[13] A. Nabae, I. Takahashi, H. Akagi, A neutral-point-clamped PWM inverter, IEEE Trans. Ind. Appl. 1A–17 (5) (1981) 518–523.

[14] S. Kouro, M. Malinowski, K. Gopakumar, J. Pou, L.G. Franquelo, B. Wu, J. Rodriguez, M.A. Perez, J.I. Leon, Recent advances and industrial applications of multilevel converters, IEEE Trans. Ind. Electron. 57 (8) (2010) 2553–2580.

[15] J. Rodríguez, S. Bernet, B. Wu, J.O. Pontt, S. Kouro, Multilevel voltage-source-converter topologies for industrial medium-voltage drives, IEEE Trans. Ind. Electron. 54 (6) (2007) 2930–2945.

[16] L.A. Serpa, P.M. Barbosa, P.K. Steimer, J.W. Kolar, in: Five-level virtual flux direct power control for the active neutral-point clamped multilevel inverter, Proc. IEEE PESC 2008, Jun. 15–19, 2008.

[17] F.Z. Peng, A generalized multilevel inverter topology with self-voltage balancing, IEEE Trans. Ind. Electron. 37 (2) (2001) 611–618.

[18] R.W. Erickson, DC-DC power converters, in: Wiley Encyclopedia of Electrical and Electronics Engineering, Wiley, Hoboken, NJ, 1999.

[19] J.J. Jozwik, M.K. Kazimierczuk, Dual sepic PWM switching-mode DC/DC power converter, IEEE Trans. Ind. Electron. 36 (1) (1989) 64–70.

[20] M. Chen, K.K. Afridi, S. Chakraborty, D.J. Perreault, Multitrack power conversion architecture, IEEE Trans. Power Electron. 32 (1) (2017) 325–340.

[21] A. Chub, D. Vinnikov, F. Blaabjerg, F.Z. Peng, A review of galvanically isolated impedance-source DC-DC converters, IEEE Trans. Power Electron. 31 (1) (2016) 2808–2828.

[22] B. Zhao, Q. Song, W. Liu, Y. Sun, Overview of dual-active bridge isolated bidirectional DC-DC converter for high-frequency-link power-conversion system, IEEE Trans. Power Electron. 29 (8) (2014) 4091–4106.

[23] P. Szcześniak, Review of AC-AC frequency converters, in: Three-Phase AC-AC Power Converters Based on Matrix Converter Topology, Springer, London, 2013, , pp. 17–86.

[24] Y. Liu, G.T. Heydt, R.F. Chu, The power quality impact of cycloconverter control strategies, IEEE Trans. Power Delivery 20 (2) (2005) 1711–1718.

[25] P.W. Wheeler, J. Rodriguez, J.C. Clare, L. Empringham, A. Weinstein, Matrix converters: a technology review, IEEE Trans. Ind. Electron. 49 (2) (2002) 276–288.

[26] A. Zuckerberger, D. Weinstock, A. Alexandrovitz, Single-phase matrix converter, IEE Proc. Electr. Power Appl. 144 (4) (1997) 235–240.

Chapter 2

Basic Control Principles in Power Electronics

Analog and Digital Control Design

Ricardo P. Aguilera*, Pablo Acuna†, Georgios Konstantinou‡, Sergio Vazquez§ and Jose I. Leon§
*University of Technology Sydney, Ultimo, NSW, Australia, †RMIT University, Melbourne, VIC, Australia, ‡University of New South Wales, Sydney, NSW, Australia, §University of Seville, Seville, Spain

2.1 INTRODUCTION

Power converters are autonomous systems. Therefore, their performance does not only depend on the hardware design but also on the control strategy used. It is for this reason that it is important to understand some basic principles related to the control of power converters and electrical drives.

The aim of this chapter is to provide an overview on basic control principles on power electronics. For that purpose, in Section 2.2, the key elements underlying the control problem in power electronics are presented. The standard approach to control power converters is to use a linear controller to define a desired closed-loop dynamic along with a modulator to finally handle the power converter switches. However, there are some control strategies that do not follow this standard approach. Therefore, a classification in terms of the use or neglect of the modulator is presented. In Section 2.3, the main modulation techniques used to handle the power switches are discussed. Then, an analysis and recommendations for designing linear controllers that include a modulator in the closed loop are presented in Section 2.4.1. In particular, the design of basic linear controllers for power converters that can be modeled as first-order systems is analyzed in detail. Finally, some insights into controllers that do not follow this standard approach are given in Sections 2.4.2 and 2.4.3.

Control of Power Electronic Converters and Systems. https://doi.org/10.1016/B978-0-12-805245-7.00002-0

2.2 BASIC CONCEPTS

2.2.1 Power Electronics Control Targets

Control targets are the keystone upon which the design of control strategies for power converters is based. First, there are some basic requirements associated with the power converter operation that need to be taken into account. The following list has been generally accepted as fundamental requirements not only related to the closed-loop response but also related to the safe operation of power converters and loads (see [1]):

- *Limited or constant switching frequency*: The importance of this control target is reflected in two main operation aspects of the power converter. On one hand, limiting the switching frequency results in known converter's power losses, which facilitates the hardware design based on the converter efficiency and thermal dissipation. On the other hand, it provides fixed current and/or voltage spectra, which facilitates the interface filtering design.
- *Reduced ripple*: This helps to reduce the size of passive components (capacitors and inductors) in a converter.
- *High dynamic response*: Power converters are normally part of larger systems (e.g., electric vehicles, renewable power plants), where a fast change in the power flow is required. Therefore, a high dynamic response is crucial to satisfy those requirements.
- *No phase and amplitude error in the steady state for AC-systems*: Standard controllers are normally designed to track constant references (regulation problem). However, in AC-systems, it is required to track sinusoidal references. If the current or voltage under control does not have the desired amplitude or phase, it can lead to undesired effects. For instance, in a rectifier with unity power factor, reactive power may be injected and/or absorbed by the converter.
- *Reduced harmonic content*: Ideally for AC or DC applications, currents or voltages under control are required to present only a desired fundamental frequency or DC component with small harmonic content, respectively. If the harmonic level is too high, some detrimental consequences can be triggered. For instance, in electrical drives, harmonics can lead to extra losses in the electric motors. In grid-connected applications, unwanted resonances in the electricity grid might be triggered.

Notice that some of these requirements contradict each other, for example, high dynamic and reduced harmonic content. For that reason, this interdependence is of great significance to select a certain class of controller to effectively satisfy these targets.

2.2.2 Power Electronics Control Classification

In this chapter, three power electronics control classes are distinguished based on the control input selection. First, the classic linear approach with modulator

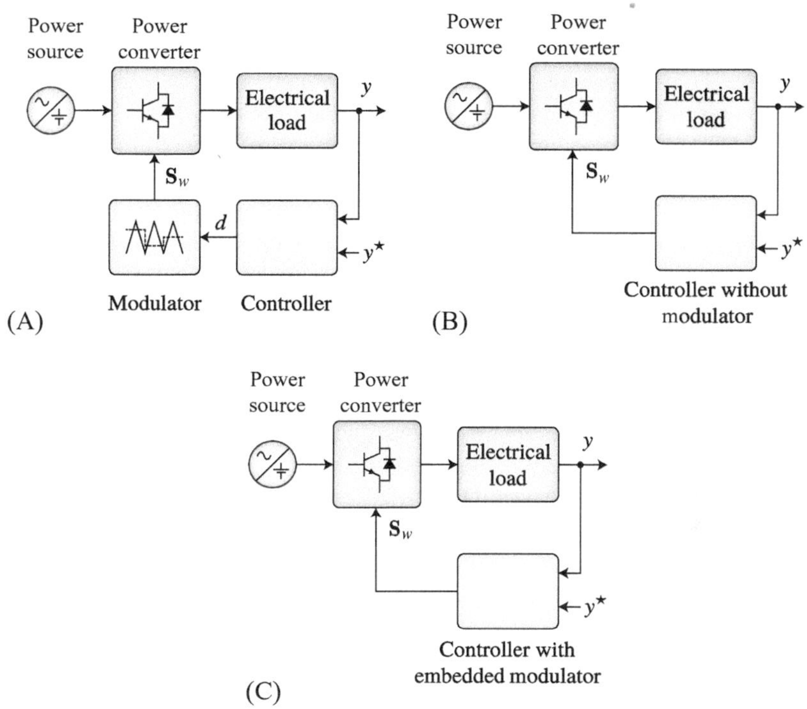

FIG. 2.1 Classification of power electronics control strategies: (A) modulation-base controllers, (B) controllers without modulator, and (C) controllers with embedded modulator.

is presented in Fig. 2.1A. For this case, the controller has to track an average value of the converter output (current, voltage, torque, power, etc.). Therefore, it will provide an average control input u. Since this average input cannot be directly implemented in the converter, this is synthesized by a modulation stage. Consequently, for this control class, the controller normally provides the required duty-cycle, $u = d$, to control the average value of y. Since the input can take a bounded continuous value, it is said that it belongs to a continuous control set (CCS) (e.g., $d \in [0, 1]$). To design a controller, an average model of the converter and load is normally used, neglecting the modulator effect. The advantage of using a modulator comes from the fact that it fixes the converter switching frequency and produces a well-defined spectrum over the controlled variables. Therefore, several power electronics control targets can be satisfied by including a modulator in the control loop.

Despite the well-known advantages of using modulators, there are some control strategies for power converters that directly consider the state of the switches (or output voltage levels) as input, $u = S_w$. Hence, a modulation stage is not needed. This second power electronics control class is depicted in Fig. 2.1B. Since in this case the control input can take only a fixed number of states, it is said that the input belongs to a finite control set (FCS) (e.g., $S_w \in \{0, 1\}$). This allows the control designer to use instantaneous models of the power converters and loads to design the controller. In that case, faster dynamic responses are generally achieved when compared to using average models. However, the output variables present a spread spectrum.

A third power electronics control class is shown in Fig. 2.1C. These control strategies explicitly consider the modulator in their formulation along with an instantaneous model of the converter and load. To do this, the instants when the switches have to commutate, t_{S_w}, within a sampling time, T_s, are considered as control input, $u = t_{S_w}$. Therefore, in this case the input also belongs to a CCS (e.g., $t_{S_w} \in [0, T_s]$). The advantage of this control class is that a fast dynamic is achieved during transients with a fixed switching frequency and well-defined spectrum.

Throughout this chapter, main emphasis will be given to the first power electronics control class, since it is the standard and accepted way to govern power converters and electrical drives. Therefore, some modulation techniques as well as standard linear control design for power converters will be provided. Nonetheless, some insights into controllers that fit into the other classes will be briefly discussed.

2.3 MODULATION METHODS

The modulator's role in controlling power electronic converters is the same throughout each and every DC/DC, DC/AC, and AC/DC power converters. Basically, the modulator decodes the controller's effort (see d in Fig. 2.1A) in the form of switching signals (see $\mathbf{S}_w$ in Fig. 2.1A) for driving the power semiconductors.

Output requirements for modulators have been developed over decades and are highly application dependant. Low-power applications, for example, usually require analog modulators that can handle several kilohertz in the frequency of the switching signals. This frequency requirement is the result of an aim to reduce weight and size of power converters by means of increasing the switching frequency. On the other hand, high-power applications, normally DC/AC or AC/DC power converters, usually require digital modulators working at lower frequency (less than kilohertz) that can handle several outputs and are generally more sophisticated.

In practice, the most used solution to implement the modulator is the well-known pulse-width modulation (PWM) method where the switched waveform is controlled by the width of the output pulses. This classic PWM technique has been the mainstream solution to implement the modulator in conventionally controlled DC/DC, DC/AC, and AC/DC power converters for more than 50 years. The basic concept of PWM is represented in Fig. 2.2 [2, 3]. In this case, the controller's effort d is decoded by generating a PWM signal. The

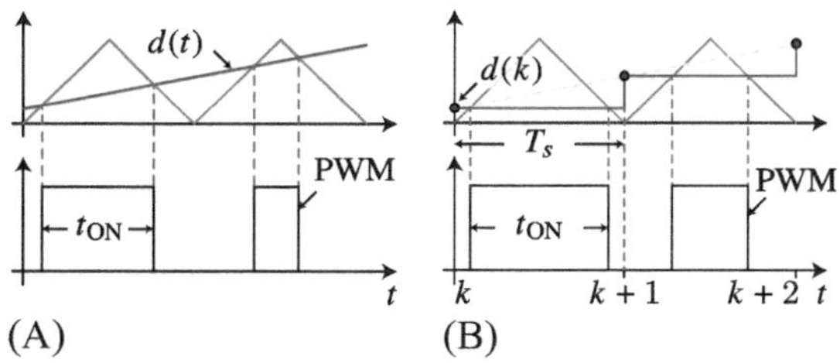

FIG. 2.2 Basic concept of modulation: (A) analog PWM and (B) digital PWM.

PWM implementation can be done either in continuous-time by using analog signals or in discrete-time as part of an analog-to-digital process.

The evolution of power electronics in the last decades has been accompanied by the rise of new modulation challenges, for example, computational cost, feasibility, optimal harmonic distortion, and so on. Those challenges are under continuous evolution due to development of new converter topologies and applications. For that reason, modifications of the classic PWM and other modulation methods have been introduced to overcome these challenges. In the next section, PWM methods for voltage source DC/AC power converters are described in more detail.

2.3.1 Classification of Modulation Methods for Voltage Source DC/AC Power Converters

A classification of the PWM methods, mainly for DC/AC power converters, is depicted in Fig. 2.3. In addition, Table 2.1 summarizes the most conventional modulation methods for two-level power converters. Finally, Table 2.2 introduces the extension of these modulation methods for multilevel power converters.

2.3.2 Conventional PWM Techniques

For voltage source DC/AC power converters, in general, the goal of the modulator is to generate a PWM voltage in which the harmonic components resemble the harmonic components of the controller's effort while keeping a maximum utilization of the DC voltage. Normally, the controller's effort waveform is sinusoidal, so it is expected to obtain a PWM waveform with harmonics in which its fundamental-frequency component matches with the fundamental-frequency component of the desired controller's effort waveform. A first approach to generate a PWM waveform from a sinusoidal waveform is to use the square-wave modulation method. This method effectively leads to a maximum utilization of the DC voltage generating a waveform with maximum modulation index m_a (ratio between the output voltage peak and the DC voltage value) at the expense of high distortion in low-order harmonics of the obtained

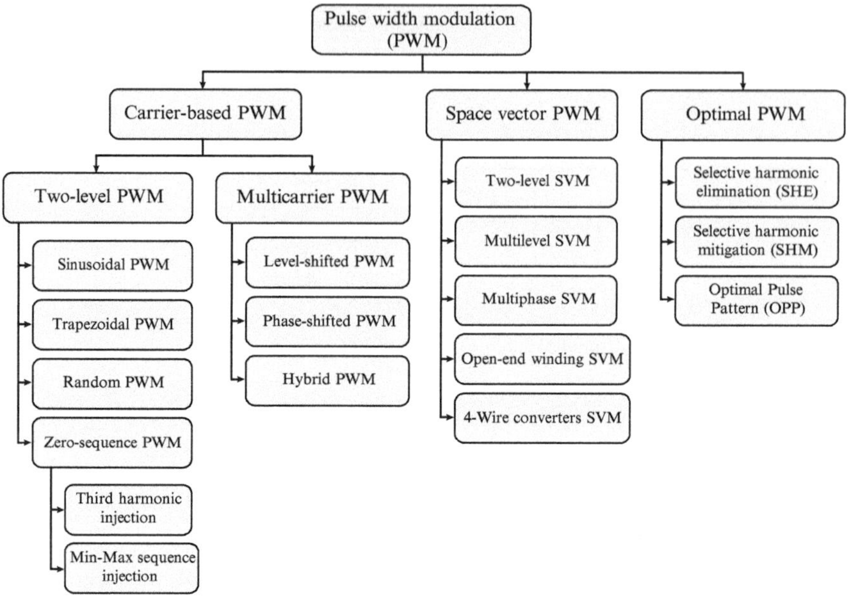

FIG. 2.3 Classification of PWM methods.

TABLE 2.1 Conventional Modulation Methods for Three-Phase Two-Level Voltage Source Inverters

Item	Operating Principle	Voltage Waveform	Harmonic Spectrum	e.g.
Square-wave	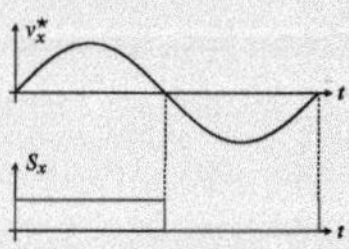	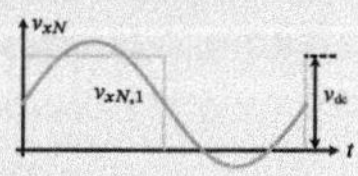	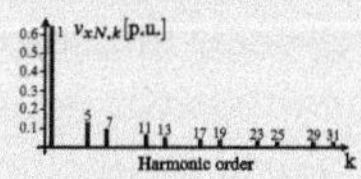	[4]
PWM	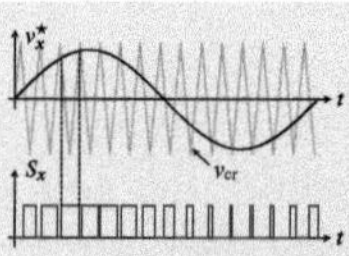	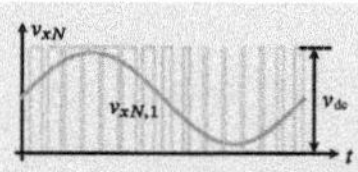	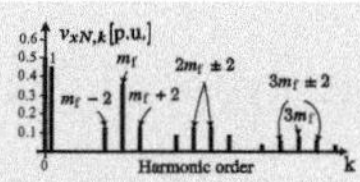	[2,4–6]
SVM	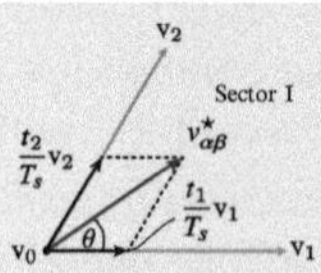	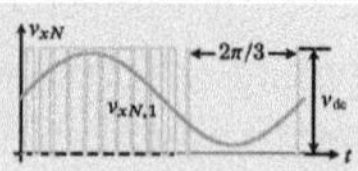	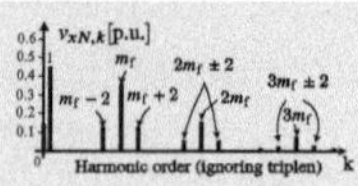	[7–9]
SHE	$m=\frac{4}{\pi}\left(1-2\sum_{i=0}^{k-1}(-1)^i\cos(\alpha_i)\right)$ $0=\frac{4}{j\pi}\left(1-2\sum_{i=0}^{k-1}(-1)^i\cos(j\alpha_i)\right)$ where $j=5,7,11,\ldots,q$	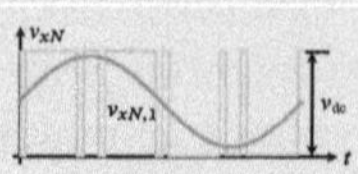	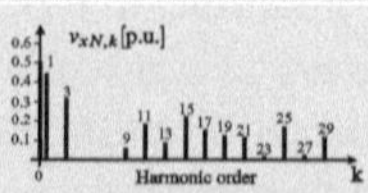	[10–12]

TABLE 2.2 Conventional Modulation Methods for Multilevel Voltage Source Inverters

Item	Operating Principle	Voltage Waveform	Harmonic Spectrum	e.g.
LS-PWM (3L-NPC example)	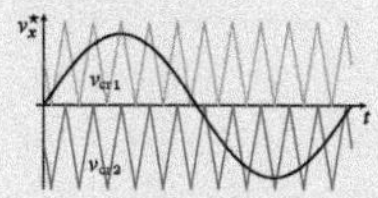	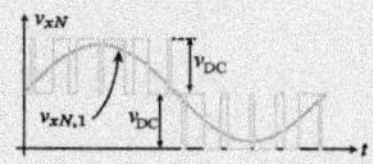	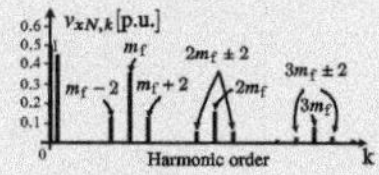	[6,13]
PS-PWM (5L-CHB example)	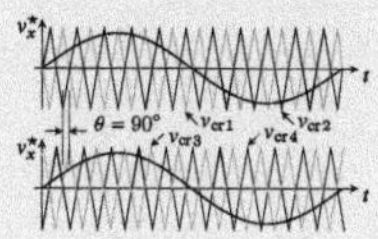	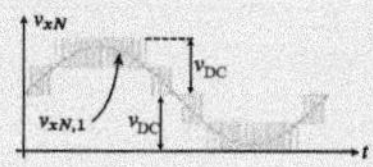	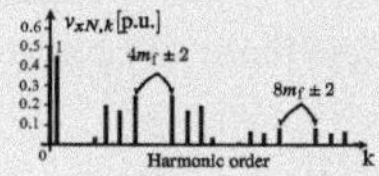	[6,13]
Multilevel SVM (3L example)	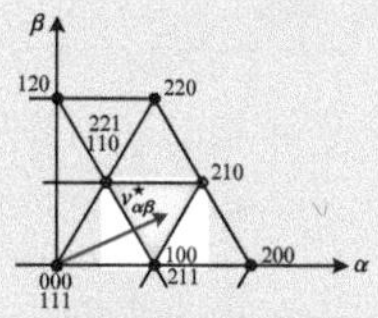	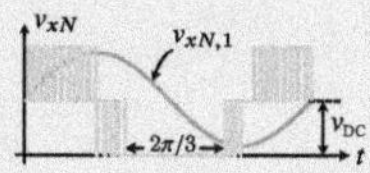	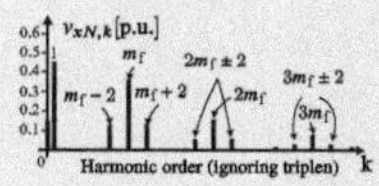	[14,15]
Multilevel SHE (3L example)	$m = \frac{4}{\pi}\sum_{i=0}^{k-1}[(-1)^i \sin(\alpha_i)]$ $0 = \frac{4}{j\pi}\sum_{i=0}^{k-1}[(-1)^i \sin(j\alpha_i)]$ where $j = 5, 7, 11, \ldots, q$	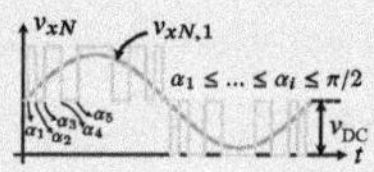	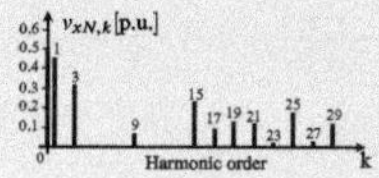	[16–18]

waveforms as shown in Table 2.1. Hence, the square-wave modulation method is very rarely used in practice. In order to improve the performance of the output waveforms, PWM-based methods can be used.

2.3.2.1 *Carrier-Based PWM Techniques: Bipolar and Unipolar PWM for Single-Phase Inverter*

A single-phase voltage source inverter (VSI) is shown in Fig. 2.4. The inverter is normally used to convert a fixed DC input voltage v_{dc} to an AC output voltage v_{AB} with variable amplitude and frequency. The fundamental principle of the PWM operation for this inverter is illustrated in Fig. 2.5, where the gating signals S_1 and S_2 are generated by comparing a sinusoidal modulating waveform v_{mA} with a periodic triangular carrier waveform v_{cr}. Since the modulating waveform is a sinusoid, this method is often referred to as sinusoidal PWM (SPWM).

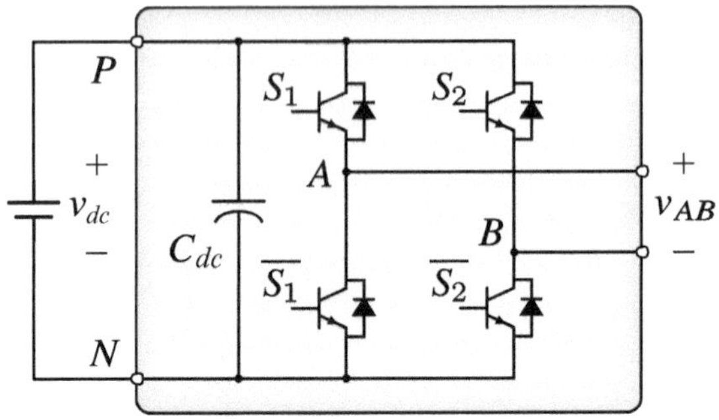

FIG. 2.4 Single-phase full-bridge inverter.

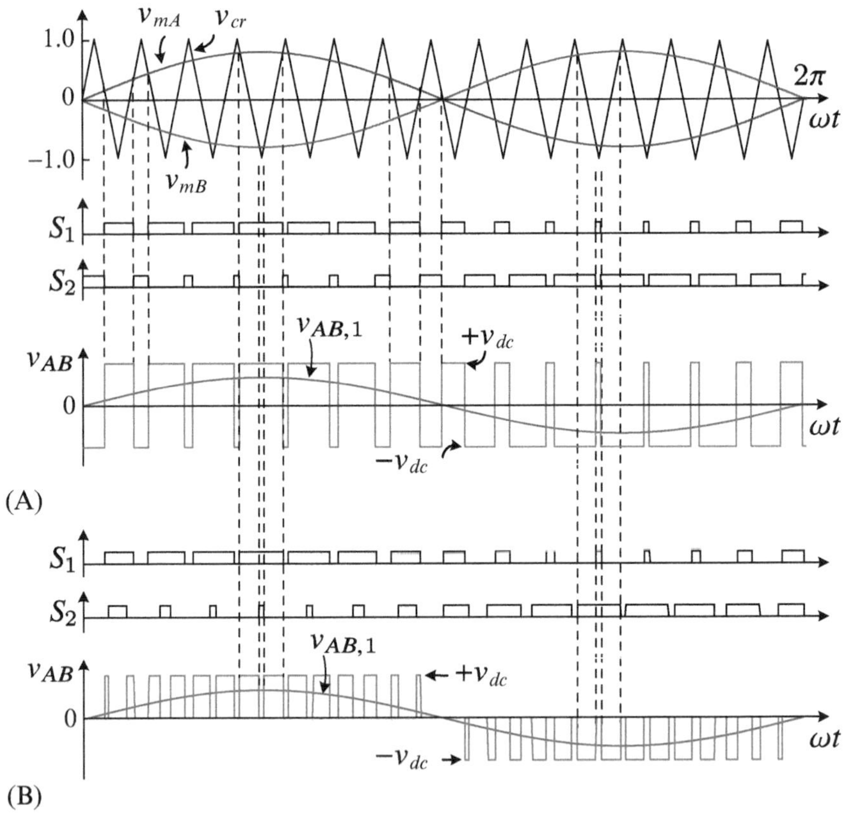

FIG. 2.5 Bipolar and unipolar PWM for single-phase inverter: (a) bipolar modulation and (b) unipolar modulation.

As a first option, the two gating signals can be complementary to each other, resulting in a bipolar voltage waveform, where the inverter output voltage v_{AB} switches between $+v_{dc}$ and $-v_{dc}$ as shown in Fig. 2.5a. Therefore, this PWM scheme is known as bipolar modulation [4]. The power switches' switching frequency f_{sw} is then equal to the carrier frequency f_{cr}.

However, by using the same power switches' switching frequency, it is possible to double the equivalent switching frequency in v_{AB}. This is done by adopting the so-called unipolar PWM scheme. Here, an additional modulating wave v_{mB} is required, which is 180 degrees out of phase with v_{mA}. The gating signals S_1 and S_2 are then generated by comparing v_{mA} and v_{mB} with the carrier wave v_{cr}. The inverter output voltage v_{AB} switches either between 0 and $+v_{dc}$ during the positive half-cycle, or between 0 and $-v_{dc}$ during the negative half-cycle of the fundamental frequency, as shown in Fig. 2.5b [4].

Unlike bipolar modulation where all the four switches are always commutated at the same time, the switches in phases A and B do not switch simultaneously under unipolar modulation. Although the output fundamental amplitude under both schemes is equivalent, and the power switches' switching frequencies are the same and equal to the carrier frequency, the inverter switching frequency under unipolar modulation is essentially doubled (i.e., $f_{sw,\ inv} = 2f_{sw}$), which facilitates a reduction in size and cost of inverter output filters, if any.

The fundamental-frequency component $v_{AB,\ 1}$ in the inverter output voltage can be adjusted by changing the amplitude of the modulation index m_a, which is defined as the ratio between the peak values of the modulating and carrier waves

$$m_a = \frac{\hat{v}_m}{\hat{v}_{cr}} \tag{2.1}$$

The modulation index m_a can normally be adjusted by varying $\hat{v}_m$ while keeping $\hat{v}_{cr}$ fixed. The frequency modulation index m_f is defined as the ratio between the frequencies of the carrier and modulating waves

$$m_f = \frac{f_{cr}}{f_m} \tag{2.2}$$

If m_f is an integer, the carrier wave is synchronized with the modulating wave and the modulation scheme is known as synchronous PWM; otherwise, it is called asynchronous PWM, in which f_{cr} is usually fixed and independent of f_m. The asynchronous PWM features a fixed switching frequency and is easy to implement with analog circuits. However, its output spectrum contains noncharacteristic harmonics whose frequency is not a multiple of f_m. The synchronous PWM scheme is more suitable for implementation with a digital processor.

2.3.2.2 *Carrier-Based PWM Techniques: PWM for Three-Phase Two-Level VSI*

The three-phase two-level VSI shown in Fig. 2.6 is arguably the most popular converter topology for DC-AC conversion with a power rating up to 750 kW without switches or converters connected in parallel or series. By comparing a set of three-phase sinusoidal modulating waveforms with a common carrier, the PWM scheme for the single-phase inverter can be readily extended to modulate the three-phase VSI. Fig. 2.7 shows the PWM for the VSI and

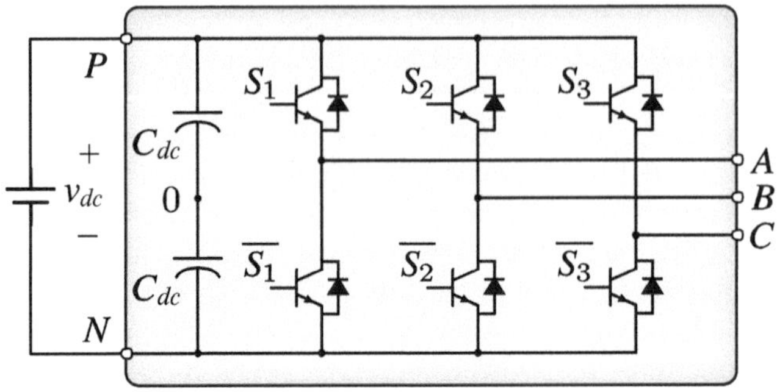

FIG. 2.6 Three-phase two-level voltage source inverter.

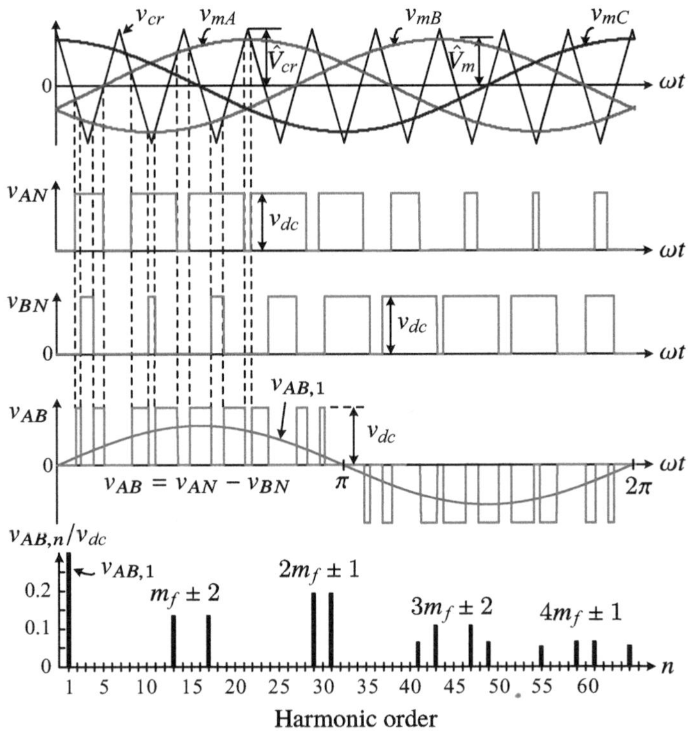

FIG. 2.7 SPWM for three-phase two-level voltage source inverters.

the harmonic spectrum of its output line-to-line voltage v_{AB}, whose fundamental-frequency component $v_{AB,1}$ has a linear relationship with the modulation index m_a.

A method to increase the maximum amplitude of the output fundamental-frequency component is to raise m_a to be greater than unity, starting from where the inverter enters the nonlinear over-modulation region [19]. The over-modulation can cause low-order harmonics and thus is rarely used in practice. To actually increase DC voltage utilization, modified modulation schemes based on injecting zero-sequence signals into the modulating waves can be employed provided that the neutral of the three-phase load is floating. Among all the possible injection methods, only a few have been proved to deliver the best results in terms of obtainable linear range and/or harmonic distortion. Some examples are the third-order harmonic injection, min-max sequence injection, discontinuous PWM schemes (DPWMs), and injection scheme equivalent to the conventional space vector modulation (SVM) [20]. Fig. 2.8A exemplifies the 1/6 coefficient third-harmonic injection, which can effectively expand the linear range of the conventional SPWM by 15.5%. However, third-harmonic injection method requires an accurate synchronization to exactly inject a third-harmonic waveform in the voltage references. This requires the introduction of phase-locked loop (PLL) systems. A simplified zero-sequence injection method is the well-known min-max injection method, which avoids the use of a PLL. In this case, as observed in Fig. 2.8B, the zero-sequence waveform to be subtracted to the reference voltage is determined by an average of the minimum

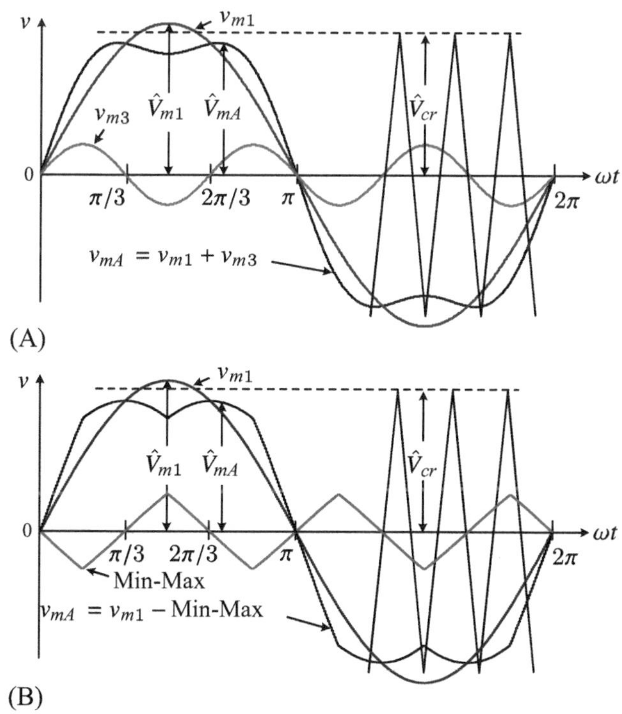

FIG. 2.8 Zero-sequence injection PWM scheme: (A) third-order harmonic injection and (B) Min-Max sequence injection.

and maximum values of the phase voltages leading to the calculation of the modified reference voltage as follows (phase A example):

$$v_{\mathrm{mA}} = v_{\mathrm{m1}} - \frac{\mathrm{MIN}[v_{\mathrm{m1}}, v_{\mathrm{m2}}, v_{\mathrm{m3}}] + \mathrm{MAX}[v_{\mathrm{m1}}, v_{\mathrm{m2}}, v_{\mathrm{m3}}]}{2} \tag{2.3}$$

2.3.2.3 Carrier-Based PWM Techniques: Multicarrier PWM for Multilevel VSI

The conventional SPWM schemes for the two-level inverters can be extended to modulate multilevel VSIs [21] by using a multiple number of carrier waves. Depending on the arrangement of the carriers, the multicarrier SPWM methods are generally divided into two categories: phase-shifted modulation (PS-PWM) and level-shifted modulation (LS-PWM) [13].

The PS-PWM scheme is particularly suited for the cascaded H-bridge (CHB), flying capacitor (FC), or modular multilevel converters (MMC) with limited number of power cells [22]. All these converter topologies are composed of modular power cells. As the name indicates, a phase shift exists between any two adjacent carriers in the PS-PWM such that they are evenly distributed in the modulation plane. The phase shift to be applied in adjacent power cells depends on the topology of the power cells. If the power cells are full bridges, the phase shift to be applied is 180 degrees divided by the number of cells per phase. If the power cells are half-bridges, the corresponding phase shift is equal to 360 degrees divided by the number of cells per phase.

Since the carriers are responsible for generating the gating signals in different power cells, and the power cells are connected and collectively supply the load voltage, the phase shifts in the carriers eventually result in the multilevel waveform shape in the inverter's output. The operation principle of PS-PWM for a three-cell CHB inverter is exemplified in Fig. 2.9. It is worth noting that due to the even distribution of the carriers, the load power is balanced among all contributing cells, and all the power switches are switched at the same frequency (f_{cr}). In addition, as the carrier phase shifting method introduces a multiplicative effect, for an m-level inverter, the switching frequency of the inverter $f_{\mathrm{sw,\ inv}}$ is $2m - 1$ times that of the power switches' switching frequency. This phenomenon can be also observed in Fig. 2.9 where the harmonic spectrum of the output voltage is represented and the harmonic distortion is located around 6 kHz with f_{cr} equal to 1 kHz.

Unlike the PS-PWM that shifts the carriers horizontally, the LS-PWM obtains its name by vertically shifting the levels of the carriers. For any m-level inverter, $m - 1$ carriers are required to compare with the sinusoidal reference. The multicarrier LS-PWM has several variations depending on the phase disposition (PD) of the carriers: (a) PD, where all carriers are in phase; (b) alternative phase opposite disposition (APOD), where all carriers are alternatively in opposite disposition; and (c) phase opposite disposition (POD),

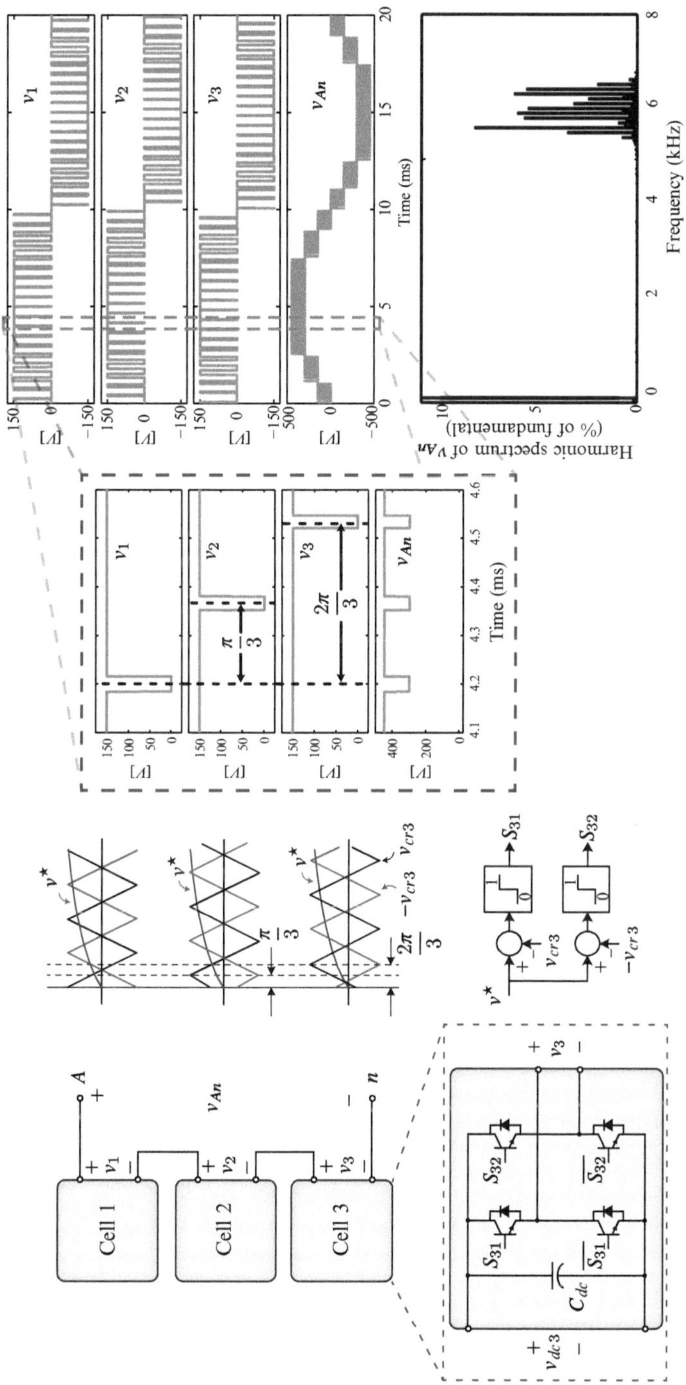

FIG. 2.9 Phase-shifted PWM scheme for a three-cell CHB converter with f_{cr} equal to 1 kHz.

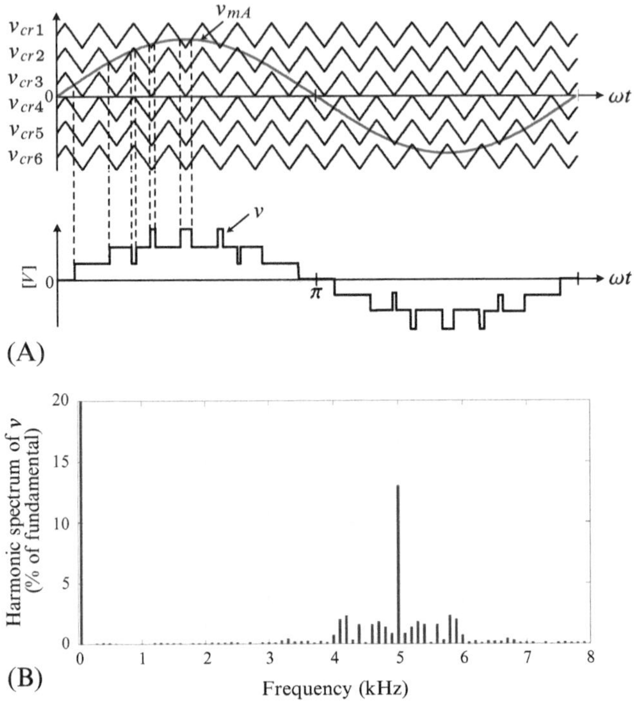

FIG. 2.10 (A) Level-shifted PWM scheme for a seven-level converter with f_{cr} equal to 750 Hz. (B) Harmonic spectrum of the obtained output voltage if f_{cr} is fixed to 5 kHz.

where all carriers above zero are in phase but in opposition with those below zero. The PD modulation scheme provides the best harmonic profile of all the three schemes [23]. The LS-PWM can be applied to any multilevel topology such as the neutral-point clamped (NPC) converter. An example of PD LS-PWM for seven-level inverter is given in Fig. 2.10.

2.3.2.4 SVM for Two-Level Three-Phase Inverters

SVM considers a complex voltage vector as the reference waveform to be generated. This reference signal is sampled with a constant frequency and the converter generates it using a linear combination of possible switching states that can be achieved by the power converter. A switching sequence formed by several switching states of the converter is performed and the average value of the output voltage must coincide with the desired reference one [7,8].

The first step to face the SVM problem and to determine the switching sequence and the time of each switching state is to make a graphical representation of the output voltages of all the discrete switching configurations of the power converter. In the two-level three-phase VSI case (represented in Fig. 2.6), only eight possible switching states are possible depending on the values of S_1, S_2, and S_3. A summary of the possible switching states and corresponding

TABLE 2.3 Switching States of the Three-Phase Two-Level Power Converter

S_1	S_2	S_3	V_{a0}/V_{DC}	V_{b0}/V_{DC}	V_{c0}/V_{DC}	$V_{\alpha n}/V_{DC}$	$V_{\beta n}/V_{DC}$	$V_{\gamma n}/V_{DC}$
0	0	0	− 1/2	− 1/2	− 1/2	0	0	0
1	0	0	1/2	− 1/2	− 1/2	$2/\sqrt{6}$	0	0
1	1	0	1/2	1/2	− 1/2	$1/\sqrt{6}$	$1/\sqrt{2}$	0
0	1	0	− 1/2	1/2	− 1/2	$-1/\sqrt{6}$	$1/\sqrt{2}$	0
0	1	1	− 1/2	1/2	1/2	$-2/\sqrt{6}$	0	0
0	0	1	− 1/2	− 1/2	1/2	$-1/\sqrt{6}$	$-1/\sqrt{2}$	0
1	0	1	1/2	− 1/2	1/2	$1/\sqrt{6}$	$-1/\sqrt{2}$	0
1	1	1	1/2	1/2	1/2	0	0	0

phase voltages (in *abc* frame) and phase to neutral voltages (in $\alpha-\beta$ frame) is shown in Table 2.3. It has to be noticed that most of SVM methods consider the well-known $\alpha-\beta$ frame to develop the modulation method but other options are also possible [14,15]. It can be also noticed that two state vectors (000 and 111) are placed in the same location. These state vectors are named redundant vectors and they are completely equal seen from the load viewpoint. The redundant vectors generate exactly the same line voltages but they can be chosen in order to improve extra features of the power converter.

The switching states of Table 2.3 can be represented in $\alpha-\beta$ frame as can be observed in Fig. 2.11, dividing the control region of the power converter in six different sectors. If a reference voltage $v^\star$ has to be generated, firstly the sector where it is located has to be found. This can be easily determined by studying

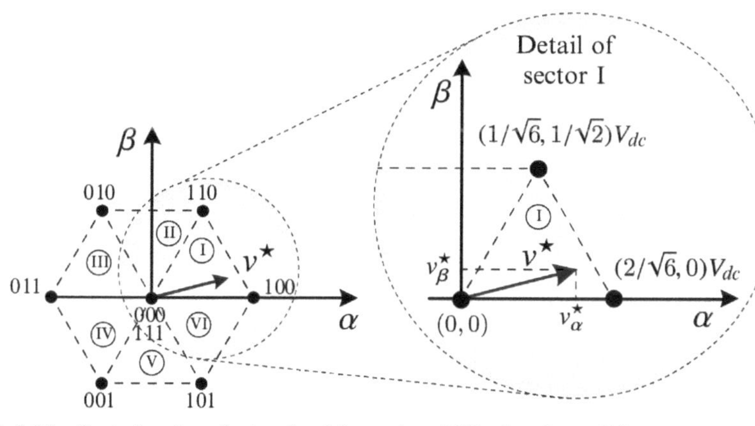

FIG. 2.11 Control region of a two-level three-phase VSI using the $\alpha-\beta$ frame.

the values of its coordinates ($v_\alpha^\star$, $v_\beta^\star$). Once the sector where $v^\star$ is located, the switching states of the switching sequence are correspondingly determined because they are the nearest switching states to $v^\star$. Then, a simple geometrical calculation has to be carried out in order to obtain the corresponding duty cycle values (d_1, d_2, and d_3).

As an example, the reference voltage $v^\star$ in Fig. 2.11 is located in sector I. Thus the geometrical calculation to obtain the duty cycles can be done as follows:

$$\begin{gathered} v_\alpha^\star = \frac{1}{T_s}\left[d_1 T_s \frac{2}{\sqrt{6}} v_{dc} + d_2 T_s \frac{1}{\sqrt{6}} v_{dc}\right] \\ v_\beta^\star = d_2 T_s \frac{1}{\sqrt{2}} v_{dc} \\ d_1 + d_2 + d_3 = 1 \end{gathered} \tag{2.4}$$

And finally, the duty cycles can be determined:

$$\begin{gathered} d_2 = \frac{\sqrt{2} v_\beta^\star}{v_{dc}} \\ d_1 = \sqrt{3/2}\frac{v_\alpha^\star}{v_{dc}} - \frac{d_2}{2} \\ d_3 = 1 - d_1 - d_2 \end{gathered} \tag{2.5}$$

An offline calculation can be repeated for all six sectors obtaining the equations that can be used to determine the duty cycles for all cases.

2.3.2.5 SVM for Multilevel Three-Phase Inverters

The control region of multilevel power converters can be determined following similar calculations that have been done for the two-level converter. For instance, the control region formed by the switching states of a three-level converter is represented in the α–β frame and the *abc* frame in Fig. 2.12. In this

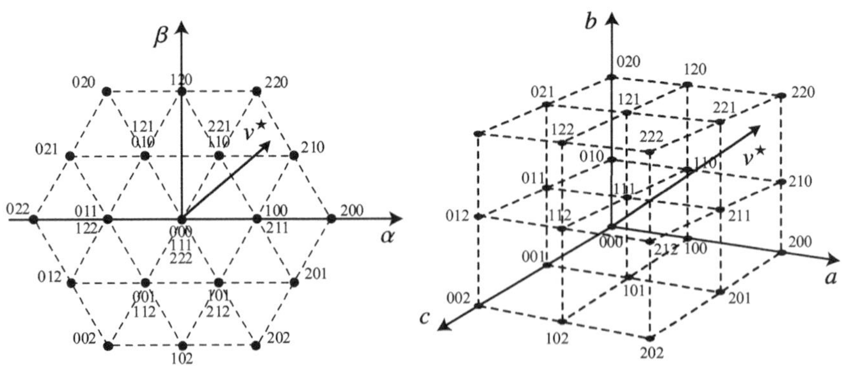

FIG. 2.12 Control region of a two-level three-phase VSI using the α–β frame and the *abc* frame.

case, the switching sequence is also formed by the nearest switching states to the reference voltage $v^{\star}$ (three states considering the α–β frame and four states considering the *abc* frame). Both methods are actually the same with different focus to face the usage of the redundant switching states. In fact, it can be noticed that SVM methods based on the *abc* frame are superior since they can be applied to systems with neutral connection (where the γ component is not necessarily zero) [24].

As the number of levels of the power converter increases, the number of possible switching states also highly increases. However, some SVM methods have been developed with much reduced computational cost in both α–β and *abc* frames. Anyway, it has to be affirmed that the choosing of the best redundant states to finally form the switching sequence is still nonstraightforward. A good design of the switching sequence and the redundant states to be used can lead to achieve minimum power losses, reduced output ripples or elimination of common-mode voltages [22].

2.3.2.6 PWM Versus SVM for Power Inverters

At first sight, PWM and SVM are different modulation methods for power converters. However, and after a deep study, it can be affirmed that they are two different ways to face the same problem that can lead to exactly the same result. Both methods can obtain the same gate signals by controlling the zero-sequence in PWM or modifying the switching sequence using the corresponding redundant switching states in SVM. Consequently, there is an unique correspondence between PWM and SVM obtaining with both techniques the same performance [25].

2.3.3 Optimal PWM

Optimal PWM techniques are typically used in high-power low-switching frequency applications such as high-voltage direct current transmission and motor drives. Such techniques use a precalculated switching pattern (Fig. 2.13A), which has been derived through the solution of a system of nonlinear trigonometrical equations in order to meet certain harmonic requirements. In its most common implementation, the switching waveforms possess quarter-wave (QW) symmetry; this assumption simplifies both the mathematical formulation and the acquisition of solutions at the cost of a restricted solution space. The target switching frequency is translated to a number of commutations per period (m) that determine the control over the output harmonic performance that can be achieved by a given waveform [16].

The most common among all optimal PWM techniques is Selective Harmonic Elimination (SHE-PWM). In SHE-PWM, the exact instant of the commutations within a fundamental period (or switching angle, α_m) is calculated so that the fundamental frequency component is regulated to a certain amplitude

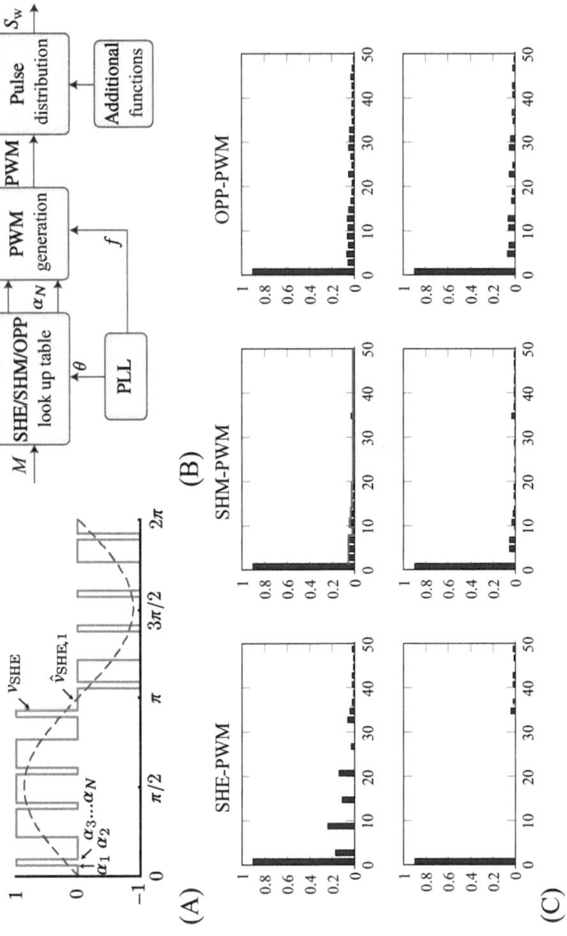

FIG. 2.13 Optimal PWM: (A) normalized three-level waveform, (B) implementation of optimal PWM, and (C) harmonic spectra of phase and line voltages for SHE-PWM, SHM-PWM, and OPP-PWM.

while $(m-1)$ harmonics (typically of low-order harmonic frequencies) are also eliminated from the output voltage spectrum. Solutions are calculated for the whole range of modulation indexes up to the overmodulation region ($M \approx 1.15$). In n-phase systems, elimination of common-mode harmonics (i.e., triplen harmonics in three-phase systems) can be omitted (Fig. 2.13C) in the problem formulation leading to a reduction in the switching frequency of any converter for a given output voltage spectrum compared with other modulation schemes.

Since optimal PWM methods allow for individual control of specific harmonics of the output voltage spectrum, the mathematical formulation can be expanded to achieve objectives other than complete elimination of harmonics, more specifically (i) compliance with specific grid codes [26], and (ii) minimization of harmonic metrics such as the total harmonic distortion (%THD or %WTHD) of the output voltage or current including minimization of peak current amplitudes [27]. The first approach is typically referred to in the literature as Selective Harmonic Minimization or SHM-PWM while the second approach is named Optimal Pulse Patterns or OPP-PWM. Their typical harmonic spectra are shown in Fig. 2.13C.

The three different approaches (SHE-PWM, SHE-PMW, or OPP-PWM) may have different goals in harmonic control but share multiple common aspects in their approach and implementation.

- The mathematical formulation is based on the Fourier decomposition of the periodical switching waveform (applicable both in two-level and multilevel converters) for the fundamental frequency and higher-order harmonics. Assuming a QW symmetrical waveform, the optimal PWM equations can be written as

$$b_n = \frac{4}{n\pi}\sum_{i=1}^{N}(-1)^k \cos(n\alpha_i)$$

  where n is the harmonic order, $i \in \{1, \ldots, N\}$ are the switching angles of the waveform, and k is defined by the pattern of the switching waveform as:

$$k = \begin{cases} 1, & \text{for rising edges} \\ 0, & \text{for falling edges} \end{cases}$$

- The assumption of QW symmetry simplifies the mathematical formulation as the even harmonics and certain Fourier coefficients become zero, but a greater number of solutions can be acquired by relaxing the symmetry requirements and considering half-wave (HW) symmetrical or nonsymmetrical waveforms. This can be particularly useful in two-level converters, becoming less practical as the number of output voltage levels increases.
- Calculation of the switching angles requires solution of a nonlinear transcendental system of equations considering a number of constraints. For simpler problems (low number of switching angles and voltage levels) this

can be done with iterative numerical methods such as Newton-Raphson, resultants theory, and Grobner algebra. However, as the problem expands these analytical methods become complicated and optimization algorithms such as Genetic Algorithms (GAs), linear programming, Differential Evolution, and Particle Swarm Optimization (PSO) [16].

- Each of the problems typically exhibits multiple solutions and a meta-analysis of the acquired solutions based on additional criteria is typically required to select the optimal pattern for a given application.
- Additional functions (e.g., voltage balancing, circulating current control for multilevel converters) may be implemented in the modulation stage either by variations in the switching angles or through generation of the gate signals as shown in Fig. 2.13C.

The main advantages of SHE-PWM lie in the tight control of the harmonic spectrum resulting in the elimination of low-order harmonics, reduction in the switching frequency, and relatively simple implementation based on precalculated switching angles and look-up tables (LUT). OPP-PWM provides optimal performance with regard to specific harmonic goals and is well suited for low-speed motor drive applications. SHM-PWM, on the other hand, is well suited for grid-connected applications due to its inherent formulation to exactly meet grid codes, which can be achieved for a greater number of harmonics compared with SHE-PWM for a given number of transitions.

As all of the optimal PWM methods rely on offline precalculation of the switching patterns, their closed-loop performance is typically slower than that of other PWM methods (SPWM, SVM, etc.). In order to improve this aspect, closed-loop SHE-PWM methods based on real-time calculation of the switching harmonics [28], extraction of the fundamental reference harmonic [18, 27], and more recently model-predictive-based SHE-PWM [17] have been demonstrated in the literature.

2.4 CONTROL OF POWER CONVERTERS

2.4.1 PWM-Based Linear Control

A widely used control structure to govern linear systems, including power converters, is the well-known linear feedback control loop, which is depicted in Fig. 2.14A. The main target is to design a feedback controller, $C(s)$, to reduce the tracking error, $e = r - y$, by applying a suitable control input, u, to the system $G(s)$. However, this *nominal* target can be affected by disturbances that might compromise the closed-loop performance. In general, the nominal system output, $\bar{y}$, may differ from the actual output, y, due to measurement noise or system uncertainties, represented by η_o. Moreover, the *nominal* input, $\bar{u}$, provided by the controller can be affected by an external disturbance or just by the way it is implemented in practice, which is represented by η_i. Therefore,

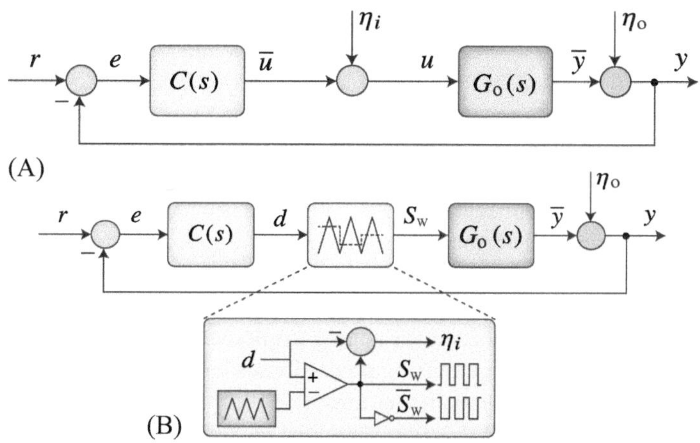

FIG. 2.14 Linear feedback controller: (A) general closed-loop representation and (B) closed-loop including modulator.

special consideration must be given to the design of a feedback controller considering both *reference tracking error performance* and *disturbance rejection capability*.

Considering the closed-loop in Fig. 2.14A, the following input-to-output relationship can be obtained:

$$y(s) = T(s)r(s) + S(s)d_o(s) + S_i(s)d_i(s) \tag{2.6}$$

where $T(s)$ is the *reference-to-output* transfer function that determines the reference tracing error performance, while $S(s)$ and $S_i(s)$ are *sensitivity* transfer functions that determine disturbance rejection. These transfer functions are given by:

$$T(s) = \frac{C(s)G_o(s)}{1 + C(s)G_o(s)} \tag{2.7}$$

$$S(s) = \frac{1}{1 + C(s)G_o(s)} \tag{2.8}$$

$$S_i(s) = G_o(s)S(s) \tag{2.9}$$

Note that $T(s)$ and $S(s)$ are complementary (i.e., $T(s) + S(s) = 1$). For a good reference tracking performance with and effective disturbance rejection, it is desired to obtain $T(s) \approx 1$ and $S(s) \approx 0$. As can be seen in Eqs. (2.7), (2.8), the controller $C(s)$ can affect both reference tracking and disturbance rejection performances. Importantly, it is not possible to influence both control targets independently for the whole frequency spectrum. However, in some cases, it is possible to achieve this ideal condition for a particular frequency range.

For the power converter control case, the output of the linear controller normally provides a duty-cycle, d, which has to be synthesized by a modulator. This is depicted in Fig. 2.14B. As explained in Section 2.3, the modulator offers constant switching frequency (known power losses) with a well-known spectrum in the steady state. However, from the control viewpoint, the actual input applied to the system, given by the state of the power switches $u = S_w$, differs from the nominal input provided by the controller, $\bar{u} = d$. This mismatch can be interpreted as an input disturbance given by $\eta_i = S_w - d$. Therefore, if the closed loop presents a high bandwidth, this input disturbance, η_i, can affect the tracking performance, since the controller will tend to compensate the modulation noise, η_i. This is particularly important for analog implementations where extra commutations can be induced, increasing power losses and harmonic distortion. On the other hand, in digital implementations, the controller sampling is normally synchronized with the PWM carrier. Thus, the closed-loop bandwidth is intrinsically limited by the PWM frequency. Moreover, due to this synchronization, the sampled measurements of the system's outputs (current, power, torque, etc.), resemble its average value. As a result, average models of power converters and loads (i.e., $G(s)$) can be used to design a linear feedback controller.

2.4.1.1 PI Controller

Despite the fact that several linear controllers have been proposed in the literature to govern power converters, the most widely used controller is the so-called proportional-integral (PI) controller. This traditional controller presents the following continuous-time expression:

$$u(t) = K_p e(t) + K_i \int e(t)\, dt \tag{2.10}$$

while its equivalent frequency domain representation is given by:

$$C_{PI}(s) = K_p + \frac{K_i}{s} = K_p\left(\frac{1 + sT_i}{sTi}\right) = K_p\left(1 + \frac{\tau_i}{s}\right) \tag{2.11}$$

where $T_i = K_p/K_i$ and $\tau_i = 1/T_i$. Based on Eq. (2.11), two standard implementations of this conventional controller are depicted in Fig. 2.15.

Basically, a proper value for K_p and T_i is required in order to achieve a desired closed-loop performance. The next part of this section details the design of such parameters by considering a common case: first-order systems.

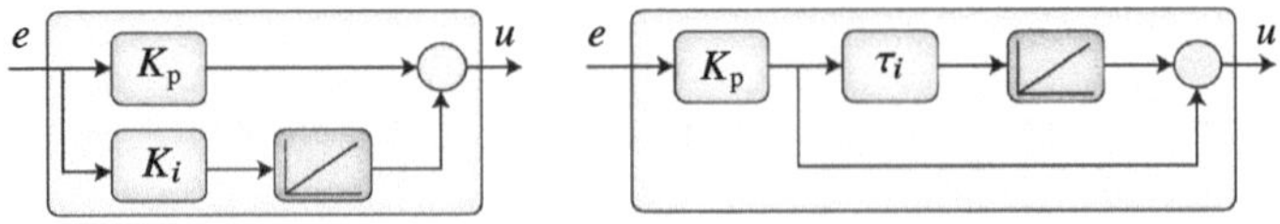

FIG. 2.15 Standard PI controller structures.

Design Criteria of PI Controllers for First-Order Systems

There are several cases where power converters and electrical drives can be controlled by considering one or more first-order systems. For instance, the AC-current of a rectifier connected to the grid through an inductor and its DC-link can be modeled as two cascaded first-order systems. Same situation happens with the speed and current in a field-oriented control strategy for induction motors. Therefore, a detailed analysis and design of a PI controller for first-order systems is given in this section. For higher order systems, the interested reader is referred to [29]. If the plant is modeled as:

$$G_o(s) = \frac{K_o}{\tau_o s + 1} \tag{2.12}$$

then, based on Eq. (2.7), the closed-loop transfer function for a first-order system governed by a PI controller is expressed as:

$$T(s) = \frac{\dfrac{K_o K_p}{\tau_o T_i}(T_i s + 1)}{s^2 + \left(\dfrac{1 + K_o K_p}{\tau_o}\right) s + \dfrac{K_o K_p}{\tau_o T_i}} \tag{2.13}$$

This closed-loop transfer function results in a second-order system with one zero. The poles of the closed-loop system can be chosen arbitrarily (pole placement) by selecting the controller parameters K_p and T_i. To facilitate this task, it is convenient to reparameterize the closed-loop characteristic equation as:

$$s^2 + 2\zeta\omega_n s + \omega_n^2 \tag{2.14}$$

where ζ is the relative damping and ω_n its natural frequency. The resulting two poles for the closed-loop transfer function, *T*(*s*), are given by:

$$p_{1,2} = -\zeta\omega_n \pm \omega_n\sqrt{\zeta^2 - 1} \tag{2.15}$$

while its closed-loop time constant is given by:

$$\tau_D = \frac{1}{\zeta\omega_n} \tag{2.16}$$

Therefore, instead of selecting the controller parameters K_p and T_i, the closed-loop poles can be chosen by selecting ζ and ω_n. By comparing Eq. (2.14) with Eq. (2.13), the following relationship can be established:

$$\omega_n^2 = \frac{K_o K_p}{\tau_o T_i} \tag{2.17}$$

$$\zeta = \frac{1 + K_o K_p}{2\omega_n \tau_o} \tag{2.18}$$

Therefore, the PI tuning parameters K_p and T_i can be determined by:

$$T_i = \frac{2\zeta\omega_n\tau_o - 1}{\omega_n^2\tau_o} = \frac{2\zeta}{\omega_n} - \frac{1}{\omega_n^2\tau_o} \tag{2.19}$$

$$K_p = \frac{2\zeta\omega_n\tau_o - 1}{K_o} = \frac{\omega_n^2\tau_o}{K_o}T_i. \tag{2.20}$$

Moreover, the closed-loop bandwidth of $T(s)$ in Eq. (2.13) is given by:

$$\omega_{BW} = \mu \cdot \omega_n \tag{2.21}$$

where

$$\mu = \sqrt{1 + 2\zeta^2 + \sqrt{4\zeta^4 + 4\zeta^2 + 2}} \tag{2.22}$$

To impose a desired closed-loop performance, it is first recommended to choose the damping factor ζ. This is normally chosen as $\zeta = 0.707$ or higher. To obtain positive PI parameters, it is necessary to constraint the natural frequency to $\omega_n > \frac{1}{2\zeta\tau_o}$. On the other hand, as per Eq. (2.21), the controller response is faster if ω_n is larger. Then, from Eq. (2.19), it follows that:

$$T_i \approx \frac{2\zeta}{\omega_n} \tag{2.23}$$

for large ω_n. This implies that T_i is independent of the system parameters, which gives the controller some robustness against parameter uncertainties. Nonetheless, for analog implementations, it is necessary to limit the closed-loop bandwidth, $f_{BW} = \omega_{BW}/(2\pi)$, to half of the PWM frequency, f_{cr}. Otherwise, the PI controller will tend to compensate the modulation error η_i, which may introduce extra switch commutations. Therefore, a good range for ω_n for analog implementations is given by:

$$\frac{1}{2\zeta\tau_o} < \omega_n \leq \frac{\pi}{\mu} f_{cr} \tag{2.24}$$

Finally, the reference-to-input and sensitivity transfer functions can be expressed in terms of ζ and ω_n as follows:

$$T(s) = \frac{(2\zeta\omega_n - 1/\tau_o)s + \omega_n^2}{s^2 + 2\zeta\omega_n s + \omega_n^2} \tag{2.25}$$

$$S(s) = \frac{s(s + 1/\tau_o)}{s^2 + 2\zeta\omega_n s + \omega_n^2} \tag{2.26}$$

Based on these transfer functions, it is possible to conclude that for $s = 0$, it follows that $T(0) = 1$ and $S(0) = 0$. This implies that the PI controller can perfectly track constant references as well as reject constant disturbances. It is important to highlight that both $T(s)$ and $S(s)$ have the same poles. Thus, the tuning of the PI controller affects both features.

Antiwindup Scheme

In power electronics applications, actuators are normally limited, for example, duty cycle in a modulator. In this case, if the controller output takes a value larger than the actuator limit, the integral action is likely to build up to a large value. Then, when the input comes back to the linear zone, a large transient will occur. This effect is called wind-up. Therefore, to avoid undesired closed-loop behaviors that could damage a power converter or electrical drive, it is necessary to take some actions against the negative effects of wind-up. In the control literature, there exist several antiwindup schemes that can help to prevent this situation. One of the simplest antiwindup schemes, known as back-calculation, is depicted in Fig. 2.16. This can be easily realized in analog implementations. First, it is possible to see that the saturation effect is taken into account by including the saturation error:

$$e_{\text{sat}}(t) = u(t) - v(t) = \text{sat}(v(t)) - v(t) \tag{2.27}$$

Then, this error is feed-backed to unwind the integral action. Notice that this antiwindup scheme has no effect over the controller output, $u(t)$, if the unsaturated PI output, $v(t)$, is within the linear range (i.e., $e(t) = 0$). For more sophisticated antiwindup schemes, see e.g., [30].

As previously mentioned, the PI controller can perfectly track constant references. However, in the case of inverters, it is required to track sinusoidal references. To address this important issue, the next part of this section will be devoted to the design of linear AC-compensators.

2.4.1.2 Resonant Controller

One accepted solution to control inverters is to work in a synchronous $d-q$ frame. Thus the sinusoidal tracking problem is translated into a regulation (constant reference) problem. Another approach is to transform the PI controller, $C_{\text{PI}}(s)$ in Eq. (2.11), into an AC-compensator working at ω_o, also known as resonant controller, $C_{\text{RE}}(s)$ [31]. As a result, the controller can work in a stationary frame without requiring transformations. To do this, the integral action is replaced by the following transformation:

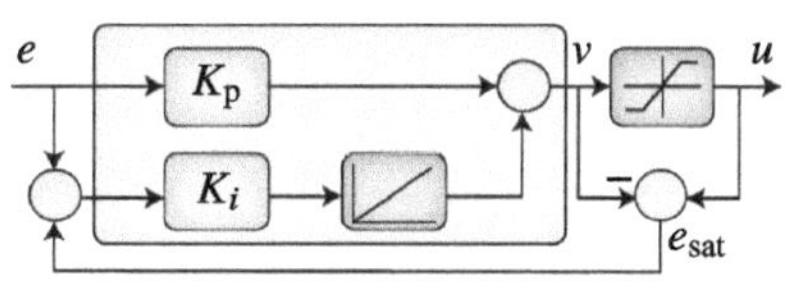

FIG. 2.16 PI controller with antiwindup strategy.

$$\frac{1}{s} \rightsquigarrow \frac{2s}{s^2+\omega_o^2} \tag{2.28}$$

Therefore, the resonant controller can be represented as:

$$C_{\mathrm{RE}}(s) = K_{\mathrm{p}} + \frac{2K_i s}{s^2+\omega_o^2} \tag{2.29}$$

Finally, the reference-to-input and sensitivity transfer functions can be expressed as follows:

$$T(s) = \frac{K_{\mathrm{p}}K_{\mathrm{o}}s^2 + 2K_iK_{\mathrm{o}}s + K_{\mathrm{p}}K_{\mathrm{o}}\omega_o^2}{(s^2+\omega_o^2)(\tau_{\mathrm{o}}s+1) + K_{\mathrm{p}}K_{\mathrm{o}}s^2 + 2K_iK_{\mathrm{o}}s + K_{\mathrm{p}}K_{\mathrm{o}}\omega_o^2} \tag{2.30}$$

$$S(s) = \frac{(s^2+\omega_o^2)(\tau_{\mathrm{o}}s+1)}{(s^2+\omega_o^2)(\tau_{\mathrm{o}}s+1) + K_{\mathrm{p}}K_{\mathrm{o}}s^2 + 2K_iK_{\mathrm{o}}s + K_{\mathrm{p}}K_{\mathrm{o}}\omega_o^2} \tag{2.31}$$

In this case, it follows that for $s = j\omega_o$, $T(j\omega_o) = 1$, and $S(j\omega_o) = 0$. This implies that the resonant controller, $C_{\mathrm{RE}}(s)$ in Eq. (2.29), can perfectly track as well as reject sinusoidal references and disturbances with frequency ω_o.

To design a resonant controller for a plant that can be modeled as a first-order system, the same design procedure presented above for the PI controller can be followed. Nevertheless, it is important to take into account is that a damping factor less than one can induce unwanted oscillations that can affect the closed-loop performance. Therefore, in this case, it is recommended to choose $\zeta \geq 1$.

2.4.1.3 Digital Implementation of PI and Resonant Controllers

Even though some applications still use analog control, most of modern power converters are currently governed by digital controllers. This comes from the fact that the cost of digital platforms has significantly decreased while the calculation power and speed has considerably increased [32]. Moreover, digital controllers are more flexible than their analog counterpart. Basically, extra features such as maximum power point track algorithms, faster and more accurate PLLs, sensorless drive strategies, etc., can be easily added to the controller in the same digital platform.

In Fig. 2.17, a standard digital control implementation of a PWM-based linear controller is depicted. Notice that both controller and PWM stages are realized inside the digital platform. Measurements are transformed into digital form by an analog-to-digital converter (ADC). In general, analog measurements are firstly passed trough a low-pass filter to reduce high-frequency noise. The most popular digital platforms for power converters are Field Programmable Gate Arrays (FPGAs) and Digital Signal Processors (DSPs). At times, a combination of both is used, where an FPGA is utilized to handle the peripherals (ADCs and modulator), while the controller is implemented in a DSP.

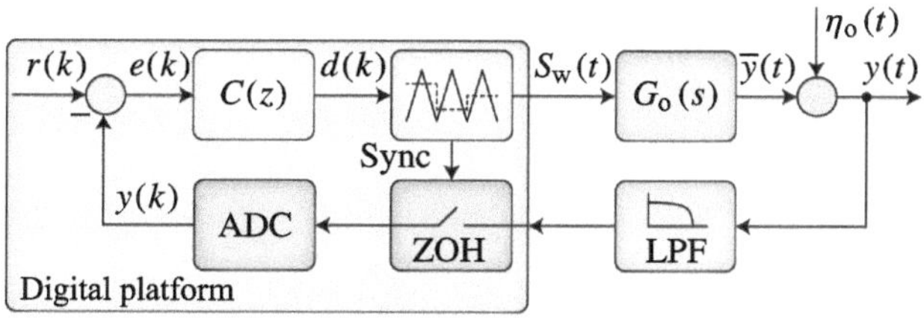

FIG. 2.17 Digital implementation of a PWM-based linear controller.

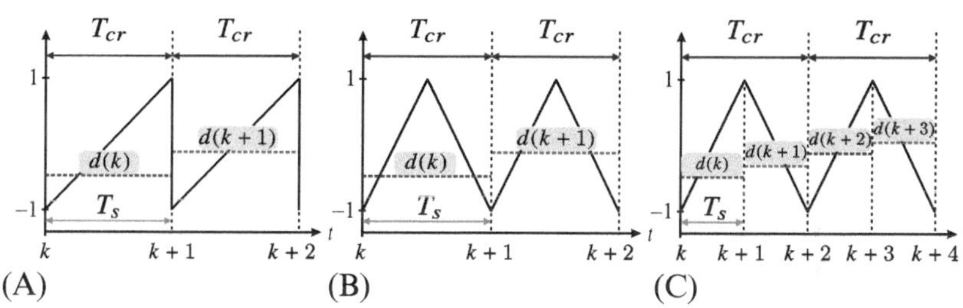

FIG. 2.18 PWM-based sampling synchronization: (A) asymmetric PWM, $f_s = f_{cr}$; (B) single update symmetric PWM, $f_s = f_{cr}$; and (C) double update symmetric PWM, $f_s = 2f_{cr}$.

In digital implementations, the PWM stage is not only used to synthesize the controller actuation, but also to synchronize the whole digital process, including measurements, control algorithm executions, and actuations. Therefore, the sampling time, T_s, depends on the PWM carrier period T_{cr}. This feature is shown in Fig. 2.18. In case of using an asymmetric carrier, both PWM and controller will have the same sampling frequency; see Fig. 2.18A. On the other hand, for symmetric carriers, the update mode can also modify the sampling frequency. In case of using a single update mode as in Fig. 2.18B, the sampling frequency will be the same as the PWM one, $f_s = f_{cr}$. Nevertheless, when a double update mode (sampling at both edges of the triangular carrier) is selected, the sampling frequency will double the carrier frequency, $f_s = 2f_{cr}$ as shown in Fig. 2.18C. Here, it is important to emphasize that duty cycles, $d(k)$, remain constant during the whole sampling time. Thus, no extra commutations are introduced.

A linear controller can be directly designed in discrete-time [33]. However, the continuous-time plant model, $G(s)$, has to be transformed into an equivalent discrete-time plant; see, e.g., [34]. Alternatively, the controller can be designed in the original s-domain obtaining, thus, $C(s)$. Later, this is transformed into an equivalent controller in z-domain to obtain $C(z)$. If the second option is adopted, then the design procedure takes advantage of the continuous-time design procedure presented above. Thus, the continuous-time PI controller, $C_{PI}(s)$, can be transformed into discrete-time by applying a zero-order hold approximation to Eq. (2.11), resulting in:

$$C_{\text{PI}}(z) = \frac{K_a z + K_b}{z - 1} = \frac{K_a + K_b z^{-1}}{1 - z^{-1}} \tag{2.32}$$

where

$$K_a = K_{\text{p}}, \quad \text{and} \quad K_b = K_i T_{\text{s}} - K_{\text{p}} \tag{2.33}$$

In addition, the PI controller can be digitally implemented by executing at each sampling instant, k, the following discrete-time polynomial equation:

$$u(k) = K_a e(k) + K_b e(k-1) + u(k-1) \tag{2.34}$$

Similarly, the continuous-time resonant controller, $C_{\text{RE}}(s)$ in Eq. (2.29), results in

$$C_{\text{RE}}(z) = \frac{a_0 z^2 + a_1 z + a_2}{z^2 - b_1 z + 1} = \frac{a_0 + a_1 z^{-1} + a_2 z^{-2}}{1 - b_1 z^{-1} + z^{-2}} \tag{2.35}$$

where

$$\begin{aligned} a_0 &= K_{\text{p}}, \quad b_1 = 2\cos(\omega_{\text{o}} T_{\text{s}}) \\ a_1 &= 2\left(\frac{K_i}{\omega_{\text{o}}}\sin(\omega_{\text{o}} T_{\text{s}}) - K_{\text{p}}\cos(\omega_{\text{o}} T_{\text{s}})\right) \\ a_2 &= K_{\text{p}} - \frac{2K_i}{\omega_{\text{o}}}\sin(\omega_{\text{o}} T_{\text{s}}) \end{aligned} \tag{2.36}$$

Then, the resonant controller can be digitally implemented by solving at each sampling instant, k, the following discrete-time polynomial equation:

$$u(k) = a_0 e(k) + a_1 e(k-1) + a_2 e(k-2) + b_1 u(k-1) - u(k-2) \tag{2.37}$$

Notice that these discretizations depend on the sampling time, T_{s}, which in turn depends on the PWM carrier period T_{cr}. To be able to obtain a good correlation between the continuous-time controller form, $C(s)$, and its discrete-time equivalent, $C(z)$, a Nyquist criterion of the sampling frequency, $f_{\text{s}} = 1/T_{\text{s}}$, has to be imposed over the closed-loop time constant, τ_{D} in Eq. (2.16), which leads to:

$$T_{\text{s}} \leq \frac{\tau_{\text{D}}}{2} = \frac{1}{2\zeta\omega_n} \tag{2.38}$$

Therefore, the range of possible values for ω_n in Eq. (2.24) needs to be replaced by the following range for the discrete-time case:

$$\frac{1}{2\zeta\tau_{\text{o}}} < \omega_n \leq \frac{f_{\text{s}}}{2\zeta} \tag{2.39}$$

Finally, both PI and resonant controllers can be digitally implemented in a DSP or FPGA by adopting Algorithms 2.1 and 2.2, respectively. Notice that a digital antiwindup scheme is also included for both cases. Here, it is considered that the control input is limited to $u \in [-1, 1]$. If the carrier runs in a different range, these limits have to be changed accordingly.

ALGORITHM 2.1 Digital PI Controller $C_{PI}(z)$

```
Initialization at k = 0 : e(k − 1) ← 0, u(k − 1) ← 0
  function u(k) = PI(r, y(k), K_a, K_b)              ▷ K_a, K_b as per Eq. (2.33)
    e(k) ← r − y(k)
    u(k) ← K_a e(k) + K_b e(k − 1) + u(k − 1)        ▷ See Eq. (2.34)
    if −1 ≤ u(k) ≤ 1 then                            ▷ Antiwindup
      u(k) ← ±1                                      ▷ Saturation
    end if
    Memory: e(k − 1) ← e(k), u(k − 1) ← u(k)
  end function
```

ALGORITHM 2.2 Digital Resonant Controller $C_{RE}(z)$

```
Initialization at k = 0 : e(k − 1) ← 0, e(k − 2) ← 0
                          u(k − 1) ← 0, u(k − 2) ← 0
  function u(k) = RE(r, y(k), a_0, a_1. a_2, b_1)    ▷ a_0, a_1. a_2, b_1 as per Eq. (2.36)
    e(k) ← r − y(k)
    u(k) ← a_0 e(k) + a_1 e(k − 1) + a_2 e(k − 2) + b_1 u(k − 1) − u(k − 2)
                                                     ▷ See Eq. (2.37)
    if −1 ≤ u(k) ≤ 1 then                            ▷ Antiwindup
        u(k) ← ±1
        e(k) ← 0
        e(k − 1) ← 0
    end if
    Memory: e(k − 2) ← e(k − 1), e(k − 1) ← e(k)
            u(k − 2) ← u(k − 1), u(k − 1) ← u(k)
  end function
```

Recommended Design Procedure

The above analysis can be summarized in the following recommended design procedure for first-order plants:

1. Select ζ according to:

$$\zeta \geq \begin{cases} 0.707 & \text{for PI controllers} \\ 1 & \text{for resonant controllers} \end{cases}$$

   Start first with one of the above-recommended values. Increase it to reduce unwanted oscillations.

2. Select ω_n according to:

$$\omega_n \leq \begin{cases} \dfrac{\pi}{\mu} f_{\text{cr}} & \text{for analog implementations} \\ \dfrac{f_{\text{s}}}{2\zeta} & \text{for digital implementations} \end{cases}$$

Start first with one of the above-recommended values to obtain a fast response. Reduce it in case of observing excessive noise. Here, $f_s = f_{cr}$ or $f_s = 2f_{cr}$, for single or double update mode, respectively.

3. Compute controller parameters

$$T_i = \frac{2\zeta}{\omega_n},\quad K_p = \frac{\omega_n^2 \tau_o}{K_o} T_i,\quad K_i = \frac{K_p}{T_i}$$

Example 2.1(Design of a Single-Phase Resonant Current Controller)
Consider the single-phase system in Fig. 2.19A. The HB power converter has the topology depicted in Fig. 2.4. The variable v_L represents a grid or back-emf voltage with frequency ω_o. The converter switches are handled by an unipolar PWM stage, as described in Fig. 2.5b, with a carrier frequency of $f_{cr} = 2$ kHz. The target is to design a digital resonant controller to provide a duty-cycle, $u = d$, to track a sinusoidal reference, $r = i_L^{\star}$. By applying simple circuit analysis and neglecting the modulation noise, the following system model in s-domain is obtained:

$$i_L(s) = \frac{v_{inv}(s) - v_L(s)}{Ls + R} \tag{2.40}$$

$$i_L(s) = \frac{V_{DC}/R}{(L/R)s + 1}\left(d(s) - \frac{v_L(s)}{V_{DC}}\right) \tag{2.41}$$

By comparing with Eq. (2.12), this plant results in a first-order system with $K_o = v_{dc}/R$ and $\tau_o = L/R$. Moreover, the voltage v_L can be considered as an input disturbance with frequency ω_o, that is, $\eta_i = v_L(s)/v_{dc}$ as shown in Fig. 2.19B. The

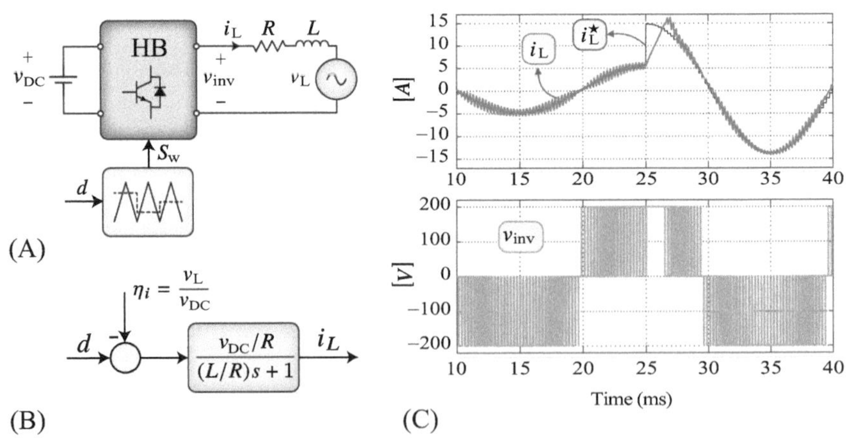

FIG. 2.19 Single-phase resonant current controller: (A) single-phase system, (B) equivalent s-domain model, and (C) closed-loop simulation result waveforms of the load current, i_L, and inverter voltage v_{inv}.

system parameters are: $L = 5$ mH, $R = 2\Omega$, and $v_L = 110$ Vrms with $\omega_o = 2\pi 50$ rad/s.

To tune the controller, the recommended design procedure is followed. First, by considering step 1, the closed-loop dumping is chosen as $\zeta = 1$. Then, a double update mode is adopted, which results in a sampling frequency of $f_s = 2f_{cr} = 4$ kHz. This leads to a natural frequency of $\omega_n = 2000$ rad/s. Following step 3, the controller parameters are chosen as $T_i = 0.001$ and $K_p = 0.1$. Finally, the resonant controller is digitally implemented by invoking Algorithm 2.2, with $a_0 = 0.1$, $a_1 = -0.1494$, $a_2 = 0.0501$, and $b_1 = 1.9938$. Notice that since the resonant controller is tuned with the same frequency than voltage v_L, it can perfectly reject the input disturbance η_i, as discussed in Section 2.4.1.2. A simulation result of a step-changed in the current reference is presented in Fig. 2.19C. Initially, the controller is tracking a sinusoidal current of 50 Hz with amplitude 5 A. Then, at the instant $t = 25$ms, the current amplitude is increased to 15A. These results show that the controller provides a perfect tracking of the sinusoidal reference with a fast response during transients.

2.4.2 Control Without Modulator

The main characteristic of the control of power converters without modulator is that the power converter's gate signals are directly generated as a output of the controller. In this case, the controller is designed to reduce the instantaneous error (sampled or continuous) between control references and control variables, which satisfies one or more particular control objectives. These control methods achieve significant improvements in transient operation at expense of a notorious random behavior in its switching frequency, mainly due to the fact that both fundamental and harmonic components (ripple) of the control variables are not decoupled for control purposes. This section covers some closed-loop control methods that do not require external modulator.

2.4.2.1 Hysteresis

Among hysteresis control methods, the most well-known is the hysteresis current control [4]. The implementation of hysteresis control methods has been done either in analog or digital form. Every time the error between control references and control variables crosses either the positive or negative hysteresis band's boundary, a significant change in the controller's output (S_w) occurs as shown in Fig. 2.20A. Thus the controller quickly reacts to any deviation from control references, which are the reason of the high gain behavior of these controllers. Notice that the digital implementation of hysteresis control methods (see Fig. 2.20B) does not guarantee the control variable ripples to be within the limits of a specified hysteresis band. This situation is illustrated in the zoomed view inside Fig. 2.20A. For that reason, in most of the cases, the

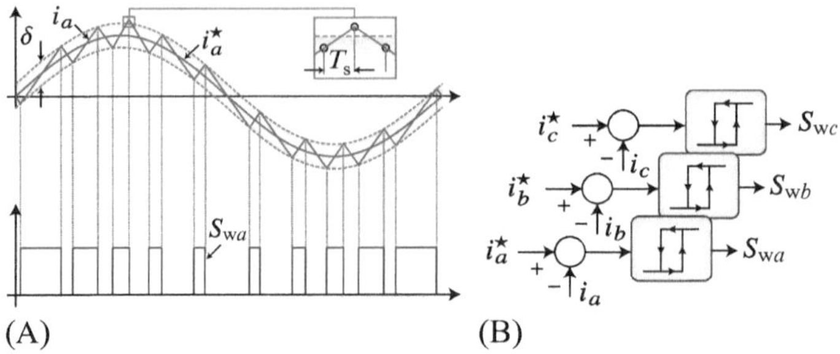

FIG. 2.20 An hysteresis current control for a three-phase two-level power converter: (A) operation principle and (B) block diagram.

operation of digital hysteresis controllers does not match the behavior of the analog ones.

2.4.2.2 Direct Torque Control/Direct Power Control

Direct torque control (DTC) for motor drive applications has been well established in both academia and industry. It offers a simple control structure, fast response, and robust operation [35]. The torque and flux references are tracked using hysteresis controllers and a switching table implemented with LUT is used for selecting the optimum converter's output. Similarly, direct power control (DPC) for grid-connected applications was developed afterward to control the instantaneous active and reactive power directly by selecting the optimum switching state of the converter [36]. Fig. 2.21A illustrates the operation principle of DTC, and Fig. 2.21B its block diagram. Basically, DPC is the analog of DTC considering the instantaneous errors of active and reactive power instead instantaneous errors of torque and flux for reading the LUT.

2.4.2.3 FCS Model Predictive Control

FCS model predictive control (FCS-MPC) belongs to the category of nonlinear control strategies. In general, they are used to enhance the control performance

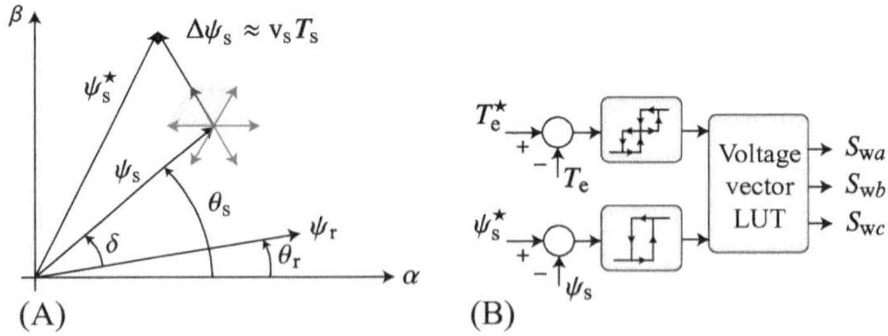

FIG. 2.21 A DTC for a three-phase two-level power converter: (A) operation principle and (B) block diagram.

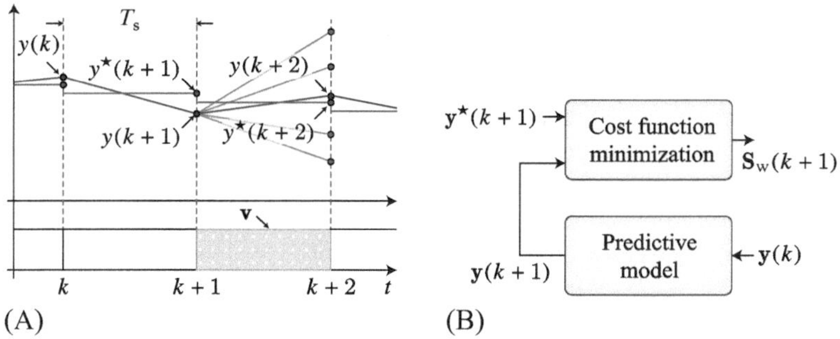

FIG. 2.22 A generalized FCS-MPC for power converters: (A) operation principle and (B) block diagram.

over whole converter's operation range but specially transient conditions. Basically, FCS-MPC needs a system's model to predict the future behavior of the variables of interest ($\mathbf{y}$) and a cost function to define the desired control goals. Each sampling period, the system's model is used to forecast the variables' values up to a certain finite prediction horizon. Then, an optimization algorithm is used to solve the optimal control problem defined by the cost function. A generalized optimization problem for a cost function ($J(k)$) over the finite prediction horizon (N) can be written as

$$\begin{aligned} J(k) &= \sum_{l=k}^{k+N-1} \left\| \mathbf{y}(l+1) - \mathbf{y}^\star(l+1) \right\|_2^2 \\ \text{s.t. } & S_w(l) \in \mathbb{S}(\text{FCS}) \ \forall l \in k,\ldots,k+N-1 \end{aligned} \tag{2.42}$$

Importantly, system's constraints can be easily included into the control problem formulation. The solution of this optimal control problem is defined as the control action to be applied to the system without an external modulator, i.e.: $\mathbf{S}_w(k) = \arg\min_{\mathbf{S}_w(k)} \{J(k)\}$. This process is repeated at each sampling instant using a receding horizon policy [37–39]. Fig. 2.22A illustrates the operation principle of FCS-MPC, and Fig. 2.22B its block diagram. For further details on long horizon FCS-MPC formulations and optimizations, see [40–43]. For stability analysis of FCS-MPC for power converters, see [44–46]

2.4.3 Control With Embedded Modulator

The advantages of control of power converters without modulator discussed earlier are the simple implementation and the enhanced dynamic performance. Without using an external modulator to generate the power switches' gate signals, the applied states of the power switches are essentially fixed between sampling instants. This makes possible to quickly achieve the control references, but unless a more sophisticated algorithm is used, for example, selective harmonic elimination MPC (SHE-MPC) [17] (see Section 2.3.3), the result is a power converter operating with variable switching frequency. This is considered a disadvantage as the design of passive components to avoid the occurrence of resonances is not straightforward. As a consequence, the research community has paid attention to develop high-performance controllers, which resemble the PWM-based output behavior but still attaining the advantages obtained by the use of controllers without modulator.

This section refers to these methods as control with embedded modulator, including some examples: Predictive DPC (P-DPC) [47] and optimal switching sequences MPC (OSS-MPC) [48]. P-DPC and OSS-MPC are within the category of MPC but in the subcategory of CCS-MPC. Basically, CCS-MPC considers the application times for each power converter's switch as control inputs. Therefore, the solutions for the optimal control problem are continuous signals (a set of application times (**t**) within T_s), inside the control region of the power converter (i.e., feasible voltage vectors of the power converter). Conceptually, explicit information of a desired evolution of the predicted trajectories of the controlled variables is embedded as part of the optimal control problem formulation to obtain optimal application times that minimizes a cost function. Normally, this cost function considers a weighted positive sum of the tracking error of the controlled variables. Finally, the conversion between those optimal application times and the gate switching signals is implemented by a set of comparators and reference signals, which are modulators in their simplest form.

2.4.3.1 Predictive Direct Power Control

P-DPC can be considered as cornerstone in the development of control methods with embedded modulator [47]. Basically, P-DPC evaluates a cost function for a predefined set of concatenated voltage vectors (**v**), which lies on the grid-voltage. This cost function is formulated taking into account the evolution of the predicted trajectories of active and reactive power from the beginning to the end of a sampling period T_s. The minimization of this cost function returns the application times ($\mathbf{t} = \{t_1, \ldots, t_i\}$ $\forall i \in \mathbb{Z}$) for these voltage vectors. These times are used to generate the gate switching signals that provide a fixed switching frequency in the converter. Since only two voltage vectors are used in the optimization process, P-DPC finds application times that are local optimal solutions, but not necessary global optimal solutions considering the whole control region of the power converter. For that reason, this concept was extended in [48] for finding the global optimum solution considering all the feasible voltage vectors of the power converter in the optimization process.

2.4.3.2 Optimal Switching Sequences Model Predictive Control

Instead of selecting a unique set of concatenated voltage vectors, OSS-MPC evaluates a cost function for each of the available switching sequences for the power converter [48]. The OSS-MPC concept has been extended for controlling the current in single-phase converters and also the inductor current

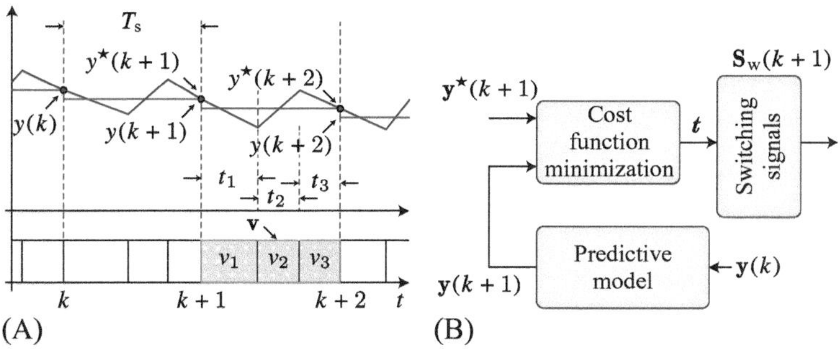

FIG. 2.23 A generalized OSS-MPC for power converters: (A) operation principle and (B) block diagram.

in DC-DC boost converters [49,50]. The main advantages are its fixed switching frequency, enhanced control performance due to the use of the optimal application times, and the straightforward capability for the inclusion of system constraints. Fig. 2.23A illustrates the operation principle of OSS-MPC, and Fig. 2.23B its block diagram. Notice that Fig. 2.23 is also valid for P-DPC.

2.5 CONCLUSION

This chapter presented an overview of basic control principles for power converters. This helps to understand the required steps not only to implement existing control strategies but also to develop new advantageous control methods for power converters. In this way, the variety of existing control strategies have been classified in three main classes: modulation-based controllers, controllers without modulator, and controllers with embedded modulator. Since power converters are autonomous systems, those control strategies have and will continue to play a key role in the past, present, and future of power electronics.

REFERENCES

[1] M.P. Kazmierkowski, M. Malinowski, M. Bech, Pulse width modulation techniques for three-phase voltage source converters, in: M.P. Kazmierkowski, R. Krishnan, F. Blaabjerg (Eds.), Control in Power Electronics, Series: Academic Press Series in Engineering, Academic Press, Burlington, 2002, pp. 89–160.

[2] A. Schönung, H. Stemmler, Static frequency changers with subharmonic control in conjunction with reversible variable speed ac drives, Brown Boveri Rev. 51 (1964) 555–577.

[3] S.R. Bowes, New sinusoidal pulsewidth-modulated inverter, IEE Proc. 122 (11) (1975) 1279–1284.

[4] N. Mohan, T.M. Undeland, W.P. Robbins, Power Electronics—Converters, Applications and Design, third ed., Media Enhanced Edition, John Wiley & Sons, Hoboken, NJ, 2003.

[5] S. Bowes, B. Bird, Novel approach to the analysis and synthesis of modulation processes in power convertors, Proc. Inst. Electr. Eng. 122 (1975) 507.

[6] D.G. Holmes, T.A. Lipo, Pulse Width Modulation for Power Converters: Principles and Practice, IEEE Press Series on Power Engineering, John Wiley & Sons, Hoboken, NJ, 2003.
[7] K.P. Kovàcs, I. Racz, Transiente Vorgänge in Wechselstrommaschinen, bd. I and II, Verlag der Ungarischen Akademie der Wissenschaften, Budapest, Hungary, 1959.
[8] K. King, A three phase transistor class-b inverter with sine wave output and high efficiency, Inst. Electron. Eng. Conf. Publ. 123 (1974) 204–209.
[9] G. Pfaff, A. Weschta, A.F. Wick, Design and experimental results of a brushless ac servo drive, IEEE Trans. Ind. Appl. IA-20 (4) (1984) 814–821.
[10] F.G. Turnbull, Selected harmonic reduction in static DC-AC inverters, IEEE Trans. Commun. Electron 83 (73) (1964) 374–378.
[11] H.S. Patel, R.G. Hoft, Generalized techniques of harmonic elimination and voltage control in thyristor inverters: part I—harmonic elimination, IEEE Trans. Ind. Appl. IA-9 (3) (1973) 310–317.
[12] H.S. Patel, R.G. Hoft, Generalized techniques of harmonic elimination and voltage control in thyristor inverters: part II—voltage control techniques, IEEE Trans. Ind. Appl. IA-10 (5) (1974) 666–673.
[13] B.P. McGrath, D.G. Holmes, Multicarrier PWM strategies for multilevel inverters, IEEE Trans. Ind. Electron. 49 (4) (2002) 858–867.
[14] N. Celanovic, D. Boroyevich, A fast space-vector modulation algorithm for multilevel three-phase converters, IEEE Trans. Ind. Appl. 37 (2) (2001) 637–641.
[15] J.I. Leon, S. Vazquez, R. Portillo, L.G. Franquelo, J.M. Carrasco, P.W. Wheeler, et al., Three-dimensional feedforward space vector modulation applied to multilevel diode-clamped converters, IEEE Trans. Ind. Electron. 56 (1) (2009) 101–109.
[16] M.S.A. Dahidah, G. Konstantinou, V.G. Agelidis, A review of multilevel selective harmonic elimination PWM: formulations, solving algorithms, implementation and applications, IEEE Trans. Power Electron. 30 (8) (2015) 4091–4106.
[17] R.P. Aguilera, P. Acuna, P. Lezana, G. Konstantinou, B. Wu, S. Bernet, et al., Selective harmonic elimination model predictive control for multilevel power converters, IEEE Trans. Power Electron. 32 (3) (2017) 2416–2426.
[18] K. Song, G. Konstantinou, W. Mingli, P. Acuna, R.P. Aguilera, V.G. Agelidis, Windowed SHE-PWM of interleaved four-quadrant converters for resonance suppression in traction power supply systems, IEEE Trans. Power Electron. 32 (10) (2017) 7870–7881.
[19] A.M. Hava, R.J. Kerkman, T.A. Lipo, Carrier-based pwm-vsi overmodulation strategies: analysis, comparison, and design, IEEE Trans. Power Electron. 13 (4) (1998) 674–689.
[20] A.M. Hava, R.J. Kerkman, T.A. Lipo, Simple analytical and graphical methods for carrier-based PWM-VSI drives, IEEE Trans. Power Electron. 14 (1) (1999) 49–61.
[21] J.I. Leon, S. Vazquez, L.G. Franquelo, Multilevel converters: control and modulation techniques for their operation and industrial applications, Proc. IEEE PP 99 (2017) 1–16.
[22] J.I. Leon, S. Kouro, L.G. Franquelo, J. Rodríguez, B. Wu, The essential role and the continuous evolution of modulation techniques for voltage-source inverters in the past, present, and future power electronics, IEEE Trans. Ind. Electron. 63 (5) (2016) 2688–2701.
[23] G. Carrara, S. Gardella, M. Marchesoni, R. Salutari, G. Sciutto, A new multilevel PWM method: a theoretical analysis, IEEE Trans. Power Electron. 7 (3) (1992) 497–505.
[24] J.I. Leon, S. Vazquez, J.A. Sanchez, R. Portillo, L.G. Franquelo, J.M. Carrasco, et al., Conventional space-vector modulation techniques versus the single-phase modulator for multilevel converters, IEEE Trans. Ind. Electron. 57 (7) (2010) 2473–2482.
[25] W. Yao, H. Hu, Z. Lu, Comparisons of space-vector modulation and carrier-based modulation of multilevel inverter, IEEE Trans. Power Electron. 23 (1) (2008) 45–51.

[26] J. Napoles, J.I. Leon, R. Portillo, L.G. Franquelo, M.A. Aguirre, Selective harmonic mitigation technique for high-power converters, IEEE Trans. Ind. Electron. 57 (7) (2010) 2315–2323.
[27] N. Oikonomou, J. Holtz, Closed-loop control of medium-voltage drives operated with synchronous optimal pulsewidth modulation, IEEE Trans. Ind. Appl. 44 (1) (2008) 115–123.
[28] H. Aggrawal, J. I. Leon, L. G. Franquelo, S. Kouro, P. Garg, and J. Rodríguez, Model predictive control based selective harmonic mitigation technique for multilevel cascaded H-bridge converters, in: Proc. IECON 2011–37th Annual Conference on IEEE Industrial Electronics Society, Oct. 2011, pp. 4427–4432.
[29] K.J. Åström, T. Hägglund, PID Controllers: Theory, Design, and Tuning, in: second ed, Instrument Society of Automation, 1995.
[30] G.C. Goodwin, S.F. Graebe, M.E. Salgado, Control System Design, Prentice Hall, Upper Saddle River, NJ, 2001.
[31] D.N. Zmood, D.G. Holmes, Stationary frame current regulation of PWM inverters with zero steady-state error, IEEE Trans. Power Electron. 18 (3) (2003) 814–822.
[32] C. Buccella, C. Cecati, H. Latafat, Digital control of power converters—a survey, IEEE Trans. Ind. Informat. 8 (3) (2012) 437–447.
[33] K.J. Åström, B. Wittenmark, Computer-Controlled Systems: Theory and Design, in: third ed, Prentice Hall, Upper Saddle River, NJ, 1997.
[34] P. Lezana, C.A. Silva, J. Rodríguez, M.A. Pérez, Zero-steady-state-error input-current controller for regenerative multilevel converters based on single-phase cells, IEEE Trans. Ind. Electron. 54 (2) (2007) 733–740.
[35] I. Takahashi, T. Noguchi, A new quick-response and high-efficiency control strategy of an induction motor, IEEE Trans. Ind. Appl. IA-22 (5) (1986) 820–827.
[36] T. Noguchi, H. Tomiki, S. Kondo, I. Takahashi, Direct power control of PWM converter without power-source voltage sensors, IEEE Trans. Ind. Appl 34 (3) (1998) 473–479.
[37] S. Kouro, P. Cortés, R. Vargas, U. Ammann, J. Rodríguez, Model predictive control a simple and powerful method to control power converters, IEEE Trans. Ind. Electron. 56 (6) (1826–1838) 2009.
[38] D.E. Quevedo, R.P. Aguilera, T. Geyer, Predictive control in power electronics and drives: basic concepts, theory, and methods, Advanced and Intelligent Control in Power Electronics and DrivesSpringer, New York, NY, USA, 2014, pp. 181–226.
[39] S. Vazquez, J. Rodríguez, M. Rivera, L.G. Franquelo, M. Norambuena, Model predictive control for power converters and drives: advances and trends, IEEE Trans. Ind. Electron. 64 (2) (2017) 935–947.
[40] T. Geyer, D.E. Quevedo, Multistep finite control set model predictive control for power electronics, IEEE Trans. Power Electron. 29 (12) (2014) 6836–6846.
[41] T. Geyer, D.E. Quevedo, Performance of Multistep Finite Control Set Model Predictive Control for Power Electronics, IEEE Trans. Power Electron. 30 (3) (Mar. 2015) 1633–1644.
[42] R. Baidya, R. P. Aguilera, P. Acuna, R. Delgado, T. Geyer, D. E. Quevedo, et al., Fast multistep finite control set model predictive control for transient operation of power converters, in: Proc. IECON 2016–42nd Annual Conference of the IEEE Industrial Electronics Society, 2016, pp. 5039–5045.
[43] R. Baidya, R.P. Aguilera, P. Acuna, S. Vazquez, H. du Toit Mouton, Multistep model predictive control for cascaded H-bridge inverters: formulation and analysis, IEEE Trans. Power Electron. 33 (1) (2018) 876–886.
[44] R. P. Aguilera and D. E. Quevedo, On stability and performance of finite control set MPC for power converters, in: Proc. 2011 Workshop on Predictive Control of Electrical Drives and Power Electronics, Oct. 2011, pp. 55–62.

[45] R.P. Aguilera, D.E. Quevedo, Stability analysis of quadratic MPC with a discrete input alphabet, IEEE Trans. Autom. Control 58 (12) (2013) 3190–3196.

[46] R.P. Aguilera, D.E. Quevedo, Predictive control of power converters: designs with guaranteed performance, IEEE Trans. Ind. Informat. 11 (1) (Feb. 2015) 53–63.

[47] S. Aurtenechea, M. A. Rodríguez, E. Oyarbide, and J. R. Torrealday, Predictive direct power control—a new control strategy for DC/AC converters, in: Proc. IEEE Industrial Electronics, IECON 2006–32nd Annual Conference on, Nov. 2006, pp. 1661–1666.

[48] S. Vazquez, A. Marquez, R.P. Aguilera, D.E. Quevedo, J.I. Leon, L.G. Franquelo, Predictive optimal switching sequence direct power control for grid-connected power converters, IEEE Trans. Ind. Electron. 62 (4) (2010–2020) 2015.

[49] S. Vazquez, R.P. Aguilera, P. Acuna, J. Pou, J.I. Leon, L.G. Franquelo, et al., Model predictive control for single-phase NPC converters based on optimal switching sequences, IEEE Trans. Ind. Electron. 63 (12) (2016) 7533–7541.

[50] L. Cheng, P. Acuna, R. P. Aguilera, M. Ciobotaru, and J. Jiang, Model predictive control for DC-DC boost converters with constant switching frequency, in: Proc. 2016 IEEE 2nd Annual Southern Power Electronics Conference (SPEC), Dec. 2016, pp. 1–6.

Chapter 3

Modeling and Control of DC-DC Converters

Yanfeng Shen*, Zian Qin† and Huai Wang*
*Aalborg University, Aalborg, Denmark, †Delft University of Technology, Delft, The Netherlands

3.1 FREQUENCY-DOMAIN VOLTAGE-MODE CONTROL OF A BUCK DC-DC CONVERTER

3.1.1 Specifications of the Buck DC-DC Converter

The circuit diagram of the Buck DC-DC converter is shown in Fig. 3.1. The input voltage $V_{in}=110\,\text{V}$, the output voltage $V_o=48\,\text{V}$, the nominal output power $P_{oN}=500\,\text{W}$, the switching frequency $f_s=100\,\text{kHz}$, the maximum current ripple in the output inductor $\Delta I_{Lo}=10\%\ I_o$, and the maximum output voltage ripple $\Delta V_o=0.5\%\ V_o$. For an output voltage of 48 V, the corresponding duty cycle $D=V_o/V_{in}=48\,\text{V}/110\,\text{V}=0.44$ and the output current $I_o=500\,\text{W}/48\,\text{V}=10.42\,\text{A}$. Based on the specifications, the required output inductance L_o and output capacitance C_o can be obtained

$$L_o \geq \frac{(1-D)T_sV_o}{\Delta I_{Lo}} = \frac{(1-0.44)\times 10\times 10^{-6}\times 48}{0.1\times 10.42} = 258\mu H \tag{3.1}$$

$$C_o \geq \frac{\Delta I_{Lo}T_s}{8\Delta V_o} = \frac{1.042\times 10\times 10^{-6}}{8\times 0.24} = 5.43\mu H \tag{3.2}$$

In practice, the output inductor $L_o=260\,\mu\text{H}$ and the output capacitor $C_o=220\,\mu\text{F}$. The equivalent series resistances (ESRs) of the output inductor and capacitor are $R_{Lo}=0.1\,\Omega$ and $R_{Co}=0.2\,\Omega$, respectively.

3.1.2 Small-Signal Modeling of Buck DC-DC Converter

The operation principle of the Buck converter is simple: when the switch S is turned-on, the input source V_{in} supplies power to the load through the output inductor L_o and output capacitor C_o, as shown in Fig. 3.2A when S is turned-off, the inductor L_o and output capacitor C_o release energy to the load via the freewheeling diode D, as shown in Fig. 3.2B. The duty cycle of S being

Control of Power Electronic Converters and Systems. https://doi.org/10.1016/B978-0-12-805245-7.00003-2

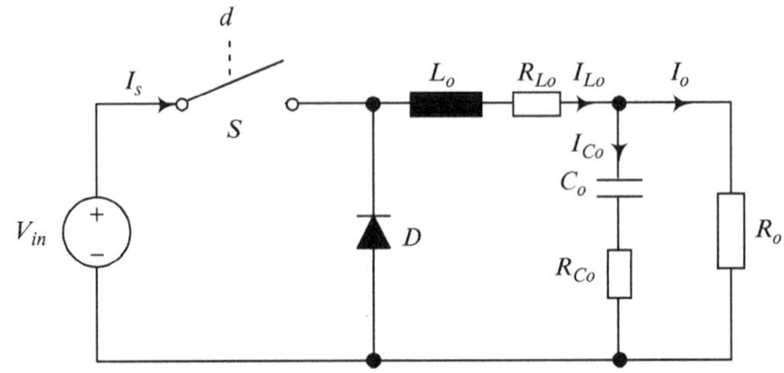

FIG. 3.1 Circuit diagram of the buck DC-DC converter.

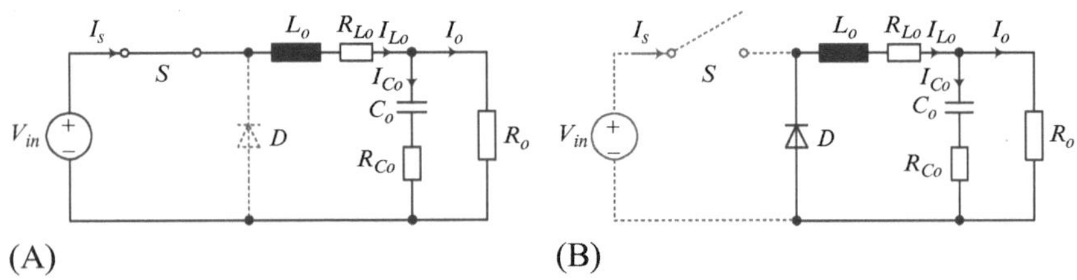

FIG. 3.2 Equivalent circuit of the Buck dc-dc converter when the switch S is (A) turned-on or (B) turned-off.

turned-on is denoted as d and then that of S being off is $1-d$ if the deadtime is neglected. According to the Kirchhoff's current and Kirchhoff's voltage laws, the basic differential equations can be derived for the two states.

State I ($t \in [0, dT_s]$), switch S is turned on:

$$\begin{cases} L_o \dfrac{di_{Lo}}{dt} = v_{in} - v_o - R_{Lo} i_{Lo} \\ C_o \dfrac{dv_{Co}}{dt} = i_{Lo} - i_o \\ v_o = v_{Co} + R_{Co}(i_{Lo} - i_o) \end{cases} \tag{3.3}$$

State II ($t \in [dT_s, T_s]$), switch S is turned off:

$$\begin{cases} L_o \dfrac{di_{Lo}}{dt} = -v_o - R_{Lo} i_{Lo} \\ C_o \dfrac{dv_{Co}}{dt} = i_{Lo} - i_o \\ v_o = v_{Co} + R_{Co}(i_{Lo} - i_o) \end{cases} \tag{3.4}$$

The time durations for states I and II are dT_s and $(1-d)T_s$, respectively. In the meanwhile, the state variables (i.e., the inductor current i_{Lo} and the capacitor voltage v_{Co}) and the output variable (i.e., the output voltage v_o) are continuous in time. Therefore, the averaged differential equations over one switching cycle can be obtained as follows

$$\begin{cases} L_o \dfrac{di_{Lo}}{dt} = dv_{in} - v_o - R_{Lo} i_{Lo} \\ C_o \dfrac{dv_{Co}}{dt} = i_{Lo} - i_o \\ v_o = v_{Co} + R_{Co}(i_{Lo} - i_o) \end{cases} \tag{3.5}$$

As can be seen from Eq. (3.5) that there is a term dv_{in} which is a product of the input variable v_{in} and the control variable d. That is the averaged model is nonlinear, and therefore the controller cannot be designed with the classical control theory. In order to obtain a linear model, perturbations at the steady-state operation (equilibrium) point should be performed. In this way, each variable contains a DC term (i.e., the steady-state operation point) and a small AC term, i.e., $i_{Lo} = I_{Lo} + \widetilde{i}_{Lo}, v_{Co} = V_{Co} + \widetilde{v}_{Co}, v_o = V_o + \widetilde{v}_o, i_o = I_o + \widetilde{i}_o$, $v_{in} = V_{in} + \widetilde{v}_{in}$, and $d = D + \widetilde{d}$. Substituting them into Eq. (3.5) yields

$$\begin{cases} L_o \dfrac{d}{dt}\left(I_{Lo} + \widetilde{i}_{Lo}\right) = \left(D + \widetilde{d}\right)(V_{in} + \widetilde{v}_{in}) - (V_o + \widetilde{v}_o) - R_{Lo}\left(i_{Lo} + \widetilde{i}_{Lo}\right) \\ C_o \dfrac{d}{dt}(V_{Co} + \widetilde{v}_{Co}) = \left(I_{Lo} + \widetilde{i}_{Lo}\right) - \left(I_o + \widetilde{i}_o\right) \\ (V_o + \widetilde{v}_o) = (V_{Co} + \widetilde{v}_{Co}) + R_{Co}\left[\left(I_{Lo} + \widetilde{i}_{Lo}\right) - \left(I_o + \widetilde{i}_o\right)\right] \end{cases} \tag{3.6}$$

Applying the steady-state differential equation and neglecting the second-order terms, Eq. (3.6) can be simplified as linear small-signal AC equations

$$\begin{cases} L_o \dfrac{d}{dt}\widetilde{i}_{Lo} = D\widetilde{v}_{in} + \widetilde{d}V_{in} - \widetilde{v}_o - R_{Lo}\widetilde{i}_{Lo} \\ C_o \dfrac{d}{dt}\widetilde{v}_{Co} = \widetilde{i}_{Lo} - \widetilde{i}_o \\ \widetilde{v}_o = \widetilde{v}_{Co} + R_{Co}\left(\widetilde{i}_{Lo} - \widetilde{i}_o\right) \end{cases} \tag{3.7}$$

Performing Laplace transform to Eq. (3.7) yields its frequency-domain form, i.e.,

$$\begin{cases} sL_o\widetilde{i}_{Lo}(s) = D\widetilde{v}_{in}(s) + \widetilde{d}(s)V_{in} - \widetilde{v}_o(s) - R_{Lo}\widetilde{i}_{Lo}(s) \\ sC_o\widetilde{v}_{Co}(s) = \widetilde{i}_{Lo}(s) - \widetilde{i}_o(s) \\ \widetilde{v}_o(s) = \widetilde{v}_{Co}(s) + R_{Co}\left[\widetilde{i}_{Lo}(s) - \widetilde{i}_o(s)\right] \end{cases} \tag{3.8}$$

Eliminate $v_{Co}(s)$ and $i_{Lo}(s)$ in Eq. (3.8), and $\widetilde{v}_o(s)$ can be obtained:

$$\widetilde{v}_o(s) = \frac{\left[D\widetilde{v}_{in}(s) + V_{in}\widetilde{d}(s) - R_{Lo}\widetilde{i}_o(s) - sL_o\widetilde{i}_o(s)\right] \cdot (1 + sR_{C_o}C_o)}{1 + sC_o(R_{Co} + R_{Lo}) + s^2C_oL_o} \tag{3.9}$$

Then different small-signal transfer functions can be derived with Eq. (3.9)

$$G_{d2v_o} = \left.\frac{\tilde{v}_o(s)}{\tilde{d}(s)}\right|_{\tilde{v}_{in}(s)=0,\tilde{i}_o(s)=0} = \frac{V_{in}(1+sR_{C_o}C_o)}{1+sC_o(R_{Co}+R_{Lo})+s^2C_oL_o}$$

$$G_{v_{in}2v_o} = \left.\frac{\tilde{v}_o(s)}{\tilde{v}_{in}(s)}\right|_{\tilde{d}(s)=0,\tilde{i}_o(s)=0} = \frac{D(1+sR_{C_o}C_o)}{1+sC_o(R_{Co}+R_{Lo})+s^2C_oL_o} \quad (3.10)$$

$$G_{i_o2v_o} = \left.\frac{\tilde{v}_o(s)}{\tilde{i}_o(s)}\right|_{\tilde{d}(s)=0,\tilde{v}_{in}(s)=0} = \frac{-(R_{Lo}+sL_o)\cdot(1+sR_{C_o}C_o)}{1+sC_o(R_{Co}+R_{Lo})+s^2C_oL_o}$$

3.1.3 Voltage-Model Controller Design of Buck DC-DC Converter

The PI controller can be expressed as

$$G_{PI}(s) = K_p\left(1+\frac{K_I}{s}\right) \quad (3.11)$$

The system switching frequency $\omega_s = 628.32$ krad/s. The bode plot of the control-to-output transfer function is shown in Fig. 3.3. As can be observed,

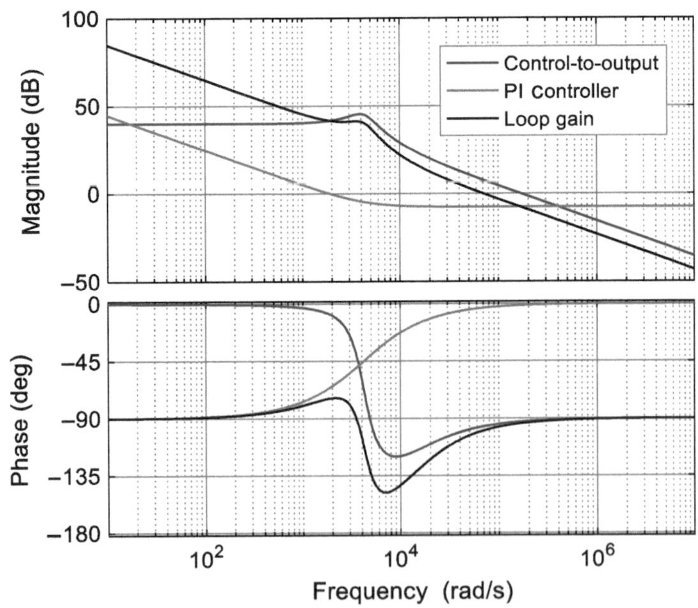

FIG. 3.3 Bode plot of the control-to-output transfer function, the PI controller, and the loop gain.

there are two corner frequencies, 4.21 and 8.35 krad/s, and the crossover frequency is 167 krad/s. In the meanwhile, the crossover frequency of the open-loop transfer function should not be higher than 1/10 to 1/5 of the switching frequency. In this case, the crossover ω_c is set as 1/9 of the switching frequency, i.e., ω_c should be corrected to be about 69.8 krad/s.

The gain of the control-to-output transfer function at $\omega_c = 69.8$ krad/s is 7.69 dB. Therefore, the required gain from the PI compensator should be $K_p = -7.69\,\text{dB} = 0.4126$. The corner frequency of the PI controller should approximately correspond to the first corner frequency of the control-to-output transfer function. Thus, $K_I = \omega_I = 4.21$ krad/s.

The transfer functions of both the PI controller and the open-loop gain are depicted in Fig. 3.3 as well. As can be seen, the open-loop transfer function has been corrected with the crossover frequency $\omega_c = 69.8$ krad/s and the phase margin being equal to 80°. The bode plots of open-loop and closed-loop transfer functions are shown in Fig. 3.4. It can be observed that the high loop gain results in a strong suppression of disturbances.

3.1.4 Simulation Results of the Case Study

The closed-loop results from the circuit simulation and the small-signal model are shown in Figs. 3.5–3.7. Both the analytical modeling and simulation results are based on the specific case study described in Section 3.1.1. As can be seen, the developed small-signal model and the designed controller can predict both the dynamic and static performances of the Buck DC-DC converter with negligible error compared to simulations. When the output voltage reference is changed from 48 to 49 V, the PI controller helps the system to track the reference tightly, as shown in Fig. 3.5. When the perturbations (i.e., the input voltage in Fig. 3.6 and the load in Fig. 3.7) appear, their impact on the output voltage can be effectively attenuated.

3.2 FREQUENCY-DOMAIN CURRENT-MODE CONTROL OF A BOOST DC-DC CONVERTER

3.2.1 Specifications of the Boost DC-DC Converter

The circuit diagram of a current mode-controlled (CMC) Boost DC-DC converter is shown in Fig. 3.8. The input voltage $v_i = 150$ V, the output voltage $v_o = 350$ V, the nominal output power $P_o = 2$ kW, the switching frequency $f_s = 50$ kHz, and the maximum current ripple in the input inductor L is limited to 25%. An inductor with inductance $L = 514\,\mu\text{H}$ and ESR $R_\text{L} = 0.02\,\Omega$ is selected. The output capacitor $C = 450\,\mu\text{F}$ and its ESR $R_\text{C} = 0.01\,\Omega$. The current sensor resistor R_s shown in Fig. 3.8 is $0.2\,\Omega$.

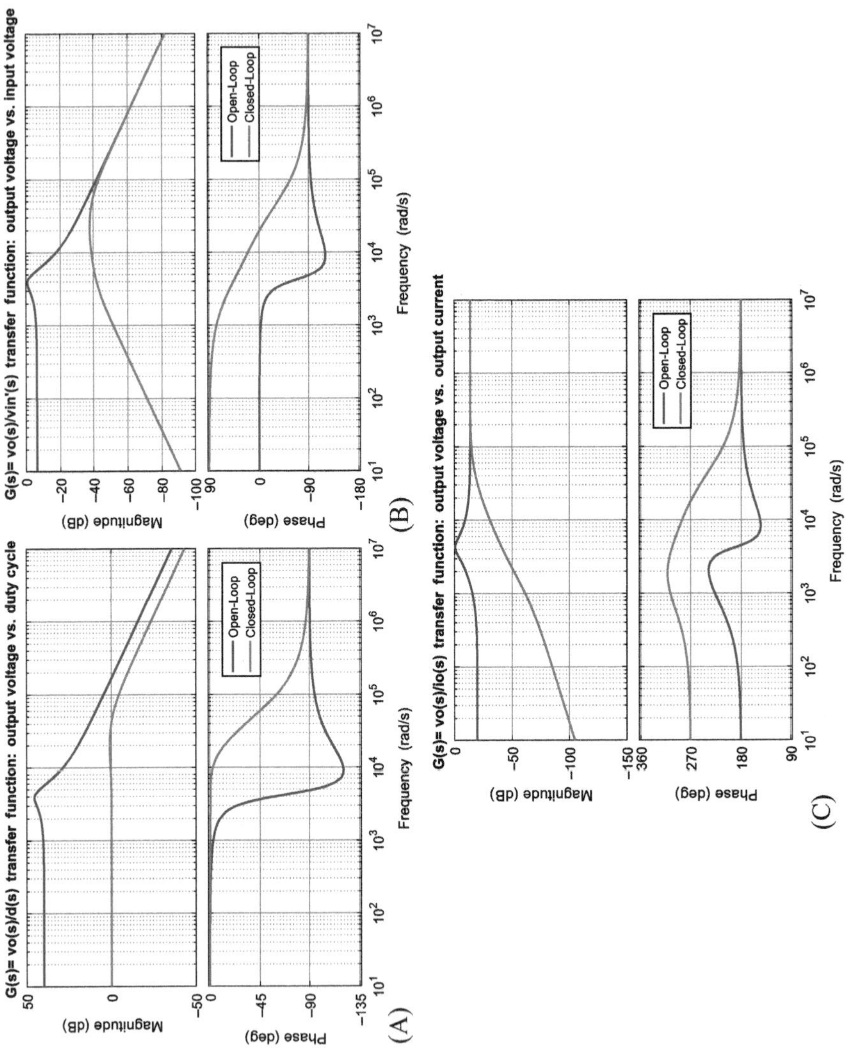

FIG. 3.4 Bode plot of open-loop and closed-loop transfer functions. (A) Control-to-output transfer function, (B) input-to-output transfer function, and (C) output impedance.

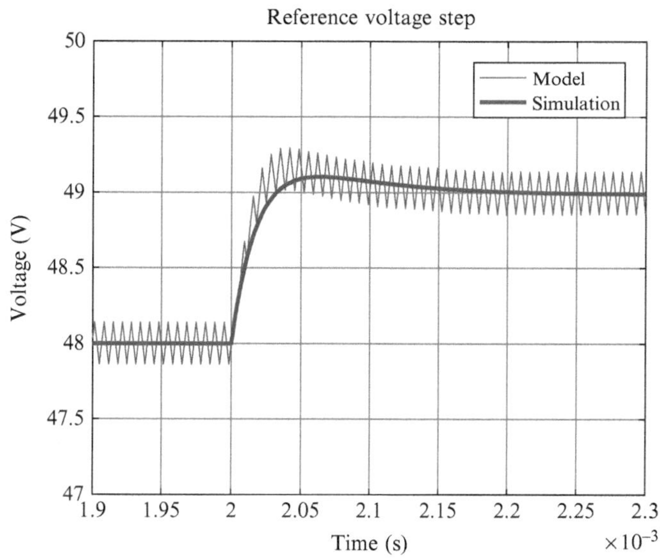

FIG. 3.5 The output voltage reference is varied from 48 to 49 V.

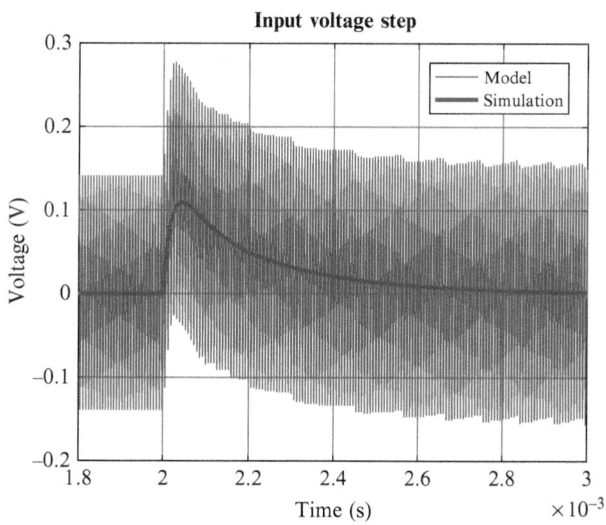

FIG. 3.6 The input voltage is step-changed from 100 to 110 V at 2 ms.

3.2.2 Small-Signal Modeling of Boost DC-DC Converter

Suppose the Boost converter is operating in the continuous current mode (CCM). When S is turned on, the state equations are

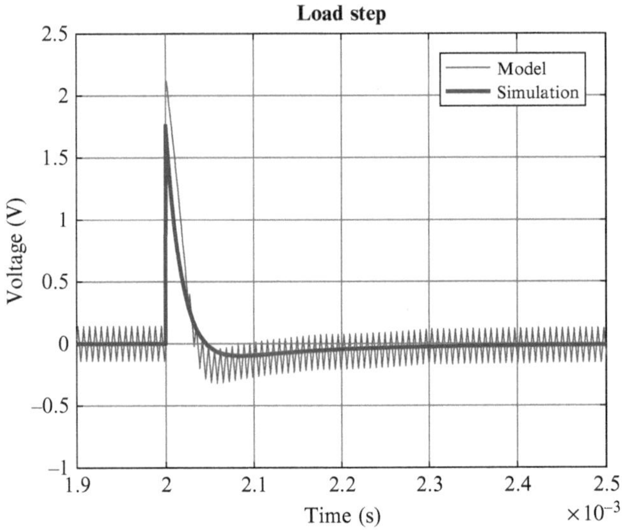

FIG. 3.7 The load is changed from 20 to 10 A at 2 ms.

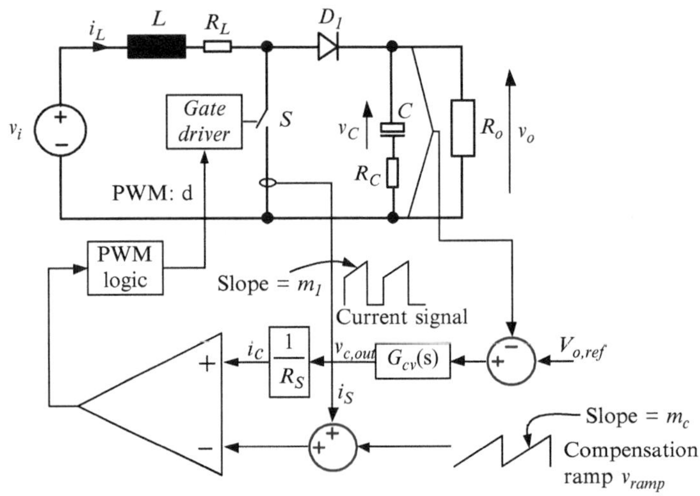

FIG. 3.8 A boost converter with current-mode control.

$$
\begin{cases} L\dfrac{d}{dt}i_L = v_i - i_L R_L \\ C\dfrac{d}{dt}v_C = -\dfrac{v_o}{R_o} \end{cases} \tag{3.12}
$$

When S is turned off, the diode D_1 is freewheeling; then the state equations are

$$\begin{cases} L\dfrac{d}{dt}i_L = v_i - i_L R_L - v_o \\ C\dfrac{d}{dt}v_C = i_L - \dfrac{v_o}{R_o} \end{cases} \tag{3.13}$$

According to Eqs. (3.12), (3.13), the average model can be obtained as,

$$\begin{cases} L\dfrac{d}{dt}\bar{i}_L = \bar{v}_i - \bar{i}_L R_L - (1-\bar{d})\bar{v}_o \\ C\dfrac{d}{dt}\bar{v}_C = (1-\bar{d})\bar{i}_L - \dfrac{\bar{v}_o}{R_o} \end{cases} \tag{3.14}$$

where d is the duty ratio of the boost converter, and $\bar{x}$ is the average value of x ($x = i_L, v_i, d, v_o, v_C$) in a switching cycle. As seen in Fig. 3.8, it is easy to acquire

$$\bar{v}_o = \bar{v}_C + R_C C\frac{d}{dt}\bar{v}_C \tag{3.15}$$

Then, by applying small signal perturbations to Eqs. (3.14), (3.15), it is gained

$$\begin{cases} L\dfrac{d}{dt}\left(\bar{I}_L + \tilde{i}_L\right) = \left(\bar{V}_i + \tilde{v}_i\right) - \left(\bar{I}_L + \tilde{i}_L\right)R_L - \left(1 - \bar{D} - \tilde{d}\right)\left(\bar{V}_o + \tilde{v}_o\right) \\ C\dfrac{d}{dt}\left(\bar{V}_C + \tilde{v}_C\right) = \left(1 - \bar{D} - \tilde{d}\right)\left(\bar{I}_L + \tilde{i}_L\right) - \dfrac{\bar{V}_o + \tilde{v}_o}{R_o} \\ \bar{V}_o + \tilde{v}_o = \bar{V}_C + \tilde{v}_C + R_C C\dfrac{d}{dt}\left(\bar{V}_C + \tilde{v}_C\right) \end{cases} \tag{3.16}$$

Extract the DC items in Eq. (3.16), it is obtained

$$\begin{cases} 0 = \bar{V}_i - \bar{I}_L R_L - (1-\bar{D})\bar{V}_o \\ 0 = (1-\bar{D})\bar{I}_L - \dfrac{\bar{V}_o}{R_o} \\ \bar{V}_o = \bar{V}_C \end{cases} \tag{3.17}$$

By removing the DC and high-order AC items in Eq. (3.16), it is acquired

$$\begin{cases} L\dfrac{d}{dt}\tilde{i}_L = \tilde{v}_i - \tilde{i}_L R_L - (1-\bar{D})\tilde{v}_o + \tilde{d}\bar{V}_o \\ C\dfrac{d}{dt}\tilde{v}_C = (1-\bar{D})\tilde{i}_L - \tilde{d}\bar{I}_L - \dfrac{\tilde{v}_o}{R_o} \\ \tilde{v}_o = \tilde{v}_C + R_C C\dfrac{d}{dt}\tilde{v}_C \end{cases} \tag{3.18}$$

By performing Laplace transform to Eq. (3.18), it is obtained

$$\begin{cases} sL\tilde{i}_L(s) = \tilde{v}_i(s) - \tilde{i}_L(s)R_L - (1-\overline{D})\tilde{v}_o(s) + \tilde{d}(s)\overline{V}_o \\ sC\tilde{v}_C(s) = (1-\overline{D})\tilde{i}_L(s) - \tilde{d}(s)\overline{I}_L - \dfrac{\tilde{v}_o(s)}{R_o} \\ \tilde{v}_o(s) = \tilde{v}_C(s) + sR_C C\tilde{v}_C(s) \end{cases} \tag{3.19}$$

Then, Eq. (3.20) can be derived from Eq. (3.19), and the control to output voltage transfer function of the open-loop boost converter is obtained as shown in Eq. (3.21).

$$\tilde{v}_o(s) = \frac{\dfrac{1+sR_CC}{1-\overline{D}}\tilde{v}_i(s) + \dfrac{1+sR_CC}{(1-\overline{D})^2}\overline{V}_i\left[1 - \dfrac{sL}{R_o(1-\overline{D})^2}\right]\tilde{d}(s)}{s^2\dfrac{LC}{(1-\overline{D})^2} + s\left[\dfrac{L+R_LR_oC}{R_o(1-\overline{D})^2} + R_CC\right] + 1} \tag{3.20}$$

$$G_{d2v_o} = \left.\frac{\tilde{v}_o(s)}{\tilde{d}(s)}\right|_{\tilde{v}_i(s)=0} = \frac{\dfrac{1+sR_CC}{(1-\overline{D})^2}\overline{V}_i\left[1 - \dfrac{sL}{R_o(1-\overline{D})^2}\right]}{s^2\dfrac{LC}{(1-\overline{D})^2} + s\left[\dfrac{L+R_LC}{(1-\overline{D})^2} + R_CC\right] + 1} \tag{3.21}$$

Fig. 3.9 shows the peak current control of the boost converter, where the switch current i_S instead of the inductor current i_L is applied to compare with the reference i_c that is generated by the voltage compensator. Moreover, in order to increase the damping and mitigate the subharmonic caused by the peak current control, a slope m_c is added into i_S before compared with i_c, as illustrated in Figs. 3.8 and 3.9. The design of m_c will be demonstrated in the following section. By solving the simple trigonometry in Fig. 3.9, it is obtained

$$\overline{i}_L = i_c - m_c\overline{d}T_s - \frac{m_1}{2}\overline{d}T_s \tag{3.22}$$

$$m_1 = \frac{v_i}{L} \tag{3.23}$$

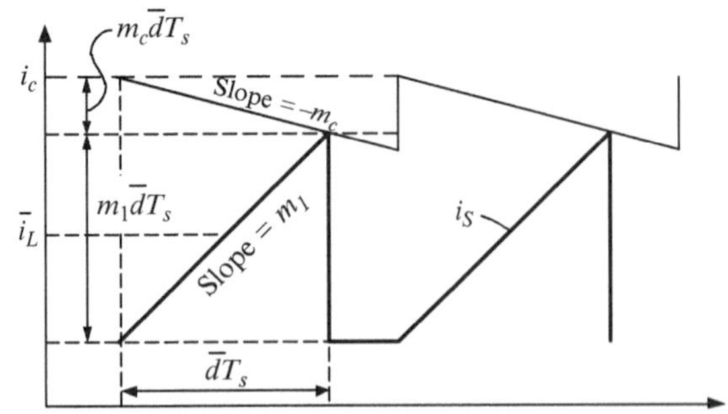

FIG. 3.9 The schematic diagram of peak current control.

By substituting Eq. (3.23) and then applying small-signal perturbation to Eq. (3.22), it is acquired

$$\overline{I}_L+\widetilde{i}_L=\overline{I}_c+\widetilde{i}_c-m_c\left(\overline{D}+\widetilde{d}\right)T_s-\frac{\overline{V}_i+\widetilde{v}_i}{2L}\left(\overline{D}+\widetilde{d}\right)T_s \tag{3.24}$$

By removing the DC and high-order AC components in Eq. (3.24), it is gained

$$\widetilde{i}_L=\widetilde{i}_c-m_c\widetilde{d}T_s-\frac{\overline{V}_i\widetilde{d}+\overline{D}\widetilde{v}_1}{2L}T_s \tag{3.25}$$

By performing Laplace transformation to Eq. (3.25), it is obtained

$$\widetilde{i}_L(s)=\widetilde{i}_c(s)-m_c\widetilde{d}(s)T_s-\frac{\overline{V}_i\widetilde{d}(s)+\overline{D}\widetilde{v}_i(s)}{2L}T_s \tag{3.26}$$

According to Eqs. (3.19), (3.26), it can be acquired

$$\widetilde{v}_o(s)=\frac{G_{NC}(s)\widetilde{v}_i(s)+G_{IC}(s)\widetilde{i}_c(s)}{\Delta(s)} \tag{3.27}$$

where

$$G_{NC}(s)=\left(1-\overline{D}\right)R_o\left[\frac{s\overline{D}T_s}{2\left(1-\overline{D}\right)^2R_o}+\frac{1}{\left(1-\overline{D}\right)^2R_o}-\frac{\overline{D}T_s}{2L}+\left(1+\frac{2m_c}{\overline{V}_i}\right)\frac{\left(1-\overline{D}\right)T_s}{2}\right] \tag{3.28}$$

$$G_{IC}(s)=\left(1-\overline{D}\right)R_o(1+sR_CC)\left[1-\frac{sL}{\left(1-\overline{D}\right)^2R_o}\right] \tag{3.29}$$

$$\begin{aligned}\Delta(s)=&\,s^2L(R_o+R_C)C\left(1-\overline{D}\right)\left(\frac{m_c}{\overline{V}_i}+\frac{1}{2L}\right)T_s\\&+s\left\{\left[L+\left(1-\overline{D}\right)^2R_oR_CC\right]\left(1-\overline{D}\right)\left(\frac{m_c}{\overline{V}_i}+\frac{1}{2L}\right)T_s+(R_o+2R_C)C\right\}\\&+2+\left(1-\overline{D}\right)^2R_o\left(1-\overline{D}\right)\left(\frac{m_c}{\overline{V}_i}+\frac{1}{2L}\right)T_s\end{aligned} \tag{3.30}$$

Assuming that the switch current is sampled via a resistor R_S, or the equivalent resistance of the current sensor is R_S, the control to output voltage transfer function of the boost converter with current mode control is obtained

$$G_{v_{c,out}2v_o}=\left.\frac{\widetilde{v}_o(s)}{\widetilde{v}_{c,out}(s)}\right|_{\widetilde{v}_i(s)=0}=\frac{G_{IC}(s)}{\Delta(s)R_S} \tag{3.31}$$

3.2.3 Current-Model Controller Design of Boost DC-DC Converter

A typical issue brought by peak current control is the subharmonics resonance, and it is indicated in Fig. 3.10. As seen, with an improperly designed m_c, the

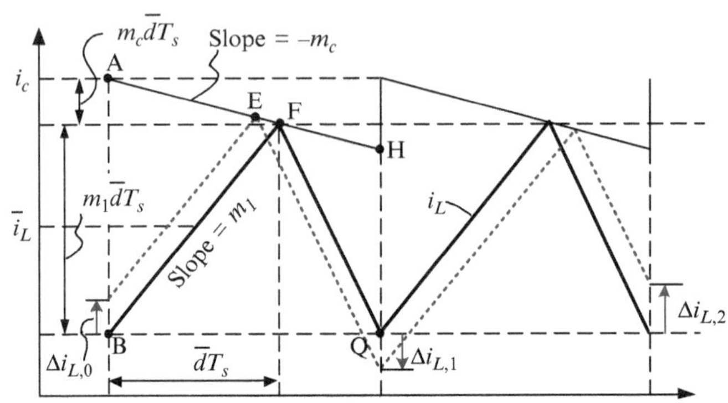

FIG. 3.10 Subharmonic resonance in current mode-controlled boost converter.

inductor current disturbance $\Delta i_{L,0}$ can be amplified after two switching cycles. The impact of m_c on the subharmonics resonance can be revealed by solving the trigonometry in Fig. 3.10, and it is obtained

$$\frac{\Delta i_{L,0}}{AB}=\frac{EF}{AF},\ \frac{EF}{FH}=\frac{-\Delta i_{L,1}}{QH} \tag{3.32}$$

Merge the two equations in Eq. (3.32), and it is acquired

$$\Delta i_{L,1}=-\frac{QH}{AB}\frac{AF}{FH}\Delta i_{L,0}=-\frac{m_1\bar{d}T_s-m_c(1-\bar{d})T_s}{(m_1+m_c)\bar{d}T_s}\frac{\bar{d}}{1-\bar{d}}\Delta i_{L,0} \tag{3.33}$$

To avoid the subharmonics resonance, it requires

$$|\Delta i_{L,1}|\leq|\Delta i_{L,0}| \tag{3.34}$$

Substitute Eq. (3.33) into Eq. (3.34), it is gained

$$\frac{m_1\bar{d}T_s-m_c(1-\bar{d})T_s}{(m_1+m_c)\bar{d}T_s}\frac{\bar{d}}{1-\bar{d}}\leq 0.5\ \Rightarrow\ m_c\geq m_1\frac{2\bar{d}-1}{2(1-\bar{d})} \tag{3.35}$$

Since the minimum value of m_c is zero, the condition to eliminate subharmonics resonance is obtained as Eq. (3.36) by substituting Eq. (3.23) into Eq. (3.35).

$$m_c\geq\begin{cases}0, & \bar{d}\leq 0.5\\ \dfrac{v_i}{L}\dfrac{2\bar{d}-1}{2(1-\bar{d})}, & \bar{d}>0.5\end{cases} \tag{3.36}$$

A voltage compensator is designed in this section for a current mode controlled boost converter with the specifications described in Section 3.2.1. Substitute the parameter values into Eq. (3.21), and the characteristics of the transfer function G_{d2v_o} is obtained as below:

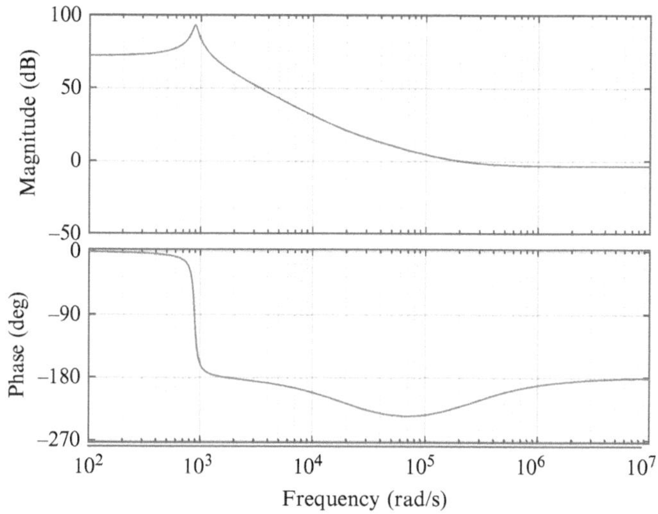

FIG. 3.11 Frequency response of the open-loop boost converter.

DC gain: 4083
Poles: $-39.4 \pm 890\text{j}$ rad/s
Zeros: -2.2×10^5 rad/s, 2.19×10^4 rad/s (Right-Half-Plane, RHP)

Fig. 3.11 shows the frequency response of the open-loop-controlled Boost converter. As seen, the open-loop system has a resonant point, where the phase is decreased sharply to $-180°$. Thus, if a voltage mode control is applied, a much lower crossover frequency than the resonant frequency is necessary for the voltage compensator to damp the resonance and make the system stable.

Substitute the parameter values into Eq. (3.31), the characteristics of the transfer function $G_{v_{c,out}2vo}$ are obtained:

DC gain: 62
Poles: -80 rad/s, -1.67×10^5 rad/s
Zeros: -2.2×10^5 rad/s, 2.19×10^4 rad/s (RHP)
Crossover frequency: 4.9×10^3 rad/s
Phase margin: 78°

where m_c is set as Eq. (3.37) according to the subharmonics resonance mitigation condition in Eq. (3.36), and a 20% safety margin is applied.

$$m_c = 1.2\frac{v_i}{L}\frac{2\bar{d}-1}{2(1-\bar{d})} \tag{3.37}$$

Fig. 3.12 shows the frequency response of the uncompensated boost converter with current-mode control. As seen, the resonant point does not exist

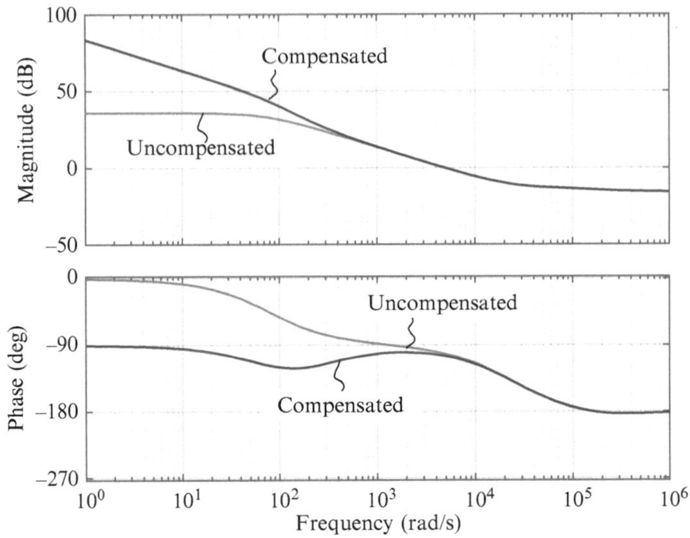

FIG. 3.12 Frequency responses of the current-model-controlled boost converter with and without compensation.

any more compared with Fig. 3.11. To realize an infinite DC gain for nonzero steady-state error tracking, a proportional + integral (PI) voltage compensator is used, as shown

$$G_{cv}(s) = K_p \frac{s + \dfrac{K_i}{K_P}}{s} \tag{3.38}$$

The crossover frequency of the uncompensated system is 4.9×10^3 rad/s which is around 1/64 of the switching frequency (50 kHz). Normally, the crossover frequency can be set to around1/10 of the switching frequency to have a reasonably high bandwidth. Therefore, the PI compensator can be designed to increase the crossover frequency of the system. But the uncompensated system has a −7 dB constant gain and −180° phase at high-frequency band due to the RHP zero, and the corner frequency to the high-frequency band is 2.19×10^4 rad/s (RHP zero) which is only four times the crossover frequency. Thus, if the crossover frequency is increased by the PI compensator, then the gain at the high-frequency band will also be increased and thereby the system may lose its stability. It is decided to keep the crossover frequency and phase margin of the system by setting a 0-dB gain at the high-frequency band in the PI compensator. Thus, we can obtain $K_p = 1$.

As known, the PI compensator will have a −45° phase at its corner frequency. To avoid a significant influence on the phase margin of the system, the corner frequency of the PI compensator is set to 1/15 of the crossover frequency of the system, i.e.,

$$K_i/K_p = 4.9/15\,\text{krad/s} \tag{3.39}$$

Thus, it is gained $K_i = 327$ and

$$G_{cv}(s) = 1 + \frac{327}{s} \tag{3.40}$$

3.2.4 Simulation Results of the Case Study

Figs. 3.13 and 3.14 show the simulation results of the specific case study. Without the slope compensations, subharmonics appear when m_c is equal to $0.8m_{c0}$. By adding a slope compensation, the subharmonics can be eliminated. The overshoot and undershoot of the Boost converter output voltage is limited to a relatively lowlevel under step load changes occur, as shown in Fig. 3.14. The results imply a proper current controller for the Boost converter is achieved.

3.3 TIME-DOMAIN CONTROL OF BASIC DC-DC CONVERTERS

The controllers of DC-DC converters are dominantly designed with small-signal linearization techniques in frequency domain, as discussed in Sections 3.1 and 3.2. However, DC-DC converters with PWM switching are highly nonlinear systems, and their large-signal characteristics will behave differently from that predicted by small-signal design approaches. To overcome the limitations, various time-domain control methods have been proposed. One major class of them is the switching surface control (i.e., boundary control). The concept is to determine the time sequence to turn on/off switches according to certain constrains, namely, switching surfaces [1]. This section is dedicated to a case study to illustrate the basic concept of a generic second-order boundary

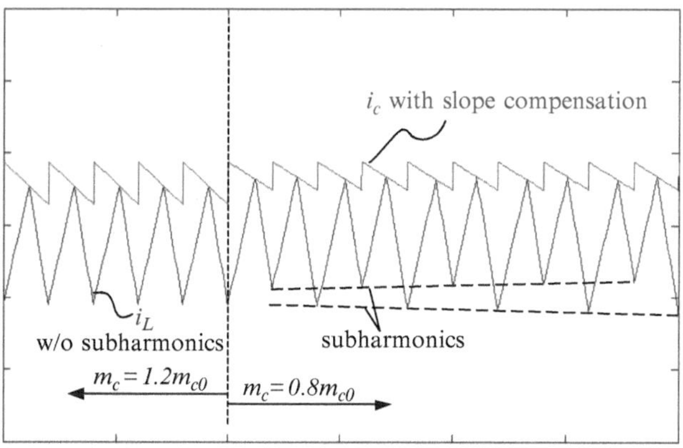

FIG. 3.13 Simulation results to show the impact of the slop compensation on the subharmonics resonance in peak current control $\left(m_{c0} = \frac{v_i}{L}\frac{2\bar{d}-1}{2(1-\bar{d})}\right)$.

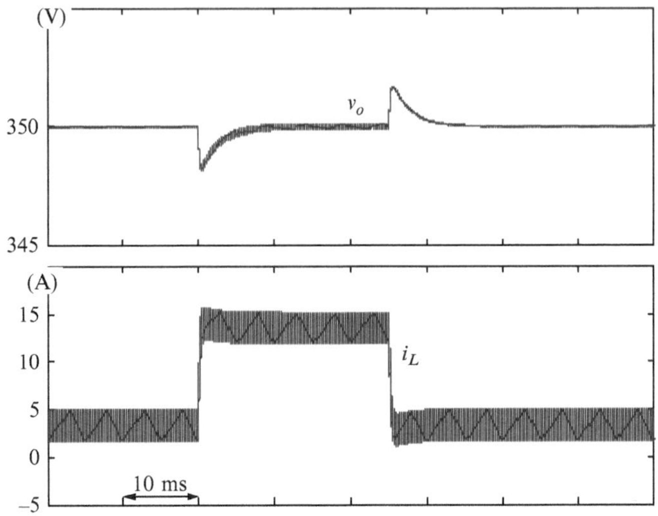

FIG. 3.14 Simulation results to show the dynamic performance of the system during load step.

control method that can be applicable to both Buck-type (i.e., minimum phase system) and Boost type (i.e., nonminimum phase system) converters. Parts of this section are based on what have been discussed in Ref. [1]. The following discussions are based on a specific example of Buck-Boost converter. It should be noted that the method is applicable to all kinds of basic DC-DC converters (i.e., Buck converter, Boost converter, Buck-Boost converter, Ćuk converter, and SEPIC).

3.3.1 Specifications of the Buck-Boost DC-DC Converter

Fig. 3.15 shows the circuit diagram of a Buck-Boost converter. In a specific case study, the parameters are as: output power 190 W, input voltage $v_{in} = 48$ V, and

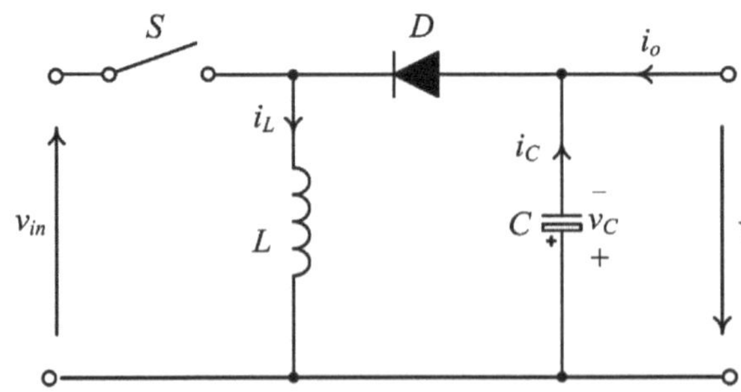

FIG. 3.15 Circuit diagram of a Buck-Boost converter.

output voltage $v_o = 48$ V (according to the polarity definition shown in Fig. 3.15). The inductor $L = 0.36$ mH and the capacitor $C = 150\,\mu$F.

3.3.2 Concept of Generic Second-Order Boundary Control

Second-order switching surfaces are proposed for buck converter [2] as shown in Fig. 3.16. The nonlinear switching surfaces approximately follow the on/off state trajectories of Buck converter. i_L and v_C are the output inductor current and output capacitor voltage of Buck converter, respectively. Therefore, the parameters used in the control law are well defined, and the converter can revert to steady state after two switching actions. A single control law is applicable for buck converter operating in both continuous conduction mode (CCM) and discontinuous conduction mode (DCM). The control methods are named as boundary control with second switching surface [2]. However, one remaining fundamental issue which poses great challenges is that the same concept cannot be easily applied to converters with nonminimum-phase characteristics [1]. It is difficult to formulate a simple switching surface on the state-plane for a converter with state trajectories in spiral shape, such as Boost converter and Buck-Boost Converter, as discussed in Ref. [1].

To handle the above fundamental issue in switching surface control and incorporate the advantages of the boundary control with second-order switching surface which has been applied on buck converter, a uniform second-order switching surface is proposed. It extends the control state variables (i.e., inductor

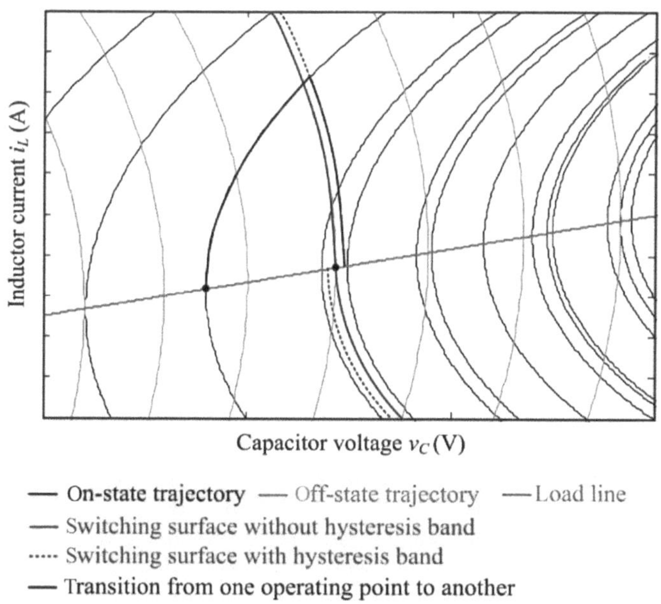

FIG. 3.16 Switching surface control method for Buck converter.

currents and capacitor voltages) in previously proposed switching surface controllers to arbitrary direct or indirect control variables, namely, x and y. Consequently, the relationships between x and y in all basic DC-DC converters are the same as that of i_L-v_C in buck converter. Therefore, the switching surfaces of different kind of converters can be easily defined on a Cartesian x-y plane. x and y have the same properties as those of inductor current and capacitor voltage, respectively, in buck converter, implying that

$$\frac{dy}{dx} = 2k_s\left(x - x_{ref}\right) \tag{3.41}$$

where k_s is state trajectory parameter related with the slope of x and x_{ref} is the reference value.

With the aid of Fig. 3.17 according to Eq. (3.41), the approximated on-state trajectory $\text{Traj}|_{on}$ and off-state CCM trajectory $\text{Traj}|_{off}$ can be expressed, respectively, as follows

$$\text{Traj}|_{on} = y(t) - y(t_0) - k_{s1}\left\{\left[x(t) - x_{ref}\right]^2 - \left[x(t_0) - x_{ref}\right]^2\right\} = 0 \tag{3.42}$$

$$\text{Traj}|_{off} = y(t) - y(t_2) - k_{s2}\left\{\left[x(t) - x_{ref}\right]^2 - \left[x(t_2) - x_{ref}\right]^2\right\} = 0 \tag{3.43}$$

where t_0 and t_2 are the turn on/off instants, respectively, k_{s1} is the trajectory parameter during on-state period, and k_{s2} is that during off-state CCM period. It is noted that $k_{s1} > 0$ and $k_{s2} < 0$. In DCM, after the x reaches zero, the trajectory moves along the y-axis, $x = 0$, and the corresponding trajectory parameter $k_{s3} = 0$ during this interval.

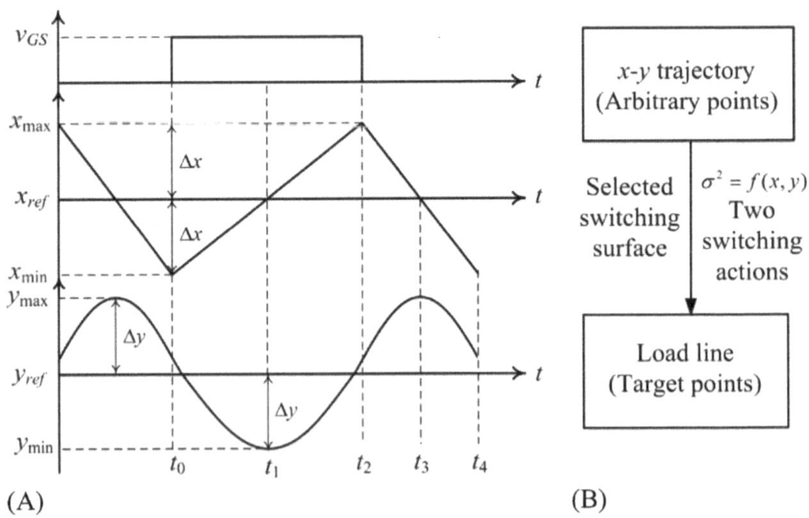

FIG. 3.17 Principle of boundary control with uniform second-order switching surface: (A) control variables and (B) control law.

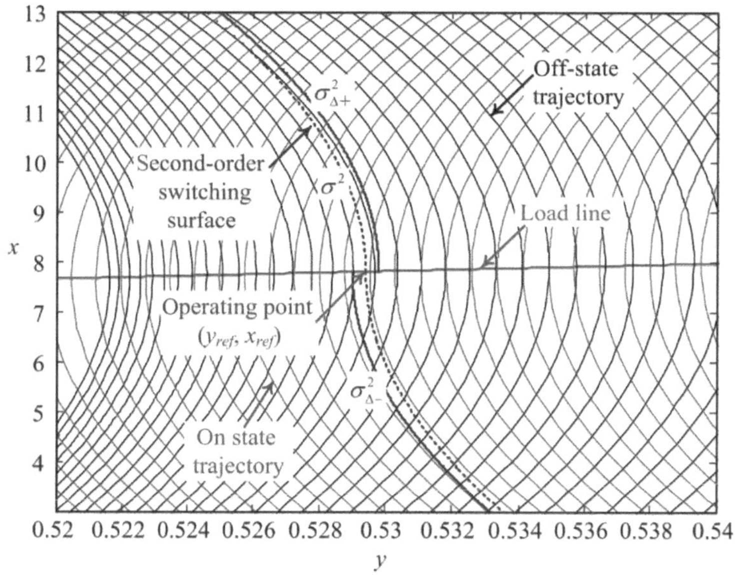

FIG. 3.18 Typical trajectories, load line, and switching surface of basic DC-DC converters on the x-y plane.

Fig. 3.18 presents the generic switching surfaces, load line, and ideal on/off state-trajectories of basic DC-DC converters on x-y plane. The trajectories are plotted by first solving on- and off-state space equations of specific converter with different initial conditions and then mapping the state variables i_L and v_C to x and y, respectively. It is the ideal state trajectories which can be approximated by Eqs. (3.42), (3.43).

As discussed in Ref. [2], the ideal second-order switching surface σ^2 should pass through the target operating point (y_{ref}, x_{ref}), and exactly along the approximated on-state trajectory when x is below the load line, and the off-state trajectory when x is above the load line. By transforming the i_L-v_C state plane to the Cartesian x-y plane, a well-defined uniform second-order switching surface σ^2 can be obtained as shown in Fig. 3.18. It approximately follows the ideal on-state trajectory and off-state trajectory when the state is below and above the load line, respectively, implying a high velocity to revert to the target operating point from arbitrary operating points.

The switching instants are determined by predicting the operating point at t_1 and t_3 as shown in Fig. 3.17. By putting $y(t)=y_{ref}$, $x(t)=x_{ref}$, $y(t_0)=y$, and $y(t_2)=y$ into Eqs. (3.42), (3.43), respectively, a general form of the switching surface is given by

$$\sigma^2 = \left(y - y_{ref}\right) - k\left(x - x_{ref}\right)^2 \tag{3.44}$$

where σ^2 is the uniform second-order switching surface and k is the control parameter and can be expressed as follows for an ideal second-order switching surface

$$k = k_{s1}\left[\frac{1-\text{sgn}\left(x-x_{ref}\right)}{2}\right] + k_{s2}\left[\frac{1+\text{sgn}\left(x-x_{ref}\right)}{2}\right] \tag{3.45}$$

k can be of other values, which affect the stability and trajectory velocity along the switching surface as discussed in Ref. [1]. In the following analysis, k_1 and k_2 are defined as general control parameters for turn-on and turn-off switching actions, respectively. Practically, to avoid chattering phenomenon, a modified switching surface σ_Δ^2 is derived by adding a hysteresis band $2\Delta y$ as shown in Fig. 3.18 into Eq. (3.44), resulting in

$$\sigma_\Delta^2 = \begin{cases} \sigma_{\Delta-}^2 = y - y_{ref} - k_1\left(x-x_{ref}\right)^2 + \Delta y & \left(x < x_{ref}\right) \\ \sigma_{\Delta+}^2 = y - y_{ref} - k_2\left(x-x_{ref}\right)^2 - \Delta y & \left(x > x_{ref}\right) \end{cases} \tag{3.46}$$

where $\sigma_{\Delta-}^2$ and $\sigma_{\Delta+}^2$ are the uniform second-order switching surface below and above load line.

Therefore, the uniform control law for the basic DC-DC converters is formulated as follows. The switch S is turned on if

$$y(t) - k_1\left(x-x_{ref}\right)^2 - \left(y_{ref}-\Delta y\right) \le 0 \quad \text{and} \quad x(t) < x_{ref} \tag{3.47}$$

S is turned off if

$$y(t) - k_2\left(x-x_{ref}\right)^2 - \left(y_{ref}+\Delta y\right) \ge 0 \quad \text{and} \quad x(t) > x_{ref} \tag{3.48}$$

For Buck converter, Boost converter, Buck-Boost converter, Ćuk converter, and SEPIC, the respective x and y can be derived, as discussed in detail in Ref. [1]. Section 3.3.3 will take Buck-Boost converter as an example to derive the x and y.

3.3.3 Second-Order Boundary Controller Design for Buck-Boost Converter

Fig. 3.19 shows the operation modes of Buck-Boost converter. x, y, k_1, and k_2 are derived by the following steps:

Step 1: Derivation of the load line. The load line is derived by considering the average steady-state inductor current $i_{L,ref}$. For the Buck-Boost converter,

$$i_{L,ref} = \frac{v_o i_o}{v_{in}} + i_o \tag{3.49}$$

where v_{in} is the input voltage, and v_o and i_o are the output voltage and current, respectively.

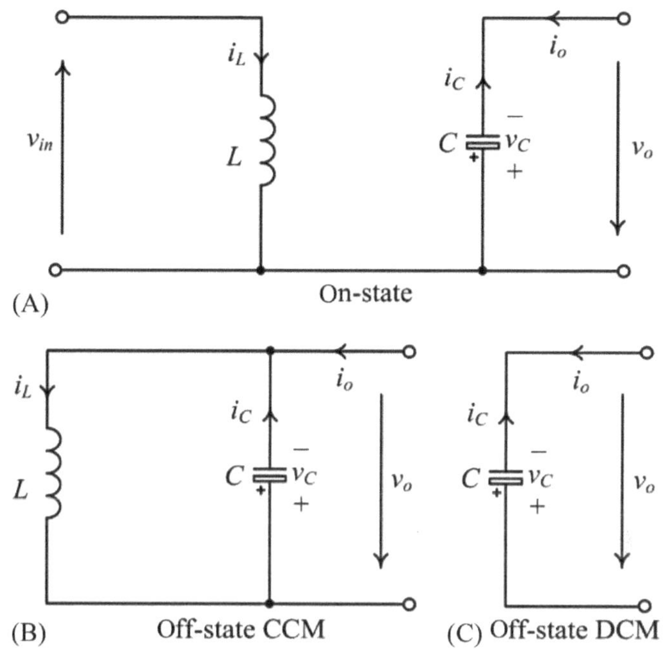

FIG. 3.19 Operation modes of buck-boost converter.

Step 2: Determination of x, y, and k_{s1}. For the Buck-Boost converter, the control variable x is first selected as the inductor current i_L. As shown in Fig. 3.19A, during on-state interval,

$$\frac{dv_o}{di_L} = -\frac{Lv_o}{RCv_{in}} \tag{3.50}$$

where L and C are the inductor and output capacitor, respectively, and R is the equivalent load resistance.

By using Eq. (3.41), it can be shown that

$$y = 2k_1 \int \left(i_L - i_{L,ref}\right) di_L = \frac{2k_1}{L}\left(\frac{1}{2}Li_L^2 + \frac{1}{2}Cv_o^2 + Cv_{in}v_o\right) \tag{3.51}$$

Hence, y can be chosen as

$$y = \frac{1}{2}Li_L^2 + \frac{1}{2}Cv_o^2 + Cv_{in}v_o \tag{3.52}$$

$$k_1 = \frac{L}{2} \tag{3.53}$$

Step 3: Determination of k_2. As shown in Fig. 3.19B, during off-state interval,

$$\frac{dy}{di_L} = \frac{dy}{dt}\frac{dt}{di_L} = -\frac{Lv_{in}}{v_o}\left(i_L - i_{L,ref}\right) \tag{3.54}$$

$$k_{s2} = -\frac{Lv_{in}}{2v_o} \tag{3.55}$$

Step 4: Calculation of y_{ref}. Based on Eqs. (3.49), (3.52),

$$y_{ref} = \frac{1}{2}L\left(\frac{v_o i_o}{v_{in}} + i_o\right)^2 + \frac{1}{2}Cv_{o,ref}^2 + Cv_{in}v_{o,ref} \tag{3.56}$$

Step 5: Formation of the control law. According to Eqs. (3.47), (3.48), the switching criterion is described as follows. The switch S is turned on if

$$y(t) - k_1\left(i_L - i_{L,ref}\right)^2 - \left(y_{ref} - \Delta y\right) \leq 0 \text{ and } i_L(t) < i_{L,ref} \tag{3.57}$$

S is turned off if

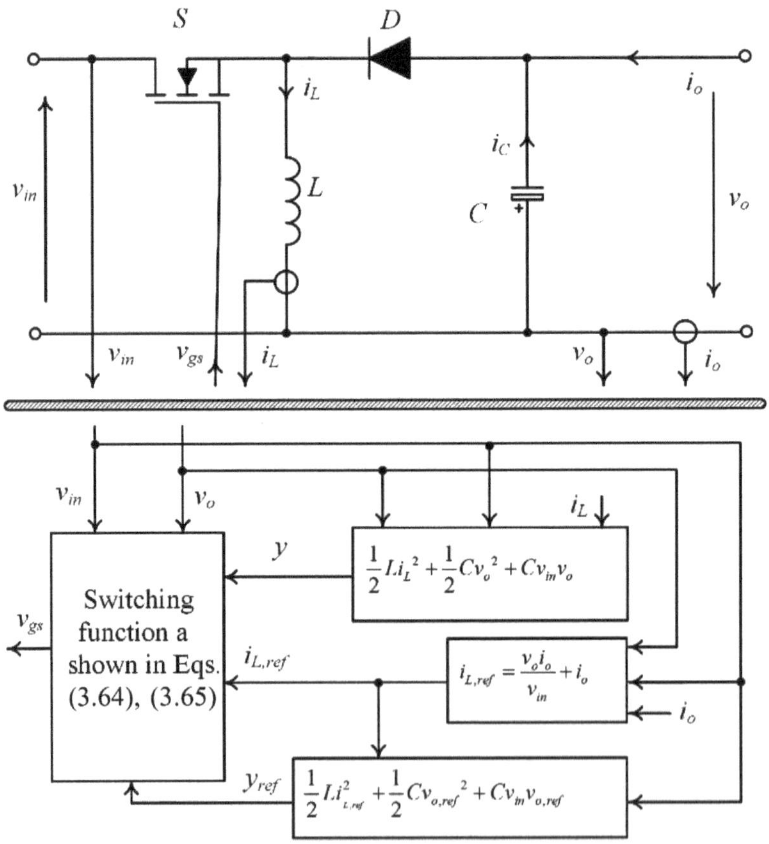

FIG. 3.20 Block diagram of the proposed controller for Buck-Boost converter.

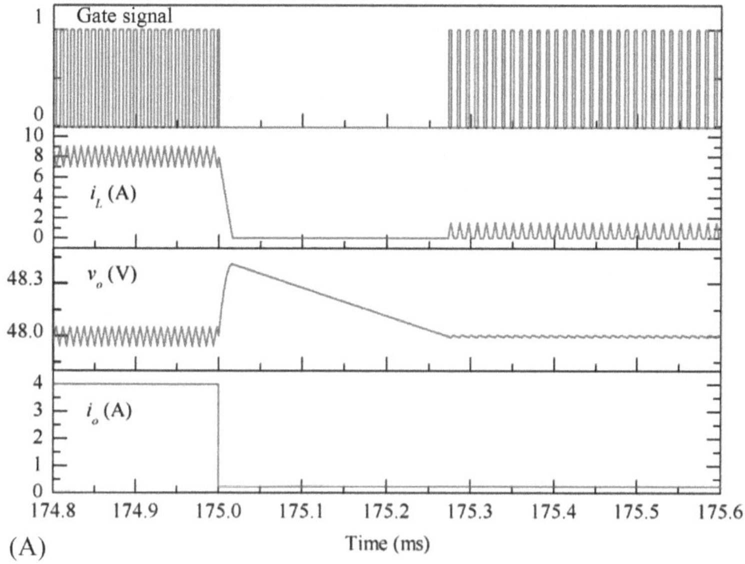

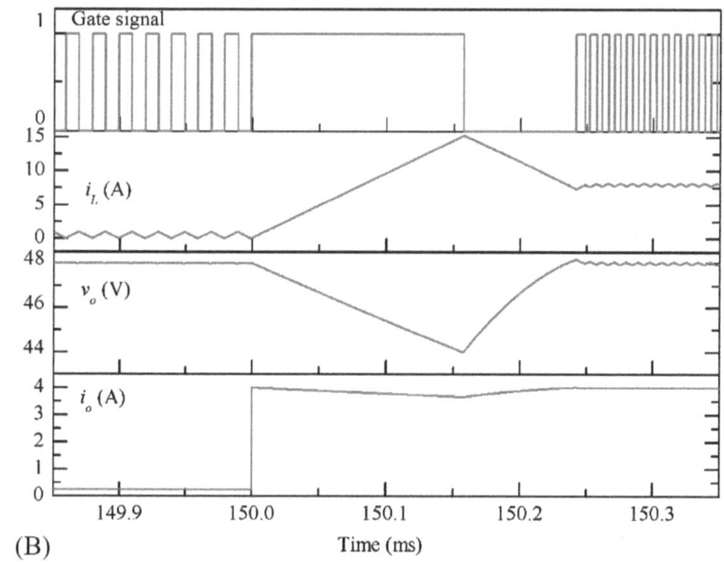

FIG. 3.21 Simulation results of Buck-Boost converters with the proposed controller. (A) Load change from 12 to 200 Ω (CCM to DCM) and (B) Load change from 12 to 200 Ω (CCM to DCM).

$$y(t) - k_2\left(i_L - i_{L,ref}\right)^2 - \left(y_{ref} + \Delta y\right) \geq 0 \text{ and } i_L(t) > i_{L,ref} \tag{3.58}$$

Step 6: Implementation of control law. Based on the above derivations, the block diagram of the controller for the Buck-Boost converter is shown in Fig. 3.20.

3.3.4 Simulation Results of the Case Study

Simulations have been performed for the Buck-Boost converter with the specifications described in Section 3.3.1. The results are shown in Fig. 3.21. The converter can revert to steady state by two switch actions after being exposed to large disturbances.

REFERENCES

[1] H. Wang, New Energy-Efficient High-Voltage DC-DC Power Conversion Technology, City University of Hong Kong, Hong Kong, Mar. 2012.

[2] K.S. Leung, H.S.H. Chung, "Derivation of a second-order switching surface in the boundary control of buck converters," IEEE Power Electron. Lett. 2 (2) (2004) 63–67.

Chapter 4

Modeling and Control of Single-Phase AC/DC Converters

Yongheng Yang* **and Keliang Zhou**[†]
[*]*Aalborg University, Aalborg, Denmark,* [†]*University of Glasgow, Glasgow, United Kingdom*

4.1 INTRODUCTION

Back to 100 years ago, there was a great "war" about currents between Thomas Edison and Nikola Tesla [1]. Although the big controversy ended with the success of Nikola Tesla and today's electricity (power generation, transmission, distribution, and utilization) is predominantly based on AC technology, many loads or systems (e.g., personal computers, electric vehicles, and high-voltage DC (HVDC) transmission systems) in practice can operate only with DC power. As a consequence of this power incompatibility (i.e., DC load and AC source), an intermediate AC/DC conversion stage is required, which is typically based on power electronics. This power processing is called AC/DC rectification, and the power electronic systems are called AC/DC power converters. Fig. 4.1 summarizes the AC/DC conversion, and the role of power electronic converters is important. It can also be seen in Fig. 4.1 that many DC loads or systems exist with a wide range of power ratings (e.g., from milliwatts to megawatts). Practically, the AC/DC converters can be categorized into (1) single-phase and (2) multi-phase [2]. This chapter will focus on single-phase AC/DC converters, which are commonly used in low-power applications, e.g., light-emitting diodes (LED) and computer adapters (chargers).

A simple way to the AC/DC conversion is to use cheap but reliable rectifier diodes. However, due to its nonlinearity and uncontrollability, considerable harmonics are emitted to the AC source when diode AC/DC converters are adopted. Additionally, with the uncontrollable power diodes, the output DC voltage level is unregulated. One possible measure to enhance voltage level adaption is to employ line-frequency transformers at the AC side or DC-DC converters (as discussed in Chapter 3). The resultant multistage AC/DC conversion system is able to change the DC output voltage upon load demands. Furthermore, harmonic distortion at the AC side can be alleviated using multi-pulse

Control of Power Electronic Converters and Systems. https://doi.org/10.1016/B978-0-12-805245-7.00004-4

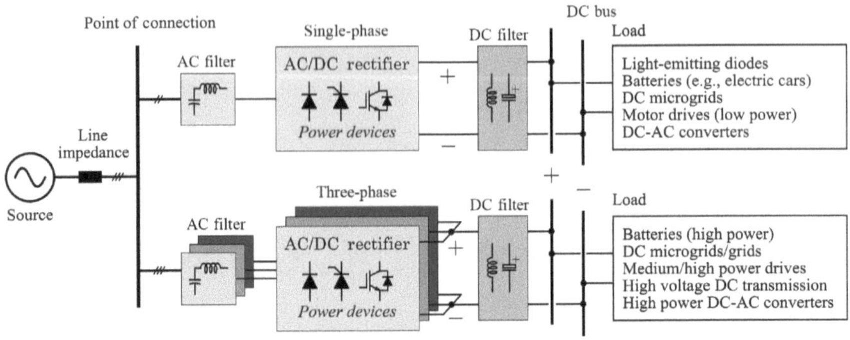

FIG. 4.1 General AC/DC system connected with various loads through power electronic converters. In the case of multiphase rectifiers, phase-shifting transformers are required.

phase-shifting transformers as the front-end unit [3,4]. However, these solutions often give rise to an increased complexity, cost, and volume.

Another approach to convert the AC power to DC power is to use thyristors instead of power diodes, being the silicon-controlled rectifier (SCR) or phase-controlled converters. By its name, the conducting phase at the AC side can be controlled. In this sense, the output DC voltage of an SCR can be adjusted upon load requirements. In spite of the fact that the controllability of the output DC voltage is increased, a large amount of low-order harmonics are still generated and appear in the currents drawn from the AC source. Nonetheless, the thyristor-based rectification converter is one active rectifier converter that can be operated in both rectification and inversion modes [2,5,6]. In many cases, SCR converters with relatively low cost and simple control are adopted to handle high voltage and high power, e.g., HVDC transmission systems, as demonstrated in Fig. 4.1.

To fully control the output DC voltage as well as to maintain good current quality at the AC power source, pulse-width modulation (PWM) techniques can be applied to rectifiers employing fully controllable power devices, e.g., IGBTs and MOSFETs. With the PWM techniques, the impedance of nonlinear rectifier loads approaches a linear resistance, which makes the rectification almost harmonic-free, being "clean" power supplies. Consequently, PWM rectifiers are increasingly used in industry to achieve high power factor, low current harmonics, and constant output DC voltage at the cost of increased control complexity [2,7,8].

Nowadays, more and more stringent rules are applied to harmonic distortions generated by AC/DC rectifiers into AC power source [9]. In this regard, the modeling of AC/DC rectifiers is important so that proper control strategies can be developed to improve the performance of AC/DC converters in terms of low total harmonic distortions (THD). In this chapter, the common single-phase AC/DC converters are modeled. It is followed by typical control schemes for single-phase AC/DC rectifiers, including the power factor correction for diode rectifiers and the current control of PWM rectifiers. More advanced control strategies for active rectifiers are discussed in Chapter 13. As the phase information of the AC input power source is important to the control of AC/DC rectifiers, phase-locked loop (PLL) techniques for single-phase converters are also discussed in this chapter.

4.2 MODELING OF SINGLE-PHASE AC/DC CONVERTERS

4.2.1 Diode Rectifiers

Power diodes conduct only when the forward voltage is positive. This makes it possible to rectify an AC voltage into a DC voltage. Fig. 4.2 shows two implementations of single-phase AC-DC converters using power diodes, i.e., half-wave diode rectifier and full-wave diode rectifier, respectively. Clearly, for the half-wave diode rectifier shown in Fig. 4.2A, only when the input AC voltage is positive, the diode will conduct. Consequently, the current at the input side will be discontinuous, which is not preferable in practical applications. Alternatively, a transformer with a center tap winding at the secondary side can be used as the front-end for two half-wave rectifiers, as illustrated in Fig. 4.2B. In this case, the positive voltage of the secondary voltage v_1 will make the upper diode to conduct, while the negative half-cycle of the secondary voltage v_2 will make the lower diode to conduct. Hence, the current of the input AC source will be continuous, when the AC voltage is rectified, as demonstrated in Fig. 4.2B. However, the full-wave rectifier requires a multiwinding transformer, leading to increased cost and system volume.

Considering those drawbacks, more power diodes are added into the circuit, forming a full-bridge diode rectifier, as shown in Fig. 4.3A. It is known from Fig. 4.3A that, in each half period of the input AC voltage, there is one pair of diodes conducting (i.e., D_1 and D_4 for the positive half-cycle; D_2 and D_3 for the negative half-cycle). Hence, the current at the input side is continuous, like the case shown in Fig. 4.2B. The full-bridge diode rectifier has relatively low volume (the transformer is not mandatory anymore), but the number of diodes is doubled in contrast to the full-wave rectifier based on a center tap

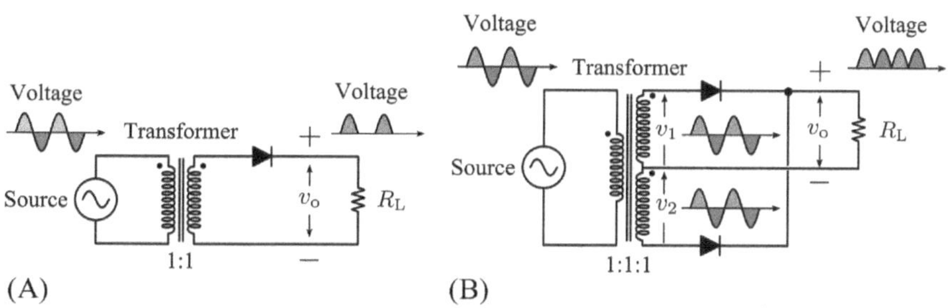

FIG. 4.2 Diode rectifiers: (A) half-wave and (B) full-wave with a center tap winding (at the secondary side) transformer. Resistive loads R_L are connected.

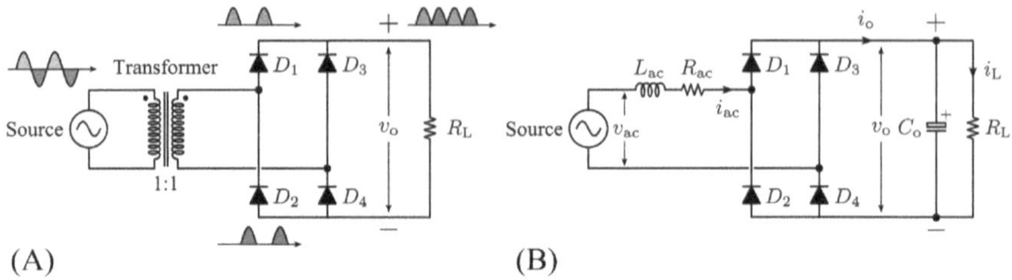

FIG. 4.3 Full-bridge diode rectifiers: (a) ideal converter and (b) practical converter with AC input impedance (L_{ac} and R_{ac}) and an output capacitor C_o. Resistive loads R_L are connected.

winding transformer (Fig. 4.2B). This full-bridge DC/AC converter is thus commonly used in practice, where however the input AC source is not ideal (with an impedance), as shown in Fig. 4.3B. Moreover, in order to smooth the DC output voltage, a large capacitor is also practically adopted at the output. Nevertheless, each pair of the diodes still conducts simultaneously in a certain period. Due to the presence of the AC inductance and the DC capacitance, the AC current will become discontinuous, resulting in a poor power quality, as shown in Fig. 4.4.

During the conducting period (e.g., t_1 to t_3), according to the Kirchhoff's Law, the following mathematical model is obtained:

$$\begin{bmatrix} \dfrac{di_o}{dt} \\ \dfrac{dv_o}{dt} \end{bmatrix} = \begin{bmatrix} -\dfrac{R_{ac}}{L_{ac}} & -\dfrac{1}{L_{ac}} \\ \dfrac{1}{C_o} & -\dfrac{1}{C_oR_L} \end{bmatrix} \begin{bmatrix} i_o \\ v_o \end{bmatrix} + \begin{bmatrix} \dfrac{1}{L_{ac}} \\ 0 \end{bmatrix} |v_{ac}| \tag{4.1}$$

where L_{ac} and R_{ac} are the inductance and resistance of the impedance for the input AC source, respectively, and C_o is the output filter capacitor. Clearly,

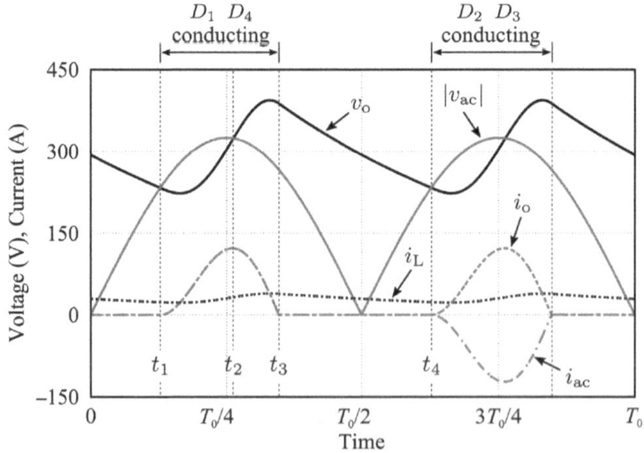

FIG. 4.4 Typical waveforms of a practical full-bridge diode rectifier (Fig. 4.3B), where the AC input voltage amplitude is 325 V and T_0 is its fundamental period.

the conducting period can be divided into two segments: inductor charging period and inductor discharging period. When the AC absolute voltage $|v_{ac}|$ is higher than the output voltage v_o, one pair of the diodes (i.e., D_1 and D_4 during the time interval from t_1 to t_2) are conducting, and the inductor L_{ac} and the capacitor C_o are being changed. This charging stops when the absolute voltage is equal to the capacitor voltage. At this moment, the diodes are not blocking, as the energy stored in the inductor must be released, and the inductor L_{ac} is discharging but the capacitor C_o is still being charged. When the capacitor charging current reaches zero, the capacitor starts to discharge in order to supply the resistive load. Within this short period, both the inductor and the capacitor are discharging. After that, when the inductor current becomes zero, the load will be supplied solely by the capacitor. That is, the diode bridge will be idle until the absolute voltage becomes equal to the capacitor voltage (e.g., at t_4, D_2 and D_3 are conducting). During this idle period, the rectifier system can be described by

$$\begin{bmatrix} \dfrac{dv_o}{dt} \\ \dfrac{dv_o}{dt} \end{bmatrix} = \begin{bmatrix} 0 & 0 \\ 0 & -\dfrac{1}{C_o R_L} \end{bmatrix} \begin{bmatrix} i_o \\ v_o \end{bmatrix} \Rightarrow \frac{dv_o}{dt} = -\frac{1}{C_o R_L} v_o \tag{4.2}$$

With the system parameters, analytical solutions to Eqs. (4.1), (4.2) can be obtained. Clearly, the inductance and the filter capacitance will affect the resultant current quality at the input side.

4.2.2 Phase-Controllable Thyristor Rectifier

Although the full-bridge rectifier is simple and reliable, the output voltage is unregulated. In certain applications, however, the rectified voltage should be controllable. One way is to use a transformer at the AC side with the frequency being the same as that of the input voltage, leading to increased cost and volume. Alternatively, phase-controllable thyristors can be adopted to replace the diodes in Fig. 4.3B, being the SCR, as shown in Fig. 4.5. Compared to the diode rectifier, it is more common to use an inductor as the DC side filter. When the input voltage is higher than the output voltage, thyristors T_1 and T_4 will conduct if positive gate currents are applied to them. This conduction leads to the current flow as what the arrows indicate in Fig. 4.5. When the input voltage becomes negative, although the thyristors T_2 and T_3 are forward biased (different from diodes), they cannot conduct if no positive gate currents are applied. Thus, the energy stored in inductors should be released, keeping the current flowing through T_1 and T_4 to the load, until the gate triggering pulses for T_2 and T_3 are enabled, and then the current will flow through T_2, the AC source, and T_3 to the load.

Considering $L_{dc} \gg L_{ac}$ and ignoring the internal resistance (i.e., R_{dc} and R_{ac}), the thyristor-based rectifier system can simply be modeled as

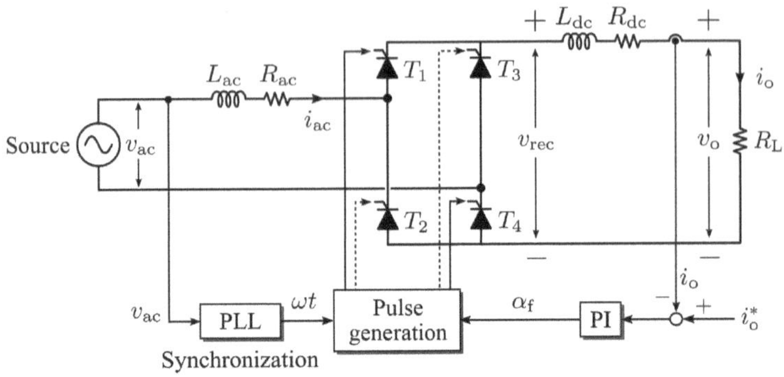

FIG. 4.5 Practical thyristor based AC/DC full-bridge rectifier and its general control block diagram (PLL—phase locked loop; PI—proportional integral), where L_{dc} is the DC-side filter inductance with R_{dc} being its internal resistance, i_o^* is the load current reference, α_f is the controlled firing angle, and ω is the fundamental frequency of the input voltage.

$$\frac{di_o}{dt} = -\frac{R_L}{L_{dc}} i_o + \frac{1}{L_{dc}} v_{rec} \tag{4.3}$$

with v_{rec} being the rectified voltage, and its average voltage can be given as

$$V_{rec} = \frac{2\sqrt{2}}{\pi} V_{ac} \cos \alpha_f \tag{4.4}$$

where V_{ac} is the root mean square (RMS) amplitude of the input voltage, and α_f is the firing angle. α_f is defined as a period from a positive zero-crossing of the input voltage (from negative to positive) to the time instant when the gate triggering pulses are applied. It is clear that the conduction period of each pair of thyristors is related to the firing angle α_f. According to Eqs. (4.3), (4.4), the load current can be controlled by regulating the firing angles of the thyristor converter. Fig. 4.5 also demonstrates the general control structure, where a synchronization unit (e.g., a phase-locked loop system) is required. The synchronization techniques for single-phase systems will be discussed in Section 4.4.

4.2.3 Active AC/DC Converters

The thyristor rectifier offers the possibility to control the output voltage by adjusting the firing angles, as exemplified in Fig. 4.5. However, the turn-off of the thyristors is strongly dependent on the circuit operation conditions. To fully control the turn-on and turn-off instants, active power semiconductor devices like insulated-gate bipolar transistors (IGBT) can be employed, forming an active AC/DC converter [10–12]. Fig. 4.6 shows the circuit diagram of a full-bridge active AC/DC converter system and its general control structure. It is indicated in Fig. 4.6 that the power conditioning is realized by switching the power devices, e.g., based on a PWM technique. With the assumption that the switching frequency is much higher than the fundamental frequency of

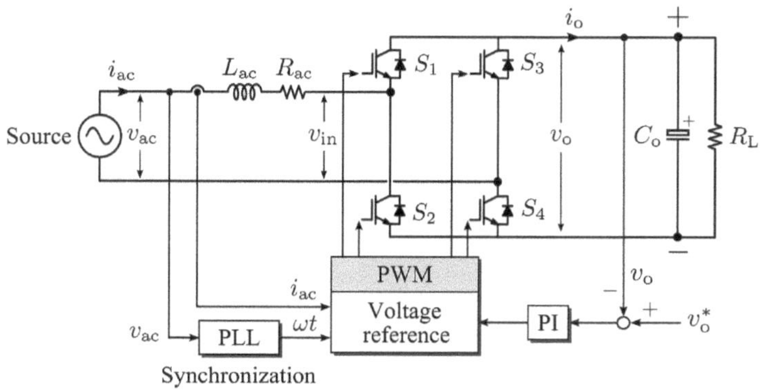

FIG. 4.6 Practical active AC/DC full-bridge rectifier and its general control block diagram (PWM—pulse-width modulation), where v_o^* is the output voltage reference, v_{in} is the rectifier input voltage, and ω is the fundamental frequency of the input voltage.

the input AC source, the rectifier input voltage v_{in} is a PWM voltage, and it can have two or three levels depending on the PWM schemes. The amplitude of the input voltage v_{in} will be equal to the DC side voltage v_o. Thus, in order to ensure a proper power flow from the AC side to the DC load, the following condition should be satisfied:

$$v_o^* > \sqrt{2}V_{ac} \tag{4.5}$$

with v_o^* being the output voltage reference.

In practice, there are mainly two modulation schemes: the bipolar modulation (S_1 and S_4 are switched simultaneously; while the other two are complementarily switched) and the unipolar modulation (each leg of the rectifier is switched separately). Namely, the bipolar modulation results in the input voltage v_{in} of two voltage levels: v_o and $-v_o$; in contrast, the unipolar modulation gives three levels: $v_o, 0$, and $-v_o$, which leads to lower requirements for the input filter. Nonetheless, according to Fig. 4.6, the mathematical model of the PWM rectifier system can be obtained as

$$\begin{cases} \dfrac{di_{ac}}{dt} = -\dfrac{R_{ac}}{L_{ac}} i_{ac} + \dfrac{1}{L_{ac}} v_{ac} - \dfrac{1}{L_{ac}} v_{in} \\ \dfrac{dv_o}{dt} = -\dfrac{1}{C_o R_L} v_o + \dfrac{1}{C_o} i_o \end{cases} \tag{4.6}$$

where it is assumed that the output capacitor is large enough to maintain the output voltage constant. Considering the PWM [11], it gives

$$\begin{cases} v_{in} = m v_o \\ i_o = m i_{ac} \end{cases} \tag{4.7}$$

with m being the modulation signal ($|m| < 1$). Submitting Eq. (4.7) into Eq. (4.6) yields.

$$\begin{bmatrix} \frac{di_{ac}}{dt} \\ \frac{dv_o}{dt} \end{bmatrix} = \begin{bmatrix} -\frac{R_{ac}}{L_{ac}} & -\frac{m}{L_{ac}} \\ \frac{m}{C_o} & -\frac{1}{C_o R_L} \end{bmatrix} \begin{bmatrix} i_{ac} \\ v_o \end{bmatrix} + \begin{bmatrix} \frac{1}{L_{ac}} \\ 0 \end{bmatrix} v_{ac} \tag{4.8}$$

which can be used to assist the controller design (e.g., the PI controller for the DC output voltage as shown in Fig. 4.6). Notably, the power conversion system shown in Fig. 4.6 can be bidirectional, i.e., AC to DC rectification as discussed in this chapter and DC to AC inversion. The DC/AC inversion operation will be presented in Chapter 6.

In addition, seen from Eq. (4.8), it is known that the control variables for the single-phase full-bridge AC/DC rectifiers are the AC current i_{ac} and the DC output voltage v_o. The output voltage can be regulated by the well-established PI controller. However, inevitably, there will be steady-state tracking errors if the PI controller is also used to directly control the AC current [13–17]. Alternatively, if a fictitious system that is in quadrature with the original system is established, two PI controllers can be adopted by transforming the AC variables into DC quantities (i.e., the Park transform). As a result, the AC current is controlled by PI controllers. Fig. 4.7 shows the relationship between the AC system (i.e., the $\alpha\beta$-stationary reference frame) and a DC system (i.e., the dq-rotating reference frame). According to Eqs. (4.6), (4.8), and Fig. 4.7, the mathematical model of the single-phase PWM rectifier (Fig. 4.6 with the created in-quadrature system) can be described in the dq-rotating reference frame as

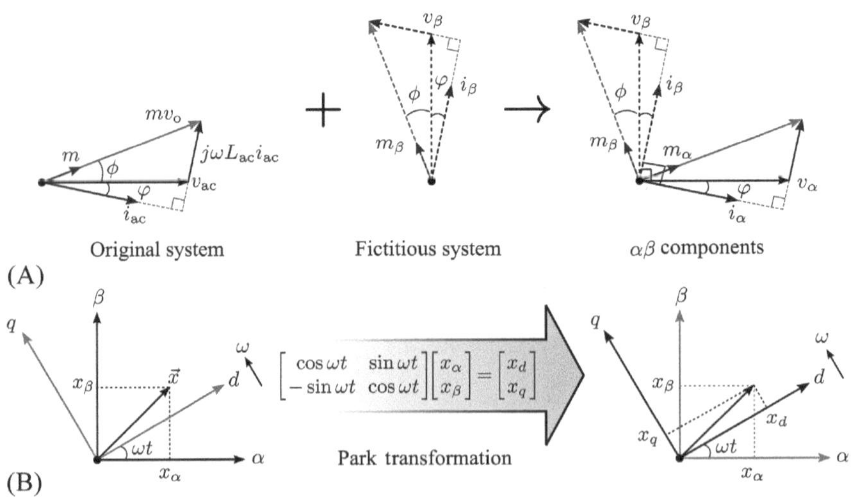

FIG. 4.7 Illustration of the Park transformation for single-phase AC systems: (A) orthogonal system generation to obtain the stationary $\alpha\beta$-components and (B) Park transformation to the synchronous dq-components ($\alpha\beta \rightarrow dq$), where ω is the frequency of the AC voltage v_{ac}, $\theta = \omega t$ becomes the phase between the d-axis and the α-axis, φ is the power angle, and ϕ is the phase difference between the AC voltage and input voltage.

$$\begin{bmatrix} \frac{di_d}{dt} \\ \frac{di_q}{dt} \end{bmatrix} = \begin{bmatrix} -\frac{R_{ac}}{L_{ac}} & \omega \\ -\omega & -\frac{R_{ac}}{L_{ac}} \end{bmatrix} \begin{bmatrix} i_d \\ i_q \end{bmatrix} + \frac{1}{L_{ac}} \begin{bmatrix} v_d \\ v_q \end{bmatrix} - \frac{1}{L_{ac}} \begin{bmatrix} m_d \\ m_q \end{bmatrix} v_o \tag{4.9}$$

where x_{dq} (x represents i, v, m) is the corresponding d- and q-component of the AC current i_{ac}, the AC voltage v_{ac}, and the modulation signal m after the Park transformation. Assuming that a large capacitor is adopted at the DC side, the variables in Eq. (4.9) will all become DC quantities. Following, the model in the s-domain is obtained as

$$\begin{cases} (L_{ac}s + R_{ac})i_d = \omega L_{ac} i_q + v_d - v_o m_d \\ (L_{ac}s + R_{ac})i_q = -\omega L_{ac} i_d + v_q - v_o m_q \end{cases} \tag{4.10}$$

which indicates that the two currents are coupled. Nevertheless, Fig. 4.8 presents the model of the rectifier system (AC side) in the dq-reference frame.

Furthermore, if the control applied to the rectifier system is proper, the modulation signal used to switch the AC/DC PWM rectifier should be sinusoidal and have the same phase with the fundamental component of the input voltage v_{in}. Without loss of generality and according to the phase diagram in Fig. 4.7A, the modulation signal and the input AC current are assumed as

$$\begin{cases} m = \sqrt{2}M\cos(\omega t + \phi) \\ i_{ac} = \sqrt{2}I_{ac}\cos(\omega t - \varphi) \end{cases} \tag{4.11}$$

with M and I_{ac} being the RMS amplitude of the modulation signal and the input AC current, respectively. Substituting Eq. (4.11) into Eq. (4.8) gives the DC output voltage differential equation as

$$\frac{d}{dt}v_o + \frac{1}{C_o R_L}v_o - \frac{1}{C_o} M I_{ac}[\cos(\phi + \varphi) + \cos(2\omega t + \phi - \varphi)] = 0 \tag{4.12}$$

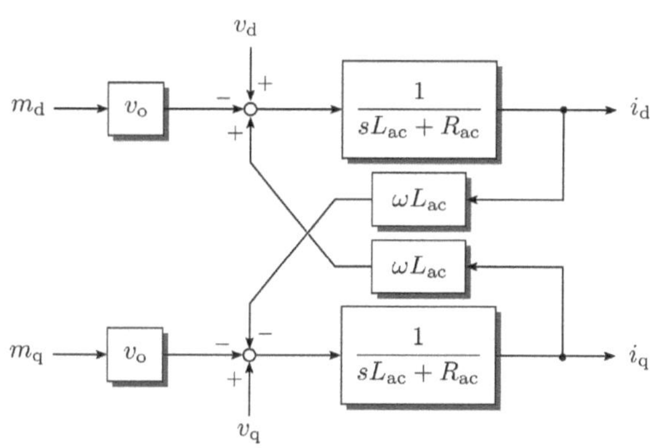

FIG. 4.8 Model of the rectifier system (AC side) in the dq-reference frame, where it is assumed that a large capacitor is adopted (i.e., the DC output voltage is constant).

Considering that the rectifier system is operating at unity power factor (i.e., $\varphi = 0$) and the DC output voltage is maintained as $v_o \simeq \sqrt{2}V_{ac}$ (resulting in $M = \sqrt{2}/2$) with a large capacitor, Eq. (4.12) can be simplified as

$$\frac{d}{dt}v_o + \frac{1}{C_o R_L}v_o - \frac{1}{\sqrt{2}C_o}I_{ac}[1 + \cos(2\omega t)] = 0 \tag{4.13}$$

where the input inductance is small (and thus $\phi \approx 0$). By applying the small signal perturbations to Eq. (4.13) and ignoring the double-frequency component, the small-signal model for the DC output voltage in the s-domain is obtained as

$$\frac{\hat{v}_o}{\hat{i}_{ac}} = \frac{\sqrt{2}}{2}\frac{\frac{1}{C_o}}{s + \frac{1}{C_o R_L}} \tag{4.14}$$

with $\hat{v}_o$ and $\hat{i}_{ac}$ being the variations of the DC output voltage and the RMS amplitude of the AC current. It is indicated in Eq. (4.14) that by properly regulating the DC output voltage using a PI controller, the amplitude of the AC current can be controlled. Notably, the DC output voltage model can also be obtained when considering the power balance. A similar case has been presented for the DC/AC inverter in Chapter 6. Hereby, the models for the AC side and the DC side of the single-phase PWM rectifier have been established, and controllers can be designed accordingly.

4.3 CONTROL OF SINGLE-PHASE AC/DC CONVERTERS

With the models established in Section 4.2, control strategies can be applied to AC/DC rectifier converters. The primary control objective is to achieve a constant DC output voltage, while the distortions at the AC side should be low. For the diode rectifiers, additional power stages or passive components should be adopted as mentioned previously. In contrast, an almost unity power factor can be achieved by properly controlling the PWM AC/DC converters. Accordingly, in the following, two cases will be demonstrated: (a) single-phase PFC system and (b) general control of the single-phase PWM rectifier in the *dq*-reference frame using the presented models.

4.3.1 Power Factor Correction

As mentioned in the last paragraph, one requirement on the diode rectifier is that the resultant AC current should maintain a satisfactory power factor that is related to the harmonic distortions. As demonstrated in Fig. 4.4, directly connecting the diode rectifier will introduce a severe power quality issue. Thus, a power factor correction (PFC) should be adopted. There are many PFC techniques for single-phase diode rectifiers [18–20]. Those solutions can be categorized into two groups: passive methods (e.g., using capacitors) and active

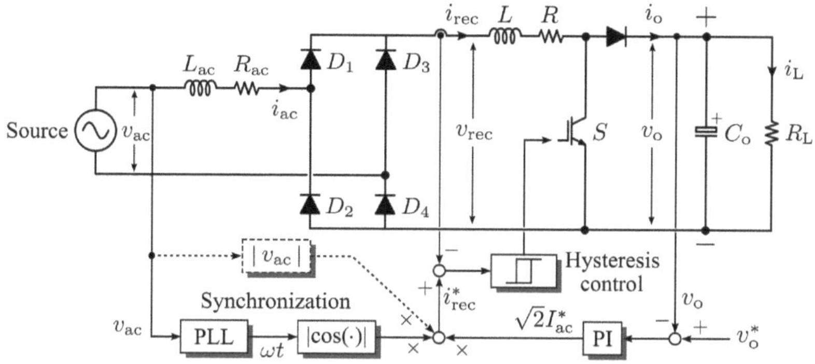

FIG. 4.9 Single-phase boost-based power factor correction system with a synchronization unit (PLL—phase locked loop; PI—proportional integral), where L and R are the boost converter inductor and its internal resistor, respectively. Notably, the AC voltage can also be directly used in the control by taking its absolute value *(dashed lines).*

methods (e.g., adding controllable power devices, stages, or converters). In the following, a case study on a boost-type PFC system is presented to demonstrate the controllability of the power factor of single-phase diode rectifiers. More details of the design of PFC rectifiers can be found in [21,22].

Fig. 4.9 shows the system structure and overall control of the single-phase AC/DC rectifier with a boost PFC stage. The parameters of this system are listed in Table 4.1. As seen in Fig. 4.9, the synchronization is mandatory, and in this case, a second-order generalized integrator (SOGI) based PLL is adopted for synchronization. Details of the SOGI PLL will be discussed in Section 4.4. Furthermore, the control involves in two loops: the DC output voltage loop and the current loop. The DC output voltage is controlled by a PI controller as

TABLE 4.1 Parameters of a Single-Phase Boost Power Factor Correction System

Parameter	Symbol	Value
Power rating	P	1.6 kW
Input voltage amplitude (RMS)	V_{ac}	230 V
Input voltage frequency	ω	314 rad/s
Source impedance	L_{ac}, R_{ac}	0.1 mH, 0.2 Ω
Boost inductor and its resistor	L, R	5 mH, 0.4 Ω
Output capacitor	C_o	2200 μF

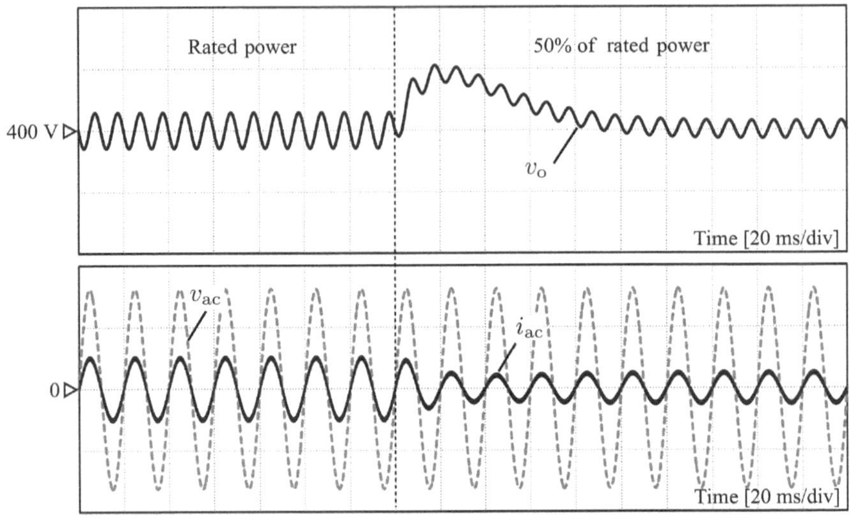

FIG. 4.10 Simulation results of the single-phase boost PFC system shown in Fig. 4.9 in the case of a load step change (rated power to 50% of the rated power): output DC voltage v_o (10 V/div), AC input voltage v_{ac} (200 V/div), and AC input current i_{ac} (20 A/div).

$$G_{PI}(s) = k_p + \frac{k_i}{s} \tag{4.15}$$

with k_p and k_i being the proportional and integral gain of the PI controller. In this case, $k_p = 0.4$ and $k_i = 10$ are selected. For the current control loop, since $i_{rec} = |i_{ac}|$ in steady state, a PI controller may not achieve zero tracking errors. Hence, a hysteresis controller is adopted as shown in Fig. 4.9 for demonstration. The hysteresis band is chosen as 0.2 A, which roughly results in an equivalent switching frequency of 50 kHz at the rated power. A load step change is applied to demonstrate the controller robustness. Simulations are done in PLECS, and Fig. 4.10 shows the simulation results of the PFC system. It is observed that the AC input current with the PFC is in phase with the AC source voltage. The distortion level of the AC current is very low. Thus, the power factor is effectively corrected.

4.3.2 Control of the Full-Bridge Active Rectifier

According to the models of the AC side (i.e., the input current), i.e., Eqs. (4.8), (4.9), the control of the AC input current of the full-bridge AC/DC can be achieved in the stationary reference frame (only the α component) and the dq synchronous rotating reference. As the PI controller is commonly used in many power converter applications in industry, the PI control of the full-bridge AC/DC rectifier shown in Fig. 4.6 is designed in this section. In both the cases, the DC output voltage is controlled by a PI controller. Regarding the controller design, the following control targets should be accomplished:

(a) Low current distortions at the AC side.
(b) The resultant AC current should be in phase with the voltage.
(c) The current control loop should be fast enough.
(d) The voltage control loop should be slow.

Referring to the AC/DC rectifier system shown in Fig. 4.6, the PI current controller in the *dq*-reference frame is thus designed. Accordingly, the PI controller for the DC output voltage can also be designed. The parameters of the AC/DC rectifier system are given in Table 4.2.

As shown in Eq. (4.10) and Fig. 4.8, the *d*- and *q*-current loops are coupled. To simplify the control, two terms are added into the reference modulation signals (m_d^* and m_q^*) in order to decouple the control loops:

$$\begin{cases} m_d^* = -m_d + \dfrac{1}{v_o}\omega L_{ac} i_q \\ m_q^* = -m_q - \dfrac{1}{v_o}\omega L_{ac} i_d \end{cases} \tag{4.16}$$

Submitting Eq. (4.16) into Eq. (4.10) and rearranging the equation, we have

$$\begin{cases} i_d = \dfrac{1}{L_{ac}s + R_{ac}} v_o m_d^* + \dfrac{1}{L_{ac}s + R_{ac}} v_d \\ i_q = \dfrac{1}{L_{ac}s + R_{ac}} v_o m_q^* + \dfrac{1}{L_{ac}s + R_{ac}} v_q \end{cases} \tag{4.17}$$

Thus, the transfer functions for the reference *d*- and *q*-axis currents can be obtained as

TABLE 4.2 Parameters of the Single-Phase Full-Bridge Active Rectifier

Parameter	Symbol	Value
Power rating (at $R_L = 50\,\Omega$)	P	3.2 kW
Input voltage amplitude (RMS)	V_{ac}	230 V
Input voltage frequency	ω	314 rad/s
Source impedance	L_{ac}, R_{ac}	2 mH, 0.2 Ω
Output capacitor	C_o	2200 μF
Carrier frequency	f_{sw}	20 kHz
Sampling frequency	f_s	10 kHz

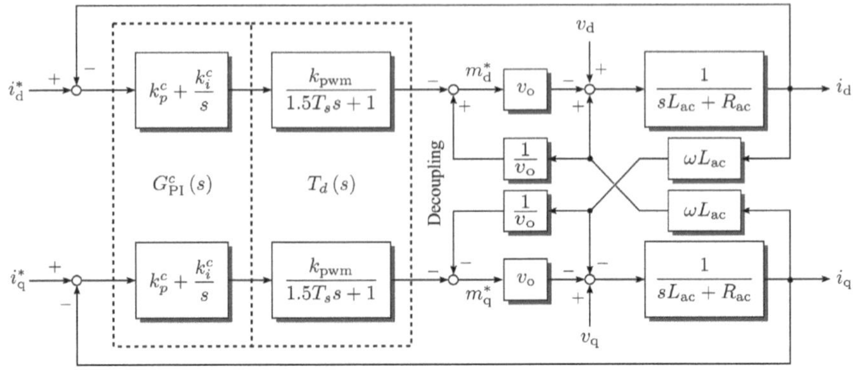

FIG. 4.11 Closed-loop current control system of the single-phase AC/DC rectifier.

$$G_p^c(s) = \left.\frac{i_d^*}{m_d^*}\right|_{v_d=0} - \left.\frac{i_q^*}{m_q^*}\right|_{v_q=0} - \frac{1}{L_{ac}s + R_{ac}} v_o \tag{4.18}$$

indicating that the two currents have the same dynamics. In Eq. (4.18), the input voltage components (i.e., v_d and v_q) are treated as disturbances. This is further illustrated in Fig. 4.11, where $G_{PI}^c(s)$ is the PI controller, and $T_d(s)$ represents the computational and PWM delays in the system. They can be expressed as

$$G_{PI}^c(s) = k_p^c + \frac{k_i^c}{s} \tag{4.19}$$

$$T_d(s) = \frac{k_{pwm}}{1.5T_s s + 1} \tag{4.20}$$

where k_p^c and k_i^c are the proportional and integral gain of the PI controller, T_s is the sampling period of the control system, and k_{pwm} is the PWM gain. According to Fig. 4.11, the coupling terms are canceled in the closed-loop control system when Eq. (4.16) is adopted. As a consequence, the closed-loop transfer function of the currents in the dq-reference frame is obtained as

$$G_{cl}^c(s) = \frac{i_d}{i_d^*} = \frac{i_q}{i_q^*} = \frac{G_{PI}^c(s) \cdot G_d(s) \cdot G_p^c(s)}{1 + G_{PI}^c(s) \cdot G_d(s) \cdot G_p^c(s)} \tag{4.21}$$

which will be a third-order system.

In practice, the PI controller is designed to cancel the pole of the current plant $G_p^c(s)$, i.e.,

$$k_i^c = \frac{R_{ac}}{L_{ac}} k_p^c \tag{4.22}$$

By doing so and also considering $k_{pwm} = 1/v_o$, the closed-loop system in Eq. (4.21) is reduced to

$$G_{cl}^c(s) = \frac{\frac{2k_p^c}{3T_sL_{ac}}}{s^2 + \frac{2}{3T_s}s + \frac{2k_p^c}{3T_sL_{ac}}} \tag{4.23}$$

which is a second-order system with the natural frequency ω_n and the damping ratio ζ being,

$$\begin{cases} \omega_n = \sqrt{\frac{2k_p^c}{3T_sL_{\text{ac}}}} \\ \zeta = \sqrt{\frac{L_{\text{ac}}}{6T_sk_p^c}} \end{cases} \tag{4.24}$$

For an optimally damped system [8,23], $\zeta = \sqrt{2}/2$, leading to

$$k_p^c = \frac{L_{\text{ac}}}{3T_s} \tag{4.25}$$

Substituting Eq. (4.25) into Eq. (4.22) gives.

$$k_i^c = \frac{R_{\text{ac}}}{3T_s} \tag{4.26}$$

Thus, the controller parameters can be determined. Notably, the resultant parameters may give a very high bandwidth of the current loop (several kilo Hz) that is related to the sampling frequency. The controller gains from Eqs. (4.25), (4.26) may not be very feasible considering system uncertainties and the assumptions made during the derivation of the model.

Alternatively, the PI controller gains can be tuned directly according to Eq. (4.21) considering the bandwidth and stability margin. Typically, the bandwidth of the current control loop is around a few hundred Hz and the phase margin should be above 45° [24,25]. With the two requirements and the system parameters shown in Table 4.2, the PI controller gains are tuned in MATLAB using the "PID Tuner" application. The obtained results are

$$k_p^c = 1.48 \quad \text{and} \quad k_i^c = 120.2 \tag{4.27}$$

which result in a bandwidth of 128 Hz and a phase margin of 85.2°.

When the controller for the currents is designed, the DC output voltage controller can be tuned. As mentioned, the DC voltage control loop should be slower compared to the current loop. Here, in this case, the DC voltage control loop is designed to be around 10 times slower. To simplify the parameter tuning, the current closed loop is approximated as a first-order system:

$$G_{cl}^c(s) = \frac{\omega_{\text{bw}}^c}{s + \omega_{\text{bw}}^c} \tag{4.28}$$

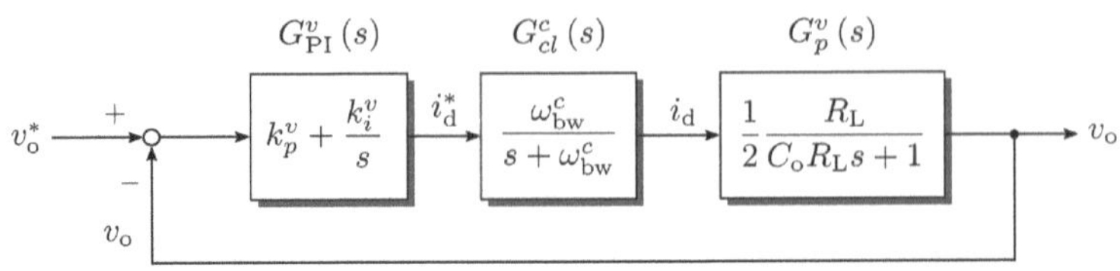

FIG. 4.12 Closed-loop DC output voltage control system of the single-phase AC/DC rectifier.

with $\omega^c{}_{bw}$ being the bandwidth of the current control loop. According to Eq. (4.14), the transfer function (plant model) from the d-axis current to the DC output voltage can be obtained as

$$G^v_p(s) = \frac{v_o}{i^*_d} = \frac{1}{2}\frac{R_L}{C_o R_L s + 1} \tag{4.29}$$

which is also shown in Fig. 4.12. Next, with Eqs. (4.28), (4.29), the closed-loop transfer function for the DC output voltage can be expressed as

$$G^v_{cl}(s) = \frac{v_o}{v^*_o} = \frac{G^v_{PI}(s) \cdot G^c_{cl}(s) \cdot G^v_p(s)}{1 + G^v_{PI}(s) \cdot G^c_{cl}(s) \cdot G^v_p(s)} \tag{4.30}$$

with $G^c_{PI}(s)$ being the PI controller as

$$G^v_{PI}(s) = k^v_p + \frac{k^v_i}{s} \tag{4.31}$$

in which k^v_p and k^v_i are the proportional and integral gain of the PI controller. Similarly, the PI gains for the DC output voltage loop are tuned in MATLAB—the "PID Tuner". Using the parameters in Table 4.2, the gains for the PI controller are designed as

$$k^v_p = 0.3 \text{ and } k^v_i = 14.5 \tag{4.32}$$

which gives a bandwidth of 18 Hz and a phase margin of 60°. The Bode plots of the designed control systems are shown in Fig. 4.13.

Simulations are then carried out in PLECS with the designed controller parameters. The SOGI PLL is adopted for synchronization (also for reference frame transformations). The results are presented in Fig. 4.14, where two cases are simulated. Fig. 4.14A shows the performance of the current control loop in the case of a reference current step change. Observations indicate that the designed PI controllers for the current loop can track the change in one cycle. However, it should be pointed out that dynamics are also dependent on the performance of the SOGI PLL, which is used to create the in-quadrature $\alpha\beta$-system. In this simulation, the settling time for the SOGI PLL was set as 100 ms. Due to the slow dynamics, the current tracking may present a relatively large overshoot or undershoot, as shown in

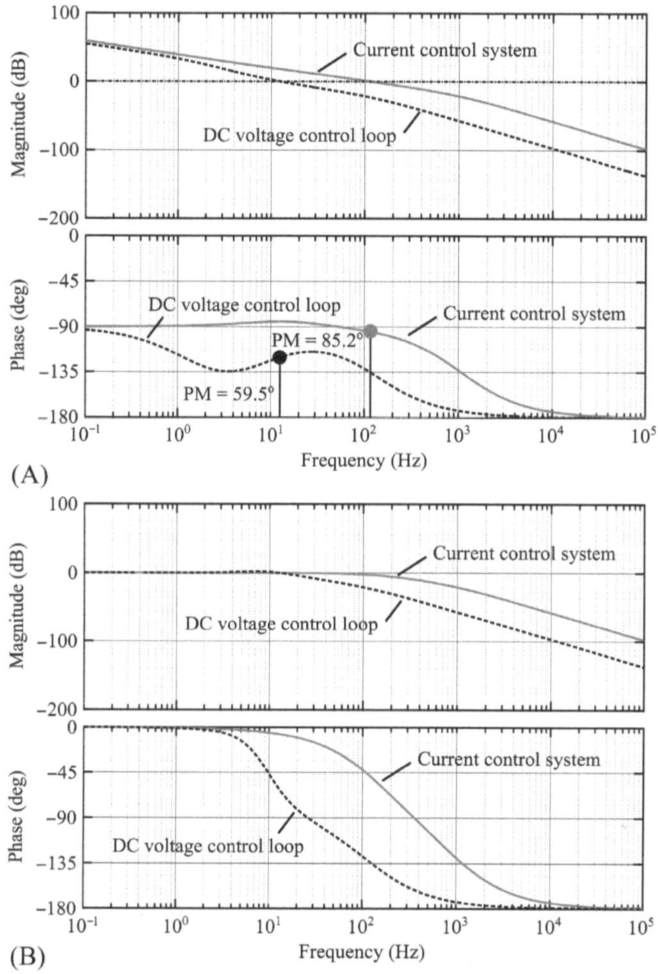

FIG. 4.13 Bode plots of the control loops with the designed parameters: (A) open-loop systems and (B) closed-loop systems (see Figs. 4.11 and 4.12).

Fig. 4.14A. Additionally, the DC output voltage control loop is simulated. It can be observed in Fig. 4.14B that the DC voltage PI controller and the current controllers enable the system to follow the change. Notably, as mentioned in the beginning of this section, the current control can also be achieved in the stationary reference frame. In that case, periodic controllers like a proportional resonant controller [13–17] and a repetitive controller [15,26] can be adopted. Additionally, the hysteresis controller is also a possibility, where it requires less parameter-tuning efforts. The audience of this book is suggested doing simulations in order to study other current controllers.

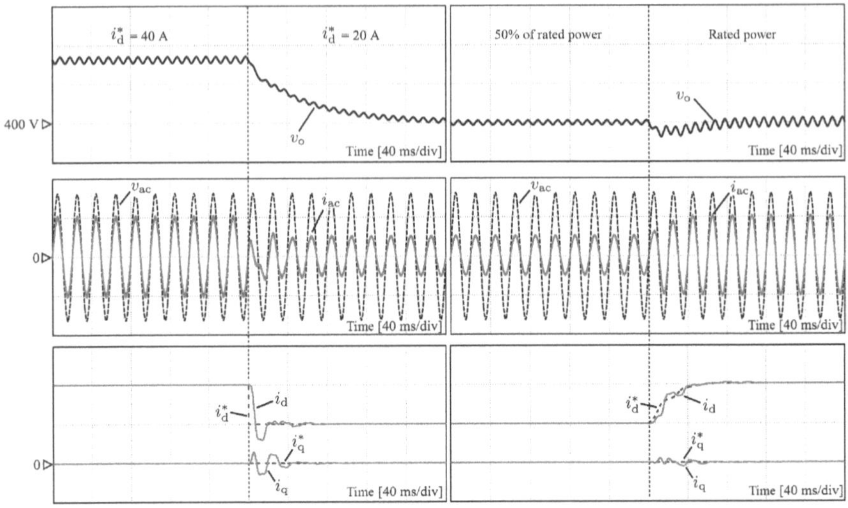

FIG. 4.14 Simulation results of the single-phase AC/DC rectifier system shown in Fig. 4.6 with the PI controllers in the dq-reference frame (output DC voltage v_o [100 V/div], AC input voltage v_{ac} [200 V/div], AC input current i_{ac} [40 A/div], and dq-currents i_d, i_q [20 A/div]): (A) step-change of the d-axis current reference and (B) load step-change.

4.4 PHASE LOCKED LOOP

It has been demonstrated earlier that the synchronization is essential in the control of AC/DC converters: to generate AC references and for reference frame transformations. In fact, the synchronization is a status-monitoring unit that may affect the dynamics and the stability of the entire control system. In three-phase systems, three voltages can be measured for synchronization, which consists of Clarke and Park transformations ($abc \rightarrow \alpha\beta \rightarrow dq$). The resultant dq-voltages are then used to generate or to estimate the phase of the input voltages, and the estimated phase is then fed back for the transformations. In this regard, the synchronization can also be taken as a feedback control system. The most popular synchronization technique is based on the phase-locked loop (PLL). However, in single-phase systems, only one voltage is available for synchronization, and thus many efforts have been devoted to advance the single-phase PLL techniques in the literature [27–33].

Nevertheless, a basic PLL system includes a phase detector, a loop filter, and a voltage-controlled oscillator, as shown in Fig. 4.15. Namely, the phase detector is used to detect the phase difference between the input AC voltage phase and the estimated. Then, a low pass filter as the loop filter compensates the phase error. In practice, a PI regulator is employed as the loop filter. Finally, the output of the loop filter is fed into the voltage-controlled oscillator, that is, an integral unit, and the phase estimation is achieved. Consequently, the small signal model of the PLL system (see Fig. 4.15) can be obtained as

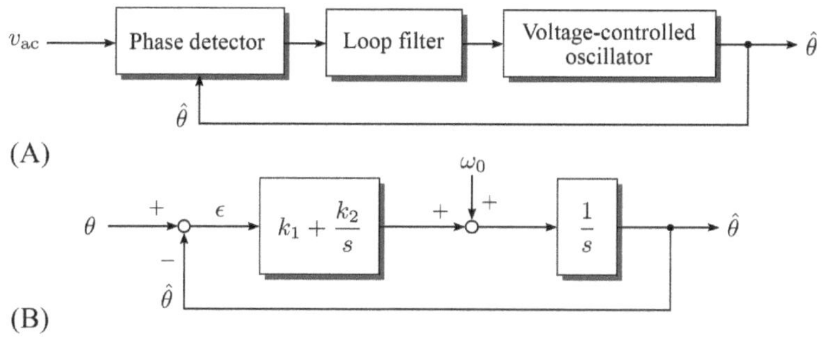

FIG. 4.15 A single-phase PLL system: (A) basic structure and (B) small signal model, where ω_0 is the nominal frequency of the input voltage and $\varepsilon = \theta - \hat{\theta}$ is the phase difference detected by the phase detector.

$$G_{\text{pll}}(s) = \frac{\hat{\theta}}{\theta} = \frac{k_1 s + k_2}{s^2 + k_1 s + k_2} \tag{4.33}$$

where a PI regulator is adopted as the loop filter with k_1 and k_2 being its proportional and integral gain, respectively, θ is the phase of the input AC voltage, and $\hat{\theta}$ is the output of the PLL system (i.e., the estimated phase).

Clearly, for the second-order system in Eq. (4.33), the damping ratio ζ_{pll} and the un-damped natural frequency ω_{pll} can be obtained as

$$\zeta_{\text{pll}} = \frac{k_1}{2\sqrt{k_2}} \quad \text{and} \quad \omega_{\text{pll}} = \sqrt{k_2} \tag{4.34}$$

When the error of the PLL system stays within 1% of the steady state, the settling time of the PLL under a step response is then approximated as

$$t_s = \frac{4.6}{\zeta_{\text{pll}}\omega_{\text{pll}}} = \frac{9.2}{k_1} \tag{4.35}$$

Considering an optimally damped system (i.e., $\zeta_{\text{pll}} = \sqrt{2}/2$), the gains for the PI-based loop filter are obtained as

$$k_1 = \frac{9.2}{t_s} \quad \text{and} \quad k_2 = \frac{42.3}{t_s^2} \tag{4.36}$$

However, in practice, the PI gains should be adjusted slightly due to the delays in the phase detection process. In addition, if the input AC voltage is not unity, the parameters in Eq. (4.36) should be divided by the amplitude of the input voltage [23]. In the case that the parameters of the PI regulator are identical for all PLL systems, the performance will then be affected by the phase detector. In the following, three phase detectors are exemplified as shown in Fig. 4.16. The PLL systems with these phase detectors are correspondingly named the T/4 Delay PLL [27], the Enhanced PLL [30], and the second-order generalized integrator (SOGI) PLL [29].

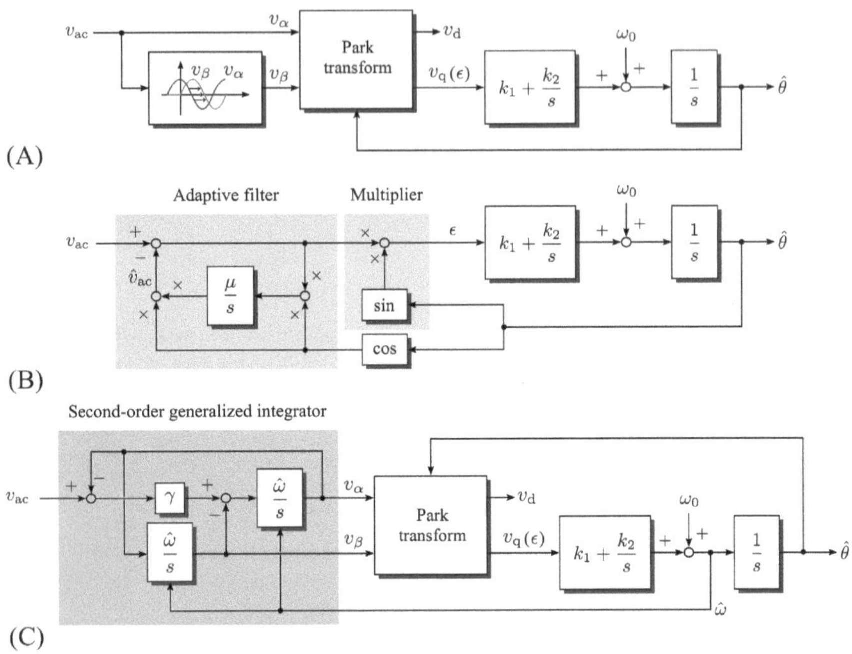

FIG. 4.16 Three single-phase PLL systems: (A) the T/4 Delay PLL, (B) the enhanced PLL (EPLL), and (C) the second-order generalized integrator PLL (SOGI PLL), where μ and γ are the control gains for the EPLL and the SOGI PLL, respectively.

As shown in Fig. 4.16, the *T*/4 Delay PLL and the SOGI PLL establish a virtual voltage that is in-quadrature with the original input AC voltage. This is beneficial to the control in the synchronous rotating reference frame, as demonstrated in Section 4.3. In contrast, the EPLL adopts an adaptive filter and then the voltage error ($v_{ac} - \hat{v}_{ac}$) is multiplied by the estimated in-quadrature signal (i.e., $\sin\left(\hat{\theta}\right)$). Hence, the phase difference is detected. With the employment of the adaptive filter, the EPLL system can reject background harmonic distortions in the AC input voltage to some extent, when compared to the T/4 Delay PLL [27]. However, the EPLL presents relatively slow dynamics. To improve this, a second-order adaptive filter can be adopted, which will act like a "generalized sinusoidal integrator" [23]. Using the second-order generalized integrator can achieve a better harmonic filtering performance. If it is employed to detect the phase difference, being the SOGI PLL as shown Fig. 4.16C, the performance in terms of dynamics and harmonic immunity will be significantly improved. Additionally, as observed in Fig. 4.16B and C, there is one control gain for the EPLL and the SOGI PLL, respectively, which will affect the entire system dynamics.

To design the two gains, it is assumed that the input AC voltage v_{ac} and the PLL-estimated AC voltage $\hat{v}_{ac}$ can be expressed as

$$\begin{cases} v_{ac} = V_m \cos(\theta) = V_m \cos(\omega t) \\ \hat{v}_{ac} = \hat{V}_m \cos(\hat{\theta}) = \hat{V}_m \cos(\hat{\omega} t) \end{cases} \tag{4.37}$$

where V_m, θ, and ω is the amplitude, phase, and frequency of the input voltage, and "$\wedge$" indicates the corresponding estimated variable. According to Fig. 4.16B, the estimated input voltage amplitude can be obtained as

$$\frac{d\hat{V}_m}{dt} = \mu (v_{ac} - \hat{v}_{ac}) \cos(\hat{\theta}) \tag{4.38}$$

which implies that the dynamics of the estimation is determined by the control parameter μ. Linearizing Eq. (4.38) yields

$$\frac{\hat{V}_m}{V_m} = \frac{1}{\tau s + 1} \tag{4.39}$$

with $\tau = 2/\mu$ being the time constant. Then, the settling time of the adaptive filter of the EPLL can be approximated as $4\tau = 8/\mu$. When the desired settling time is set, the control gain of the EPLL can be obtained accordingly. For the SOGI PLL shown in Fig. 4.16C, the phase detector can be described in the s-domain as

$$\begin{bmatrix} v_\alpha \\ v_\beta \end{bmatrix} = \begin{bmatrix} \dfrac{\gamma \hat{\omega} s}{s^2 + \gamma \hat{\omega} s + \hat{\omega}^2} \\ \dfrac{\gamma \hat{\omega}^2}{s^2 + \gamma \hat{\omega} s + \hat{\omega}^2} \end{bmatrix} v_{ac} \tag{4.40}$$

which is a typical second-order system. Hence, the control gain of the SOGI in-quadrature generation system can be designed as $\gamma = \sqrt{2}$ for an optimally damped system in terms of settling time and overshoots.

Notably, in practical applications, the AC input voltage may experience various disturbances (e.g., voltage sags, frequency variations, and harmonic distortions). These eventualities challenge the synchronization. As an exercise, the audience can benchmark the three PLL systems with identical parameters (designed according to the earlier discussion) in order to find the suitable synchronization technique. In general, the SOGI PLL stands out in terms of fast dynamics, good accuracy, and high immunity to harmonic distortions. This is the reason why it has been employed in the control of AC/DC converters demonstrated in this chapter.

4.5 SUMMARY

In this chapter, single-phase AC/DC rectifier converters have been briefly overviewed. Modeling of the full-bridge diode rectifier, thyristor-based phase-controllable rectifier, and the fully controllable active AC/DC converter have been presented. The models enable proper control of the single-phase AC/DC converters to achieve a satisfactory power quality at the AC side. The controllability of the diode rectifier has been demonstrated on a single-phase power factor correction system, where an almost unity power factor is achieved. Furthermore, with the presented models of the active PWM rectifier, the design of proportional integral controllers for the DC output voltage and the current has been exemplified. The designed parameters were validated by simulations. Additionally, the control of AC/DC rectifiers requires the information of the AC input voltage, known as synchronization. Hence, this chapter has also presented the basics of phase locked loop (PLL) based synchronization techniques, and discussed three PLL single-phase PLL systems in terms of overall performance and design.

REFERENCES

[1] U.S. Department of Energy, The war of the currents: ac vs. dc power, https://energy.gov/articles/war-currents-ac-vs-dc-power, Nov. 14, 2014. Accessed 30 July 2017.

[2] N. Mohan, T.M. Undeland, W.P. Robbins, Power Electronics: Converters, Applications, and Design, third ed, Wiley, Hoboken, NJ, USA., 2002.

[3] J. Sun, Z. Bing, K.J. Karimi, Input impedance modeling of multipulse rectifiers by harmonic linearization, IEEE Trans. Power Electron. 24 (12) (2009) 2812–2820.

[4] F.J. Chivite-Zabalza, A.J. Forsyth, D.R. Trainer, A simple, passive 24-pulse ac-dc converter with inherent load balancing, IEEE Trans. Power Electron. 21 (2) (2006) 430–439.

[5] H.S. Patel, R.G. Hoft, Generalized techniques of harmonic elimination and voltage control in thyristor inverters: part I—harmonic elimination, IEEE Trans. Ind. Appl. IA-9 (3) (1973) 310–317.

[6] M.A. Cardiel-Alvarez, J.L. Rodriguez-Amenedo, S. Arnaltes, M. Montilla-DJesus, Modeling and control of LCC rectifiers for offshore wind farms connected by HVDC links, IEEE Trans. Energy Convers. (2017) early access online. https://doi.org/10.1109/TEC.2017.2696261.

[7] B. Singh, B.N. Singh, A. Chandra, K. Al-Haddad, A. Pandey, D.P. Kothari, A review of single-phase improved power quality AC-DC converters, IEEE Trans. Ind. Electron. 50 (5) (2003) 962–981.

[8] M.P. Kazmierkowski, R. Krishnan, F. Blaabjerg (Eds.), Control in Power Electronics: Selected Problems, Academic Press, San Diego, CA, USA, 2002.

[9] International Electrotechnical Commission (IEC), Electromagnetic Compatibility (EMC)—Part 3-2: Limits-Limits for Harmonic Current Emissions (Equipment Input Current $\leq$ 16 A Per Phase), IEC Standards, 2014.

[10] O. Stihi, B.-T. Ooi, A single-phase controlled-current PWM rectifier, IEEE Trans. Power Electron. 3 (4) (1988) 453–459.

[11] O. Kukrer, H. Komurcugil, Control strategy for single-phase PWM rectifiers, Electron. Lett. 33 (21) (1997) 1745–1746.

[12] J.R. Rodriguez, J.W. Dixon, J.R. Espinoza, J. Pontt, P. Lezana, PWM regenerative rectifiers: state of the art, IEEE Trans. Ind. Electron. 52 (1) (2005) 5–22.

[13] X. Yuan, W. Merk, H. Stemmler, J. Allmeling, Stationary-frame generalized integrators for current control of active power filters with zero steady-state error for current harmonics of concern under unbalanced and distorted operating conditions, IEEE Trans. Ind. Appl. 38 (2) (2002) 523–532.

[14] D.N. Zmood, D.G. Holmes, Stationary frame current regulation of PWM inverters with zero steady-state error, IEEE Trans. Power Electron. 18 (3) (2003) 814–822.
[15] K. Zhou, D. Wang, Y. Yang, F. Blaabjerg, Periodic Control of Power Electronic Converters, IET Publisher, London, UK, 2016.
[16] A.G. Yepes, F.D. Freijedo, Ó. Lopez, J. Doval-Gandoy, High-performance digital resonant controllers implemented with two integrators, IEEE Trans. Power Electron. 26 (2) (2011) 563–576.
[17] A.G. Yepes, F.D. Freijedo, Ó. Lopez, J. Doval-Gandoy, Analysis and design of resonant current controllers for voltage-source converters by means of Nyquist diagrams and sensitivity function, IEEE Trans. Ind. Electron. 58 (11) (2011) 5231–5250.
[18] O. Garcia, J.A. Cobos, R. Prieto, P. Alou, J. Uceda, Single phase power factor correction: a survey, IEEE Trans. Power Electron. 18 (3) (2003) 749–755.
[19] B. Singh, B.N. Singh, A. Chandra, K. Al-Haddad, A. Pandey, D.P. Kothari, A review of single-phase improved power quality AC-DC converters, IEEE Trans. Ind. Electron. 50 (5) (2003) 962–981.
[20] H. Akagi, Active harmonic filters, Proc. IEEE 93 (12) (2005) 2128–2141.
[21] ON Semiconductor, Power factor correction (PI) handbook, Tech. Rep. No. HBD853/D, http://www.onsemi.com/pub/Collateral/HBD853-D.PDF, 2014. Accessed 27 August 2017.
[22] Infineon, PFC boost converter design guide, Application Note, https://www.infineon.com/dgdl/Infineon-ApplicationNote_PFCCCMBoostConverterDesignGuide-AN-v02_00-EN.pdf?fileId=5546d4624a56eed8014a62c75a923b05, 2016. Accessed 27 August 2017.
[23] R. Teodorescu, M. Liserre, P. Rodriguez, Grid Converters for Photovoltaic and Wind Power Systems, John Wiley & Sons, West Sussex, UK, 2011.
[24] D. Zhou, F. Blaabjerg, Bandwidth oriented proportional-integral controller design for back-to-back power converters in DFIG wind turbine system, IET Renew. Power Gener. 11 (7) (2017) 941–951.
[25] H. Mao, D. Boroyevich, F.C.Y. Lee, Novel reduced-order small-signal model of a three-phase PWM rectifier and its application in control design and system analysis, IEEE Trans. Power Electron. 13 (3) (1998) 511–521.
[26] K. Zhou, D. Wang, Digital repetitive controlled three-phase PWM rectifier, IEEE Trans. Power Electron. 18 (1) (2003) 309–316.
[27] Y. Yang, L. Hadjidemetriou, F. Blaabjerg, E. Kyriakides, in: Benchmarking of phase locked loop based synchronization techniques for grid-connected inverter systems, Proc. of ICPE-ECCE Asia, 2015, pp. 2167–2174.
[28] Y. Han, M. Luo, X. Zhao, J. Guerrero, L. Xu, Comparative performance evaluation of orthogonal-signal-generators based single-phase PLL algorithms, IEEE Trans. Power Electron. 31 (5) (2016) 3932–3944.
[29] M. Ciobotaru, R. Teodorescu, F. Blaabjerg, in: A new single-phase PLL structure based on second order generalized integrator, Proc. of PESC, 2006, , pp. 1–6.
[30] M. Karimi-Ghartemani, M.R. Iravani, A nonlinear adaptive filter for online signal analysis in power systems: applications, IEEE Trans. Power Delivery 17 (2) (2002) 617–622.
[31] J. Wang, J. Liang, F. Gao, L. Zhang, Z. Wang, A method to improve the dynamic performance of moving average filter based PLL, IEEE Trans. Power Electron. 30 (10) (2015) 5978–5990.
[32] L. Hadjidemetriou, E. Kyriakides, Y. Yang, F. Blaabjerg, A synchronization method for single-phase grid-tied inverters, IEEE Trans. Power Electron. 31 (3) (2016) 2139–2149.
[33] Y.F. Wang, Y.W. Li, Analysis and digital implementation of cascaded delayed-signal-cancellation PLL, IEEE Trans. Power Electron. 26 (4) (2011) 1067–1080.

Chapter 5

Modeling and Control of Three-Phase AC/DC Converter Including Phase-Locked Loop

Dao Zhou, Yipeng Song and Frede Blaabjerg
Aalborg University, Aalborg, Denmark

5.1 MODELING AND ANALYSIS OF GRID-TIED CONVERTER

Due to the existence of power electronic semiconductor in the grid-tied converter, featuring as the switching device, the circuit inevitably becomes discontinuous. Similar with the dynamic modeling of the DC/DC converter, a state-average approach can be used to enable the circuit model continuous. However, different from the DC/DC converter, an AC voltage in the input side is always time-varying, a quiescent operating point can hardly be found without the help of the reference frame transformation (Clarke transformation and Park transformation), which changes a static three-phase system to a rotating two-phase system. The *dq* circuits are still nonlinear because of the multiple disturbances from the control objective and control plant, whereas the classical controller design (like root locus, Bode plots, and Nyquist plot) normally relies on the linear system [1,2]. Due to the fact that circuit variables are linear with a small disturbance around the quiescent operation point, a small-signal model can be established to implement the classical procedure of the controller design.

5.1.1 Mathematical Model Under *abc* Reference Frame

With functions of the fixed DC-link voltage and the unity power factor, the topology of the three-phase grid-tied converter is shown in Fig. 5.1. Although an LCL filter is widely adopted in modern industry applications, as a single L filter has the similar characteristics compared with the LCL filter within low frequency [3], the L filter is applied for simplicity.

According to Kirchhoff voltage law (KVL) in the AC-side and the Kirchhoff current law (KCL) in the DC-side, following equations can be obtained,

Control of Power Electronic Converters and Systems. https://doi.org/10.1016/B978-0-12-805245-7.00005-6

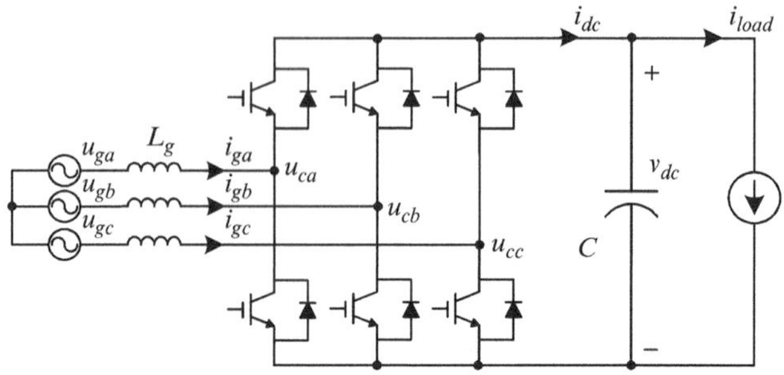

FIG. 5.1 Typical structure of three-phase grid-tied converter using single inductor filter.

$$\begin{cases} u_{ga} - L_g \dfrac{di_{ga}}{dt} - S_a v_{dc} = u_{gb} - L_g \dfrac{di_{gb}}{dt} - S_b v_{dc} = u_{gc} - L_g \dfrac{di_{gc}}{dt} - S_c v_{dc} \\ C \dfrac{dv_{dc}}{dt} = S_a i_{ga} + S_b i_{gb} + S_c i_{gc} - i_{load} \end{cases} \tag{5.1}$$

where S_X denotes the switching function, whose value becomes 1 or 0 with the turn-on or turn-off status of the power switch; u_{gx} denotes the grid voltage; i_{gx} denotes the current of the AC-side; and subscripts *abc* denote the three-phase system. v_{dc} is the DC voltage; i_{load} is the loading current; and L_g and C are the boost inductor and the DC capacitor, respectively.

In the symmetrical three-phase system with $u_{ga}+u_{gb}+u_{gc}=0$, together with the $i_{ga}+i_{gb}+i_{gc}=0$ in the three-wire system, Eq. (5.1) can be deduced as,

$$\begin{cases} \begin{bmatrix} u_{ga} \\ u_{gb} \\ u_{gc} \end{bmatrix} = L_g \dfrac{d}{dt} \begin{bmatrix} i_{ga} \\ i_{gb} \\ i_{gc} \end{bmatrix} + \begin{bmatrix} S_a - \dfrac{S_a + S_b + S_c}{3} \\ S_b - \dfrac{S_a + S_b + S_c}{3} \\ S_c - \dfrac{S_a + S_b + S_c}{3} \end{bmatrix} v_{dc} \\ C \dfrac{d}{dt} v_{dc} = [S_a \ S_b \ S_c] \begin{bmatrix} i_{ga} \\ i_{gb} \\ i_{gc} \end{bmatrix} - i_{load} \end{cases} \tag{5.2}$$

where $\frac{S_a+S_b+S_c}{3} v_{dc}$ is the zero-sequence voltage. It is noted that Eq. (5.2) behaves not only complicated and coupled, but also discontinuous and nonlinear.

5.1.2 State-Average Model Under *abc* Reference Frame

To ensure the control circuit continuous, the state-average approach can be used to obtain the integral of the variable within one switching period T_s, which yields,

$$\langle x \rangle_{T_s} = \frac{1}{T_s}\int_t^{t+T_s} x(\tau)d\tau \tag{5.3}$$

where $\langle x \rangle_{T_s}$ denotes the state-average value.

By using this approach, the switching function is the same with duty cycle, and state equations of inductor current and the capacitor voltage of Eq. (5.2) become,

$$\begin{cases} \begin{bmatrix} \langle u_{ga} \rangle_{T_s} \\ \langle u_{gb} \rangle_{T_s} \\ \langle u_{gc} \rangle_{T_s} \end{bmatrix} = L_g \dfrac{d}{dt} \begin{bmatrix} \langle i_{ga} \rangle_{T_s} \\ \langle i_{gb} \rangle_{T_s} \\ \langle i_{gc} \rangle_{T_s} \end{bmatrix} + \begin{bmatrix} d_a - \dfrac{d_a + d_b + d_c}{3} \\ d_b - \dfrac{d_a + d_b + d_c}{3} \\ d_c - \dfrac{d_a + d_b + d_c}{3} \end{bmatrix} \langle v_{dc} \rangle_{T_s} \\ C\dfrac{d}{dt}\langle v_{dc} \rangle_{T_s} = [d_a \;\; d_b \;\; d_c] \begin{bmatrix} \langle i_{ga} \rangle_{T_s} \\ \langle i_{gb} \rangle_{T_s} \\ \langle i_{gc} \rangle_{T_s} \end{bmatrix} - \langle i_{load} \rangle_{T_s} \end{cases} \tag{5.4}$$

where d denotes the duty cycle. In this way, the circuit becomes continuous, while the switching ripple is removed compared with the real system.

5.1.3 Coordinate Transformation and Space Vector

With Clarke transformation, the three-phase static abc reference frame can be converted into the two-phase static $\alpha\beta$ reference frame, and vice versa,

$$C_{32} = \frac{2}{3}\begin{bmatrix} 1 & -\dfrac{1}{2} & -\dfrac{1}{2} \\ 0 & \dfrac{\sqrt{3}}{2} & -\dfrac{\sqrt{3}}{2} \end{bmatrix} \quad C_{23} = \begin{bmatrix} 1 & 0 \\ -\dfrac{1}{2} & \dfrac{\sqrt{3}}{2} \\ -\dfrac{1}{2} & -\dfrac{\sqrt{3}}{2} \end{bmatrix} \tag{5.5}$$

As shown in Fig. 5.2A, Eq. (5.5) is obtained by the projection of abc coordinate to $\alpha\beta$ coordinate. Meanwhile, a gain of 2/3 is used in order to keep the same amplitude.

Park transformation changes the static $\alpha\beta$ reference frame into the rotating dq reference frame, and vice versa,

$$C_{2s/2r} = \begin{bmatrix} \cos\omega t & \sin\omega t \\ -\sin\omega t & \cos\omega t \end{bmatrix} \quad C_{2r/2s} = \begin{bmatrix} \cos\omega t & -\sin\omega t \\ \sin\omega t & \cos\omega t \end{bmatrix} \tag{5.6}$$

where θ is the angle between the d-axis and α-axis, and ω is the angular speed of the grid. As shown in Fig. 5.2B, Eq. (5.6) is obtained by the projection of $\alpha\beta$ coordinate to dq coordinate

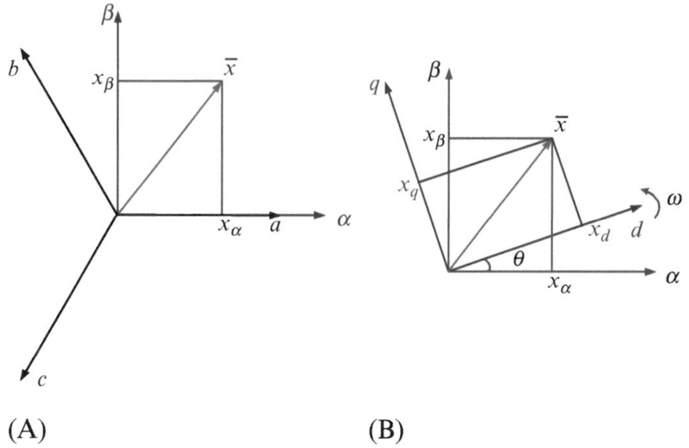

FIG. 5.2 Coordinate transformation. (A) Three-phase static *abc* to two-phase static *αβ* reference frame; (B) static *αβ* reference frame to rotating *dq* reference frame.

To express three scalars of three-phase in terms of one vector, the concept of the space vector can be introduced,

$$\vec{x} = x_a + \gamma x_b + \gamma^2 x_c \tag{5.7}$$

where $\gamma = \exp\left(j\frac{2\pi}{3}\right)$. It is noted that this space vector includes all the information of the three-phase, and the three-phase can be simplified to single-phase.

In the three-phase system, compared to phase a, the electrical values in phase b and phase c lag $2\pi/3$ and $4\pi/3$, respectively.

$$x_a = X\cos(\omega t);\;\; x_b = X\cos(\omega t - 2\pi/3);\;\; x_c = X\cos(\omega t + 2\pi/3) \tag{5.8}$$

$$\vec{X} = x_\alpha + jx_\beta = X\exp(j\omega t) \tag{5.9}$$

It can be seen that trace of the space vector completely becomes a circle, whose radium is the same as the phase value. It rotates counter-clockwise with the rotating speed of grid. Actually, the expression of the space vector is the same with Clarke transformation. Combining Eqs. (5.5) and (5.8),

$$\begin{bmatrix} x_\alpha \\ x_\beta \end{bmatrix} = C_{32}\begin{bmatrix} x_a \\ x_b \\ x_c \end{bmatrix} = \begin{bmatrix} X\cos(\omega t) \\ X\sin(\omega t) \end{bmatrix} \tag{5.10}$$

Furthermore, if Park transformation is applied,

$$\begin{bmatrix} x_d \\ x_q \end{bmatrix} = C_{2s/2r}\begin{bmatrix} x_\alpha \\ x_\beta \end{bmatrix} = \begin{bmatrix} X \\ 0 \end{bmatrix} \tag{5.11}$$

It can be seen that the advantage of the coordinate transformation is to simplified three-phase system into two-phase system. Besides, the AC values become DC values under dq rotating reference frame.

5.1.4 Small-Signal Model Under *dq* Reference Frame

Substituting Eq. (5.5) into Eq. (5.4), the large-signal model under $\alpha\beta$ reference frame can be obtained,

$$\begin{cases} \begin{bmatrix} \langle u_{g\alpha}\rangle_{T_s} \\ \langle u_{g\beta}\rangle_{T_s} \end{bmatrix} = L_g \dfrac{d}{dt} \begin{bmatrix} \langle i_{g\alpha}\rangle_{T_s} \\ \langle i_{g\beta}\rangle_{T_s} \end{bmatrix} + \begin{bmatrix} d_\alpha \\ d_\beta \end{bmatrix} \langle v_{dc}\rangle_{T_s} \\ C\dfrac{d}{dt}\langle v_{dc}\rangle_{T_s} = \dfrac{3}{2}\begin{bmatrix} d_\alpha & d_\beta \end{bmatrix} \begin{bmatrix} \langle i_{g\alpha}\rangle_{T_s} \\ \langle i_{g\beta}\rangle_{T_s} \end{bmatrix} - \langle i_{load}\rangle_{T_s} \end{cases} \tag{5.12}$$

Substituting Eq. (5.6) into Eq. (5.12), the large-signal model under dq reference frame can be obtained,

$$\begin{cases} \begin{bmatrix} \langle u_{gd}\rangle_{T_s} \\ \langle u_{gq}\rangle_{T_s} \end{bmatrix} = L_g \dfrac{d}{dt} \begin{bmatrix} \langle i_{gd}\rangle_{T_s} \\ \langle i_{gq}\rangle_{T_s} \end{bmatrix} + L\begin{bmatrix} 0 & -\omega \\ \omega & 0 \end{bmatrix} \begin{bmatrix} \langle i_{gd}\rangle_{T_s} \\ \langle i_{gq}\rangle_{T_s} \end{bmatrix} + \begin{bmatrix} d_d \\ d_q \end{bmatrix} \langle v_{dc}\rangle_{T_s} \\ C\dfrac{d}{dt}\langle v_{dc}\rangle_{T_s} = \dfrac{3}{2}\begin{bmatrix} d_d \\ d_q \end{bmatrix}^T \begin{bmatrix} \langle i_{gd}\rangle_{T_s} \\ \langle i_{gq}\rangle_{T_s} \end{bmatrix} - \langle i_{load}\rangle_{T_s} \end{cases} \tag{5.13}$$

It is noted that, the coupling components in the first equation of Eq. (5.13) comes from the Park transformation, while the gain of 1.5 appears in the second equation of Eq. (5.13) because an equal amplitude transformation results in a different transformed power. Compared with Eq. (5.4), the continuous state-average model of the grid-tied converter under two-phase synchronous rotating frame becomes simplified and continuous.

The small-signal dynamic model can be achieved by adding the disturbance around the quiescent operation point,

$$\langle x\rangle_{T_s} = X + \hat{x} \tag{5.14}$$

where X is the DC component, while $\hat{x}$ is the disturbance.

As the disturbances may come from the AC grid, the loading current, and the duty cycle, substituting Eq. (5.14) into Eq. (5.13), neglecting the second-order terms of the disturbance, the equation at the quiescent operation point yields,

$$\begin{cases} U_{gd} = -\omega L_g I_{gq} + D_d V_{dc} \\ U_{gq} = \omega L_g I_{gd} + D_q V_{dc} \\ I_{load} = \dfrac{3}{2}\left(D_d I_{gd} + D_q I_{gq}\right) \end{cases} \tag{5.15}$$

Meanwhile, the small-signal model can be obtained,

$$\begin{cases} \hat{u}_{gd} = L_g \dfrac{d\hat{i}_{gd}}{dt} - \omega L_g \hat{i}_{gq} + D_d \hat{v}_{dc} + V_{dc}\hat{d}_d \\ \hat{u}_{gq} = L_g \dfrac{d\hat{i}_{gq}}{dt} + \omega L_g \hat{i}_{gd} + D_q \hat{v}_{dc} + V_{dc}\hat{d}_q \\ C\dfrac{d}{dt}\hat{v}_{dc} = \dfrac{3}{2}\left(D_d \hat{i}_{gd} + I_{gd}\hat{d}_d + D_q \hat{i}_{gq} + I_{gq}\hat{d}_q\right) - \hat{i}_{load} \end{cases} \tag{5.16}$$

The small-signal model of the grid-side converter is shown in Fig. 5.3. It is equivalent to two DC/DC boost converters with paralleled output. On the basis of the small-signal model, the transfer function can be obtained, and its linear characteristic is suitable for the classical controller.

5.2 GRID SYNCHRONIZATION

As commonly known, the phase angle of the grid voltage is critical for the control performance of the grid-connected power converters, since the popular control strategies (e.g., the vector oriented control, the direct power control) strongly rely on the phase angle of the grid voltage to implement the related control units such as the *dq* reference frame transformation, the *dq* variable extraction, and the decoupled current control under the *dq* axis. Thus, the grid synchronization technique serves as the fundamental part for the control strategy of the grid-connected converter, and it is quite essential to ensure satisfactory performance of the converter.

The grid synchronization technique has been under intensive investigation, and a large number of different approaches [4,5] have been proposed in order to achieve precise and fast synchronization. It should be noted that the grid voltage during grid synchronization may be in either normal or abnormal conditions (i.e., the single-phase grid voltage, the three-phase unbalanced grid voltage, the three-phase distorted grid voltage), as well as the circumstance of the grid voltage fault, phase angle jump, etc.

The phase-locked loop (PLL) is the most popular grid synchronization technique, and it has been already adopted for a few decades. This section starts with

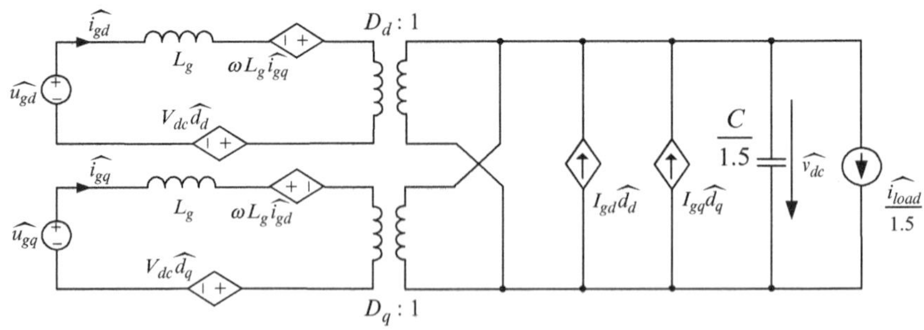

FIG. 5.3 The small-signal model of grid-tied converter under the rotating reference frame.

the description of the basic structure of the PLL. Then, the synchronous reference frame PLL (SRF-PLL) [6,7], which can be considered as a fundamental PLL technique implemented in the dq synchronous reference frame, will be introduced. Thereafter, another type of advanced PLL implemented in the stationary frame including the dual second-order generalized integrator, named as DSOGI-PLL [4,8,9], will be addressed. It is worth noting the three-phase PLL is on focus of this section.

5.2.1 SRF-PLL

The basic structure of a standard PLL contains the following parts, i.e., phase detector (PD), low-pass filter (LPF), and voltage controlled oscillator (VCO) [4]. This kind of PLL is applicable for different types of PLLs, and the schematics of the SRF-PLL, which can be served as the foundation for the following description, is firstly presented in Fig. 5.4 [4].

As shown in Fig. 5.4, the SRF-PLL can be divided into three parts as described in the following,

(1) Phase Detector: the three-phase grid voltage u_a, u_b, and u_c are delivered into the Clarke transformation $T_{abc/\alpha\beta}$, which transforms the three-phase voltage into the two-phase voltage u_α and u_β in the stationary frame. Then, based on the sine and cosine value of the grid phase angle θ_1, the Park transformation $T_{\alpha\beta/dq}$ is applied to obtain the two-phase voltage u_d and u_q in the synchronous frame.

(2) Low-Pass Filter: the PI controller is adopted as a low-pass filter. The grid voltage q-axis component u_q is selected as the input of the low-pass filter, while the PI controller outputs the angular speed of the grid voltage ω_1. Due to the operation of the PI controller, the q-axis component of the grid voltage u_q is regulated to zero, therefore the grid voltage can be aligned with the d-axis, and the dq axis decoupling can consequently be achieved.

(3) Voltage Controlled Oscillator: Once the angular speed of the grid voltage is obtained, the integrator is implemented to obtain the phase angle, and then the sine and cosine of the grid voltage phase angle can be calculated and delivered as the input of the Park transformation.

Based on the earlier explanation, the operation principle of the SRF-PLL can be summarized as following, firstly transform the three-phase grid voltage u_{abc} to

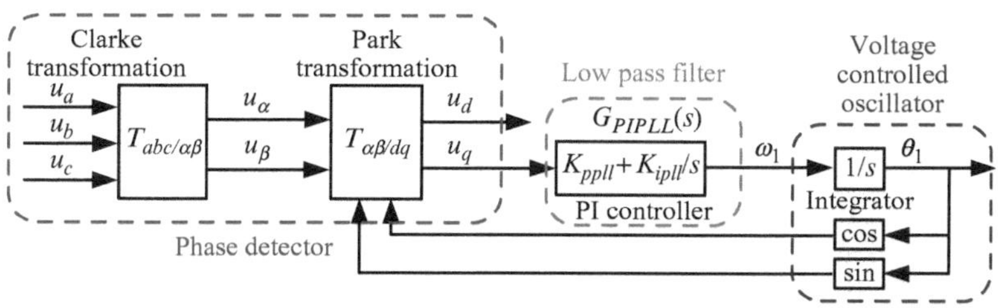

FIG. 5.4 Schematics of the SRF-PLL for grid synchronization.

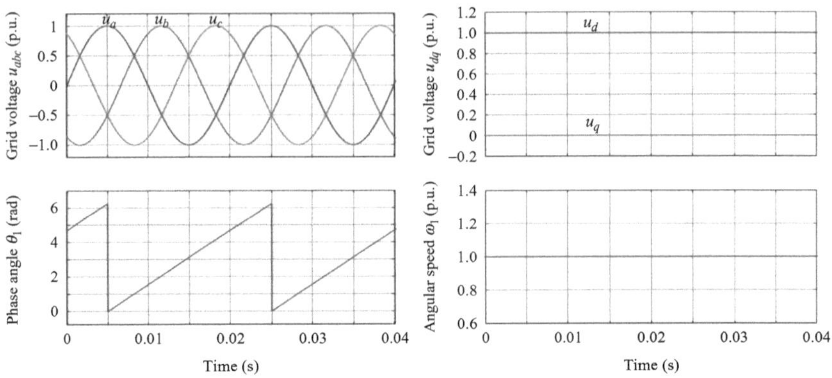

FIG. 5.5 Simulation results of SRF-PLL with the ideal (balanced and sinusoidal) three-phase grid voltage.

the u_{dq} in the synchronous frame, then the PI controller is adopted to suppress the u_q to zero, and consequently align the grid voltage to *d*-axis. The output of the PI controller together with the integrator finally produces the grid voltage phase angle, which is also delivered back to the Park transformation. Thus, it can be observed that the SRF-PLL can be considered as a closed-loop control with the control target of aligning the grid voltage to *d*-axis, and simultaneously obtain the grid voltage phase angle.

In order to better understand the performance of the SRF-PLL, the simulation results of SRF-PLL under the ideal three-phase grid voltage are shown in Fig. 5.5. As it can be seen, the three-phase grid voltage in the stationary frame is ideally balanced and sinusoidal, the phase angle increases linearly and periodically from 0 to 2π rad within 0.02 s (due to the fundamental frequency of 50 Hz). Moreover, the phase angle precisely follows the u_a, that is, the phase angle $= 2\pi$ rad at 0.005 s when the u_a is at positive peak, and $\pi/2$ rad at 0.01 s when the u_a is at 0 value, and π rad at 0.015 s when the u_a is at negative peak, and $3\pi/2$ rad at 0.02 s when the u_a is at 0 value, and again 2π rad at 0.025 s when the u_a is at positive peak.

Furthermore, as the u_{dq} are kept constant as 1.0 p.u. and 0.0 p.u., respectively, this indicates the precise alignment of the grid voltage along with the *d*-axis. The angular speed of the grid voltage is also kept constant as 100π rad/s (50 Hz), which also indicates the good synchronization.

Besides the simulation results of the SRF-PLL in the circumstance of the ideal grid voltage, it is also interesting to discuss its performance when the three-phase grid voltage is unbalanced or harmonic distorted as shown in Figs. 5.6 and 5.7, respectively.

As seen from Fig. 5.6, the negative sequence, due to the unbalanced three-phase grid voltage, causes deteriorated performance for the SRF-PLL. The phase angle no longer increases linearly, but contains fluctuating component.

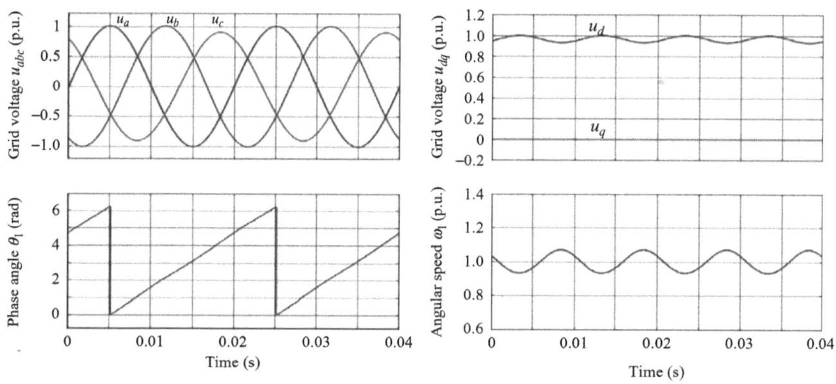

FIG. 5.6 Simulation results of SRF-PLL in the three-phase unbalanced grid voltage.

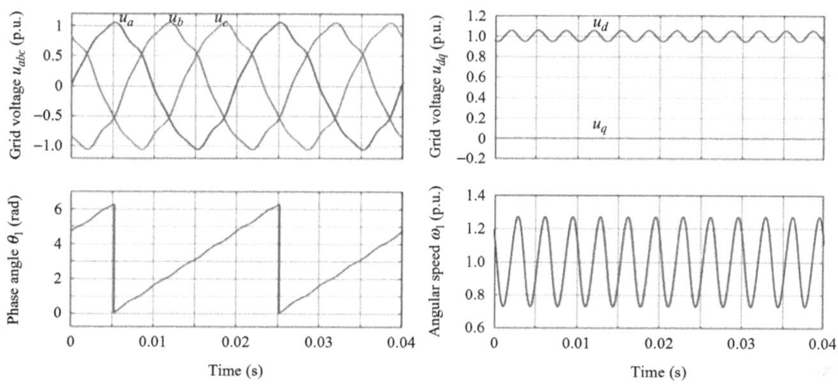

FIG. 5.7 Simulation results of SRF-PLL in the three-phase harmonic distorted grid voltage.

Also, the d-axis component of the grid voltage u_d contains the fluctuation of 100 Hz at twice the fundamental frequency, and the angular speed also inevitably contains the 100 Hz pulsating component.

For the case of the grid voltage with harmonic distortion as shown in Fig. 5.7, the similar simulation results as Fig. 5.6 can be obtained. Due to the existence of both 5th and 7th low-order harmonic sequences in the three-phase grid voltage, both the angular speed and the grid voltage d-axis component u_d contain the sixfold fundamental frequency at 300 Hz, and the phase angle is no longer linear.

Thus, based on these simulation results, it can be concluded that the conventional SRF-PLL can achieve satisfactory grid synchronization only under the circumstance of the ideal (balanced and sinusoidal) grid voltage. Once the three-phase grid voltage becomes unbalanced or harmonic distorted, the SRF-PLL fails to give the accurate phase angle, but contains unexpected 100 Hz or 300 Hz fluctuation. As a consequence, the entire performance of the grid-connected converter may be jeopardized due to the inaccurate grid synchronization.

5.2.2 DSOGI-PLL

In order to solve these problems of the SRF-PLL implemented in the nonideal grid voltage conditions (three-phase unbalanced, harmonic distorted, or low voltage fault), several improvements have been proposed during the last decades. One of the most popular improved PLL, i.e., dual second-order generalized integrator-phase locked loop (DSOGI-PLL), will be introduced in this section. To certain extent, the DSOGI-PLL can be regarded as the combination of both the DSOGI unit and the conventional SRF-PLL unit. Its main principle of excellent performance against grid voltage unbalance and harmonic distortion is able to eliminate the negative and harmonic sequence in the u_α and u_β, this indicates that the pure balanced and sinusoidal grid voltage is delivered into the conventional SRF-PLL.

Prior to the description of the DSOGI-PLL, it is necessary to introduce the second-order generalized integrator (SOGI), with its schematic diagram shown in Fig. 5.8. It can be seen that the input of the SOGI is the grid voltage u (which can either be the grid voltage α-axis or β-axis components $u_{\alpha\beta}$) as well as the estimated angular speed ω_1 of the grid voltage. On the other hand, the output of the SOGI is the filtered pure sinusoidal grid voltage u' and its quadrature voltage component qu', thus the SOGI also has the characteristics called quadrature signal generator (QSG).

The filtered sinusoidal direct-axis u' and quadrature axis qu' can be validated by conducting the mathematical deduction of Fig. 5.8 and the following transfer functions can consequently be obtained

$$\frac{u'}{u} = \frac{k\omega_1 s}{s^2 + k\omega_1 s + \omega_1^2} \tag{5.17}$$

$$\frac{qu'}{u} = \frac{\omega_1}{s}\frac{u'}{u} = \frac{k\omega_1^2}{s^2 + k\omega_1 s + \omega_1^2} \tag{5.18}$$

where k is the gain parameter of the SOGI.

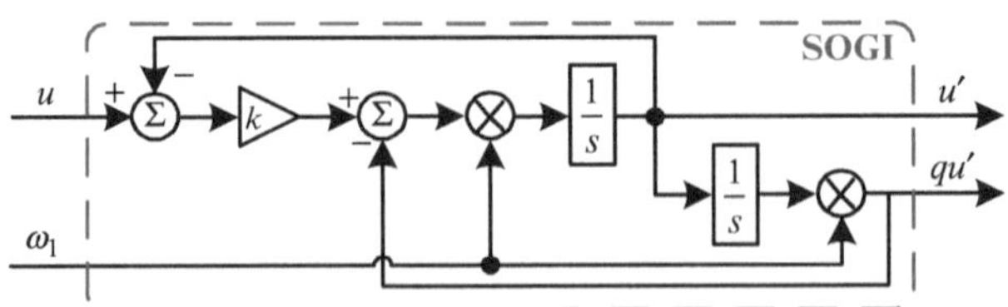

FIG. 5.8 Schematic diagram of the SOGI.

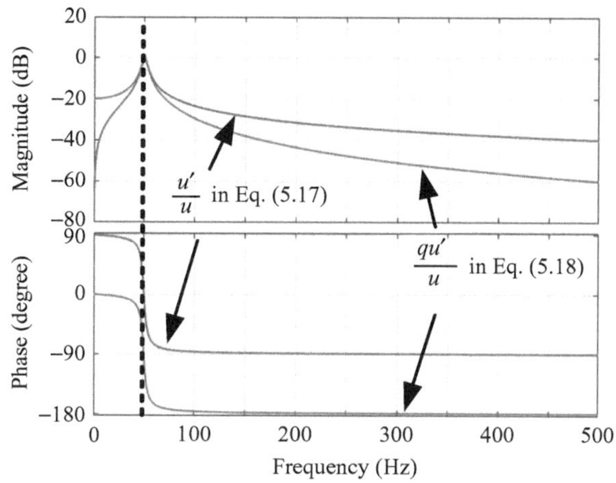

FIG. 5.9 Bode diagram of the transfer function in Eqs. (5.17) and (5.18), $k=0.1$, $\omega_1=100\pi$ rad/s.

Fig. 5.9 shows the Bode diagram of the transfer function in Eqs. (5.17) and (5.18), $k=0.1$, $\omega_1=100\pi$ rad/s. It can be seen from Fig. 5.8 that both two curves are able to maintain 0 dB at the fundamental frequency of 50 Hz. In addition, the significant amplitude attenuation can be seen for both curves at the harmonic frequency of 250 and 350 Hz. Moreover, the phase responses of both curves at 50 Hz are 0° and −90°, respectively, for Eqs. (5.17) and (5.18), which indicates that the direct voltage component u' is 90° phase leading compared with the quadrature component qu'.

According to Eqs. (5.17) and (5.18), together with Fig. 5.9, both the two transfer functions can be considered as second-order bandpass filters, which are able to extract only the AC signals at the fundamental angular speed ω_1, while the other AC sequences (for instance 5th and 7th harmonic components) can be eliminated. However, it is worth mentioning that by using the single SOGI to either u_α or u_β, the harmonics can be eliminated, but the negative sequence still unfortunately remains.

In order to eliminate the negative sequence simultaneously and aim for the symmetry purpose, the dual SOGI (DSOGI) can be used for both the u_α and u_β, and Fig. 5.10 shows the schematics of the proposed DSOGI, including the block diagram for eliminating the grid voltage negative components.

Fig. 5.11 shows the phasor diagram of grid voltage $\alpha\beta$ direct/quadrature and positive/negative components. The blue phasors (the first diagram) are positive α direct and quadrature components, while the green phasors (the second diagram) are positive β direct and quadrature components. Similarly, the red phasors (the third diagram) are negative α direct and quadrature components, while the yellow phasors (the fourth diagram) are negative β direct and quadrature components.

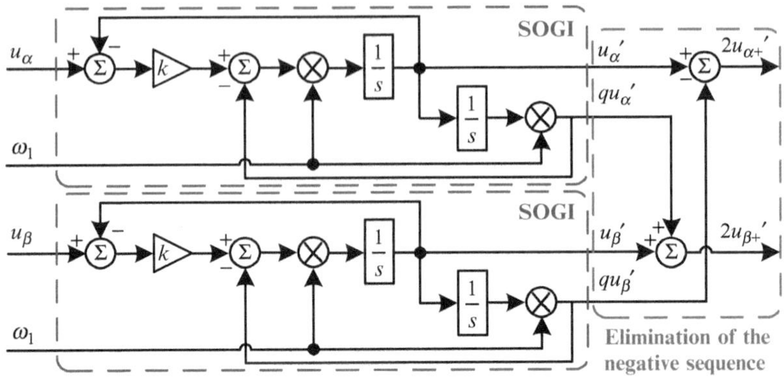

FIG. 5.10 Schematic diagram of the DSOGI.

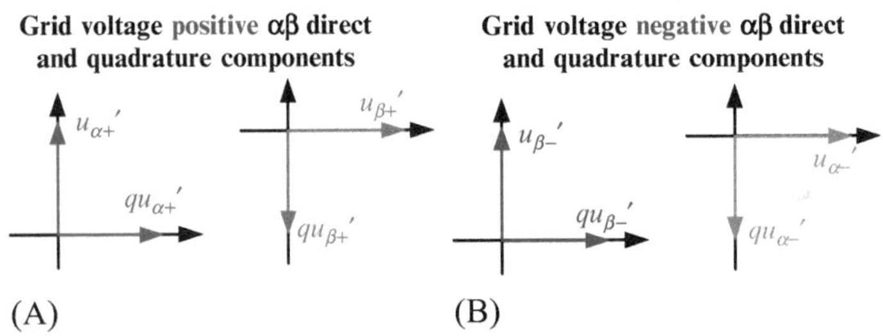

FIG. 5.11 Phasor diagram of (A) grid voltage positive $\alpha\beta$ direct/quadrature components and (B) negative $\alpha\beta$ direct/quadrature components.

It needs to point out that in all four phasors, the direct phasors are always 90° phase leading compared with the quadrature phasors since this is determined by the transfer functions in Eqs. (5.17) and (5.18). However, due to the Clarke transformation, the α components are 90° phase leading compared with the β components for the positive components shown in Fig. 5.11A, while the α components are 90° phase lagging compared with the β components for the negative components shown in Fig. 5.11B since the negative components are rotating at the opposite direction of −50 Hz.

Therefore, by combining Figs. 5.10 and 5.11, the following equations can be obtained as,

$$u_\alpha' - qu_\beta' = (u_{\alpha+}' + u_{\alpha-}') - (qu_{\beta+}' + qu_{\beta-}') = u_{\alpha+}' - qu_{\beta+}' = 2u_{\alpha+}' \quad (5.19)$$

$$qu_\alpha' + u_\beta' = (qu_{\alpha+}' + qu_{\alpha-}') + (u_{\beta+}' + u_{\beta-}') = qu_{\alpha+}' + u_{\beta+}' = 2u_{\beta+}' \quad (5.20)$$

Thus, according to Eqs. (5.19) and (5.20), it is evident that the DSOGI schematics are able to extract the $\alpha\beta$ positive components, which contain only the pure balanced and sinusoidal components without any negative or harmonic distorted components. Obviously, by using the ideal AC grid voltage signals

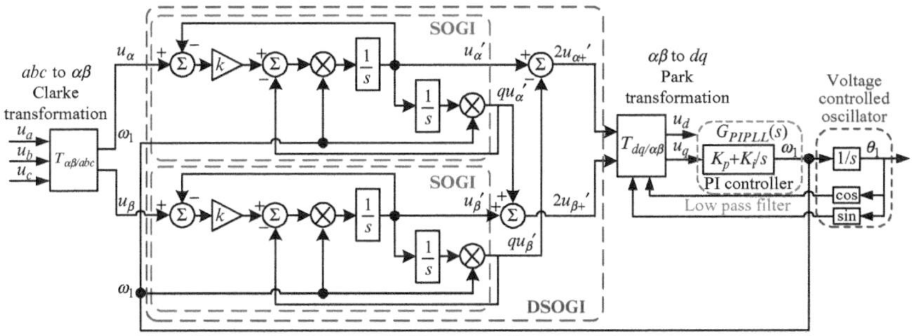

FIG. 5.12 Schematic diagram of the DSOGI-PLL.

into the conventional SRF-PLL, the accurate tracking of the grid voltage phase angle under the circumstance of the unbalance and distortion can be accomplished.

Fig. 5.12 shows the schematics diagram of the DSOGI-PLL, which can be considered as the combination of both DSOGI and conventional SRF-PLL. Comparing with the schematics of SRF-PLL in Fig. 5.4 and the DSOGI-PLL in Fig. 5.10, it can be found that the DSOGI-PLL is actually the special example of the SRF-PLL with the inclusion of the DSOGI, which is responsible for eliminating the negative and harmonic sequences.

In order to verify the effectiveness of the DSOGI-PLL in the nonideal grid voltage condition, the simulation results are shown in Fig. 5.13 where the grid voltage contains both the unbalanced and harmonic distorted components.

As it can be seen from Fig. 5.13A, the three-phase grid voltage u_{abc} obviously contains the harmonic distorted and unbalanced components, and the grid voltage $\alpha\beta$ components $u_{\alpha\beta}$ in the stationary frame shown in Fig. 5.13B are also unbalanced and distorted by using simple Clarke transformation.

By implementing the DSOGI to u_α and u_β, the fundamental components, including both the positive fundamental +50 Hz and the negative fundamental −50 Hz, can be extracted simultaneously, while the harmonic distorted components can be eliminated, and finally the u_α' and u_β' can be obtained as shown in Fig. 5.13C. Note that both u_α' and u_β' are sinusoidal, but they have different amplitudes, which indicates that the negative fundamental components still exist and result in the different amplitude.

Moreover, the corresponding quadrature axis $\alpha\beta$ voltages qu_α' and qu_β' can also be obtained in this control scheme as shown in Fig. 5.13E and F. It can be observed that both quadrature axis components qu_α' and qu_β' are phase lagging by 90° compared with the direct-axis components u_α' and u_β'. The quadrature axis components qu_α' and qu_β' can later be adopted to remove the negative fundamental component.

In order to remove the negative fundamental components in u_α' and u_β', the calculation in Eqs. (5.19) and (5.20) can be applied. As a consequence,

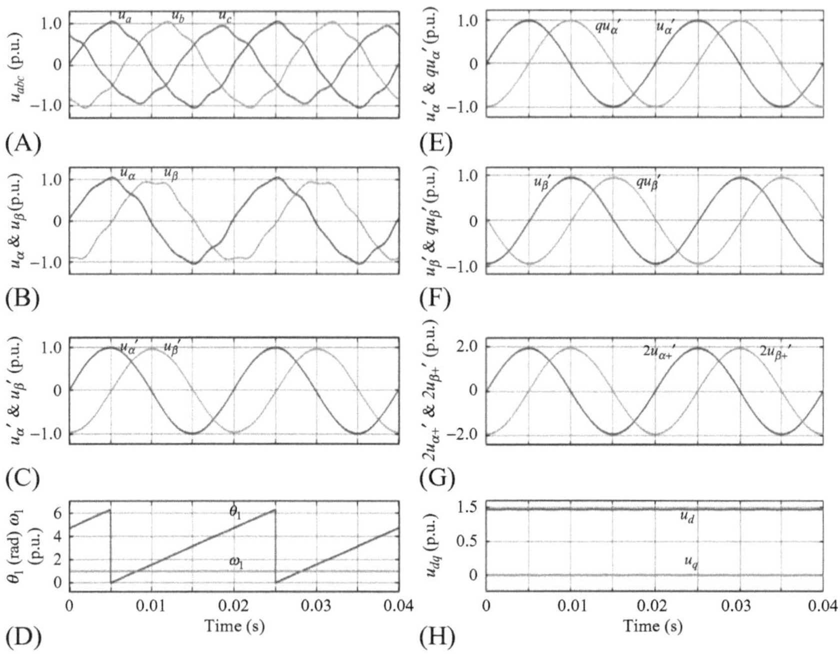

FIG. 5.13 Simulation results of the DSOGI-PLL under the three-phase unbalanced and harmonic distorted grid voltage. (A) Three-phase grid voltage u_{abc}. (B) Grid voltage α_β components $u_{\alpha\beta}$. (C) Grid voltage α_β components $u_{\alpha\beta}$ (sinusoidal). (D) Grid voltage angular speed ω_1 and phase angle θ_1. (E) Grid voltage α' and quadrature $q_{u\alpha}{}'$ components. (F) Grid voltage β' and quadrature $q_{u\beta}{}'$ components. (G) Sinusoidal and balanced grid voltage α_β components $2u_{\alpha+}{}'$ and $2u_{\beta+}{}'$. (H) Grid voltage direct and quadrature-axis components u_{dq}.

the sinusoidal and balanced grid voltage $\alpha\beta$ components $2u_{\alpha+}{}'$ and $2u_{\beta+}{}'$ in Fig. 5.13G can be obtained, and their amplitude can be easily corrected by multiplying by 0.5. These pure sinusoidal and balanced AC grid voltage signals $u_{\alpha+}{}'$ and $u_{\beta+}{}'$ can then be delivered as the input of the conventional SRF-PLL.

Finally, as shown in Fig. 5.13D, the grid voltage angular speed ω_1 is precisely calculated as 1.0 p.u. (50 Hz or 100π rad/s), and consequently the grid voltage phase angle θ_1 increases linearly. Besides, as shown in Fig. 5.13H, even under the circumstance of the grid voltage unbalance and distortion, the grid voltage direct-axis component u_d is kept constant as the peak amplitude value of the three-phase grid voltage, while the quadrature axis component u_q is kept zero due to the effective operation of the DSOGI-PLL.

In order to further compare the conventional SRF-PLL and the DSOGI-PLL, the simulation results under the grid voltage unbalance and distortion using both the two grid synchronization methods are shown in Fig. 5.14. As it can be seen, the phase angle θ_1 using the DSOGI-PLL increases linearly, while the phase angle θ_1 using the SRF-PLL unfortunately twists due to the fluctuating

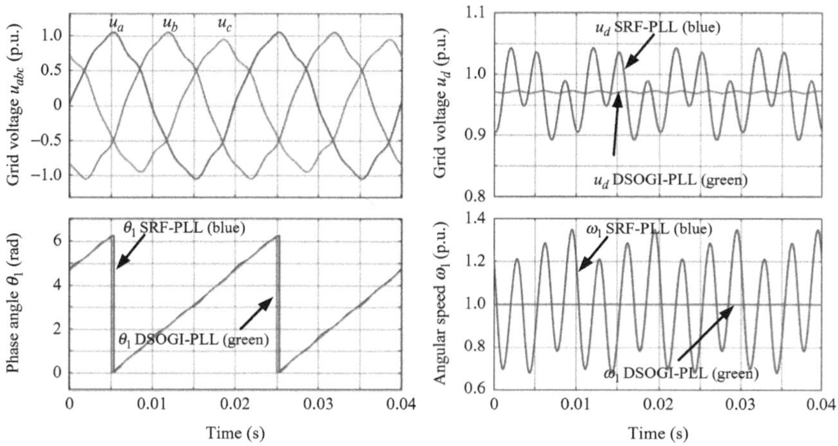

FIG. 5.14 Simulation results of both SRF-PLL and DSOGI-PLL in the three-phase harmonic distorted grid voltage.

components caused by the grid voltage unbalance and distortion. Similarly, the grid voltage d-axis component u_d using the DSOGI-PLL is almost constant, while the large fluctuation components at 100 and 300 Hz exist in the u_d using SRF-PLL. Furthermore, the similar performance regarding the angular speed ω_1 can also be obtained.

According to the earlier theoretical illustrations and the simulation results, it can be concluded that,

(1) The conventional SRF-PLL performs well only under the circumstance of three-phase balanced and sinusoidal grid voltage, but unsatisfactory grid synchronization results can be expected in the case of any nonideal grid voltage happens.
(2) Due to the effective operation of the DSOGI, which is capable of eliminating both the negative and harmonic distorted components in the grid voltage, the DSOGI-PLL is able to track precisely the grid voltage under nonideal grid voltage conditions.

5.3 GRID CURRENT CONTROL BASED ON PI CONTROLLER

As well-understood, a three-phase PWM power converter normally employs the cascade control—the inner-current loop and the outer-voltage/power/speed/flux loop. In the case of the grid-tied converter used in renewable power application, two grid current controllers are employed to realize the fast response during the changing load, and one DC-link voltage controller are applied to keep the independent control of the grid-tied converter.

5.3.1 Controller Design of Inner-Current Loop

In the situation of a stable power grid and loading condition, the disturbance from the grid voltage and the loading current can be neglected. Moreover, as the response time of the inner-current loop is much faster than the outer-voltage loop, the DC-link voltage can be considered constant without any disturbance. As a result, the first two equations in Eq. (5.13) can be simplified as,

$$\frac{d}{dt}\begin{bmatrix}\hat{i}_{gd}\\\hat{i}_{gq}\end{bmatrix}=-\begin{bmatrix}0 & -\omega\\\omega & 0\end{bmatrix}\begin{bmatrix}\hat{i}_{gd}\\\hat{i}_{gq}\end{bmatrix}-\frac{1}{L_g}\begin{bmatrix}\hat{d}_d\\\hat{d}_q\end{bmatrix}V_{dc} \tag{5.21}$$

It is obvious from Eq. (5.21) that it is a two-input and two-output system, where the inputs are $\begin{bmatrix}\hat{d}_d\\\hat{d}_q\end{bmatrix}$, and the outputs are $\begin{bmatrix}\hat{i}_{gd}\\\hat{i}_{gq}\end{bmatrix}$. The control of $\hat{d}_d$ not only affects $\hat{i}_{gd}$, but also changes $\hat{i}_{gq}$. Similarly, due to the coupling of dq circuit, the control of $\hat{d}_q$ not only affects $\hat{i}_{gq}$, but also changes $\hat{i}_{gd}$. In order to simplify the controller design to the single-input and single-output, the decoupling control is necessary to be introduced.

If the current controller outputs of the d-axis and the q-axis modify from $\hat{d}_d=G_{PI}\left(\hat{i}^*_{gd}-\hat{i}_{gd}\right)$ and $\hat{d}_q=G_{PI}\left(\hat{i}^*_{gq}-\hat{i}_{gq}\right)$ to,

$$\begin{cases}\hat{d}_d=G_{PI}\left(\hat{i}^*_{gd}-\hat{i}_{gd}\right)-\omega L_g\hat{i}_{gq}/V_{dc}\\\hat{d}_q=G_{PI}\left(\hat{i}^*_{gq}-\hat{i}_{gq}\right)+\omega L_g\hat{i}_{gd}/V_{dc}\end{cases} \tag{5.22}$$

Eq. (5.21) can be simplified as,

$$\frac{d}{dt}\begin{bmatrix}\hat{i}_{gd}\\\hat{i}_{gq}\end{bmatrix}=-\frac{1}{L_g}\begin{bmatrix}\hat{d}_d\\\hat{d}_q\end{bmatrix}V_{dc} \tag{5.23}$$

By using these modifications of the d-axis and q-axis current controller, the decouple control of the d-axis and q-axis can be achieved, whose control block diagram is shown in Fig. 5.15.

As inferred from Eq. (5.23), the control plant of the d-axis and q-axis current loops perform as an integrator, and the transfer functions from the duty cycle to the current can be deduced as,

$$\begin{cases}\dfrac{\hat{i}_{gd}(s)}{\hat{d}_d(s)}=-\dfrac{V_{dc}}{L_g s}\\[2ex]\dfrac{\hat{i}_{gq}(s)}{\hat{d}_q(s)}=-\dfrac{V_{dc}}{L_g s}\end{cases} \tag{5.24}$$

Considering the outer loop of the DC-link voltage, based on Eq. (5.13), the disturbance of the DC voltage is linked to the changing duty cycle and the grid current,

$$\begin{cases}\dfrac{\widehat{i_{gd}}(s)}{\widehat{d_d}(s)} = -\dfrac{V_{dc}}{L_g s} \\ \dfrac{\widehat{i_{gq}}(s)}{\widehat{d_q}(s)} = -\dfrac{V_{dc}}{L_g s}\end{cases}$$

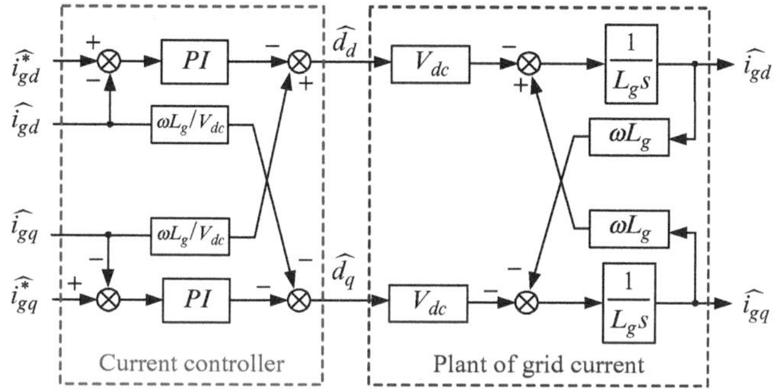

FIG. 5.15 Decoupling control block diagram of d-axis and q-axis current.

$$\frac{d}{dt}\hat{v}_{dc} = \frac{3}{2C}\left[\hat{d}_d \;\; \hat{d}_q\right]\begin{bmatrix} I_{gd} \\ I_{gq}\end{bmatrix} + \frac{3}{2C}\left[D_d \;\; D_q\right]\begin{bmatrix}\hat{i}_{gd} \\ \hat{i}_{gq}\end{bmatrix} \tag{5.25}$$

By using the grid voltage orientation control, the d-axis is aligned to the grid voltage vector,

$$\begin{cases} u_{gd} = U_g \\ u_{gq} = 0 \end{cases} \tag{5.26}$$

where U_g denotes the phase peak voltage of the power grid.

As the active power is determined by the d-axis grid current, the DC voltage is only related by the d-axis grid current due to the balance power between the AC-side and the DC-side of the power converter by neglecting the power loss. As a result, Eq. (5.25) can be further simplified as,

$$\frac{d}{dt}\hat{v}_{dc} = \frac{3}{2C}I_d\hat{d}_d + \frac{3}{2C}D_d\hat{i}_d \tag{5.27}$$

Due to the high inertia of the DC capacitor, the first term in the right side of Eq. (5.27) can be regarded as the disturbance of the DC voltage. In the unity power factor operation, the quiescent q-axis current always keeps zero, the transfer function from the d-axis current to the DC voltage can be obtained and simplified by using Eqs. (5.15) and (5.26),

$$\frac{\hat{v}_{dc}(s)}{\hat{i}_d(s)} = \frac{3D_d}{2Cs} = \frac{3U_g}{2V_{dc}Cs} \tag{5.28}$$

which performs also as an integrator. With the help of the transfer function of the inner-current loop and the outer-voltage loop, the compensation network can be applied to the controllers design.

As aforementioned, the transfer function from the duty cycle to converter output is considered as a proportion unit. In reality, due to the transport delay induced by the PWM technique and the sampling delay caused by the digital control [10,11], a delay unit is more acceptable. Different with the DC/DC converter, where the saw-tooth carrier is commonly applied, a triangular counter is used as the carrier signal for sampling the average inductor current.

As shown in Fig. 5.16, the symmetrical PWM and the asymmetrical PWM can be achieved by single sample and doubly sample within a switching period T_{sw}. In respect to the transport delay T_{d_tr}, it is the time interval between the data sample and PWM toggle. With the triangular carrier, the PWM signal can minimally respond without time interval, and can maximally react with half of the switching period, which results in the average delay of quarter switching period. In respect to the sampling delay T_{d_sa}, it is related to the sample period T_{sa}, where a time delay of one switching period can be found in the symmetrical PWM, while it is reduced to half of the switching frequency using the asymmetrical PWM. Above all, the total delay time of the symmetrical and asymmetrical PWM can be considered as 5/4 and 3/4 of the switching period, respectively. Moreover, as the delay time is around hundreds of microsecond, it can be roughly regarded as an inertia unit $G_{PWM}(s)$ [1],

$$G_{PWM}(s) = \frac{1}{T_d s + 1} \tag{5.29}$$

where T_d is the inertia of the PWM converter. It becomes $1.25T_{sw}$ in single sample but $0.75T_{sw}$ in doubly sample.

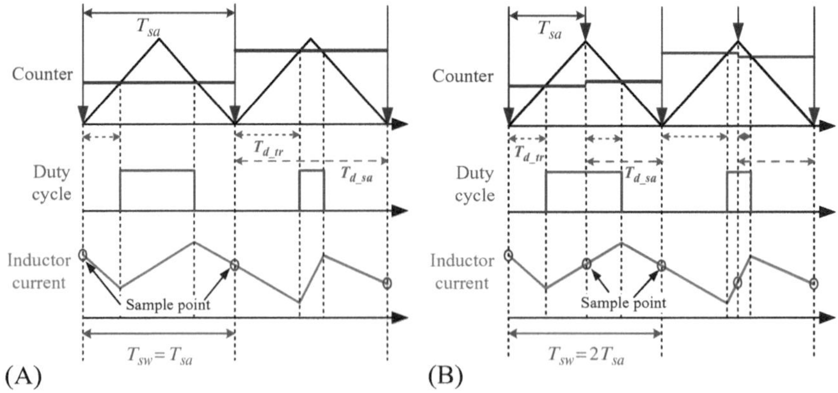

FIG. 5.16 Transport delay and sampling delay existed in digital (A) symmetrical PWM; and (B) asymmetrical PWM.

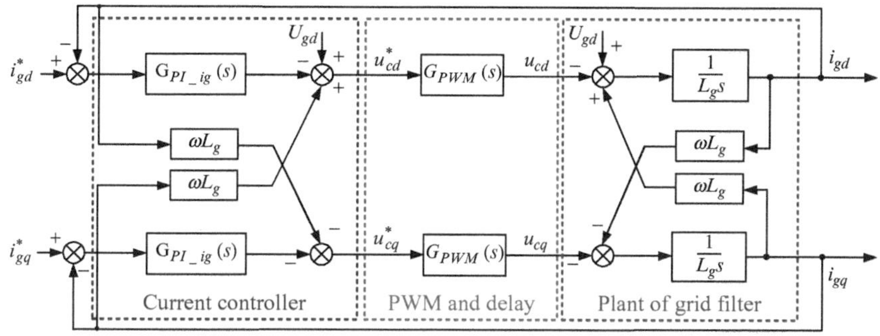

FIG. 5.17 Overall block diagram of the grid current loop control.

With the state-average model of the grid-tied converter as described in Fig. 5.15, the block diagram of the filter plant is ready. Together with the aforementioned PWM model as well as the PI controller, the overall block diagram of the grid current loop control is shown in Fig. 5.17. It is worth mentioning that the output of the PI controller is negative because of the negative transfer function of the control plant as listed in Eq. (5.24). Moreover, the DC bias and decoupling components are also feedforward to cancel out components existing in the control plant for the fast dynamic response.

The transfer function of the grid current controller $G_{PI\_ig}(s)$, and the plant of grid filter $G_{pl\_ig}(s)$ can be expressed by,

$$G_{PI\_ig}(s) = K_{p\_ig} + \frac{K_{i\_ig}}{s} \tag{5.30}$$

$$G_{pl\_ig}(s) = \frac{1}{L_g s} \tag{5.31}$$

where K_{p_ig} and K_{i_ig} are the proportional and integral coefficients of the grid current controller.

Neglecting the DC component, and decoupling components, the transfer function of the open-loop grid current can be simplified as shown in Fig. 5.18, and it can be found as,

$$G_{ol\_ig}(s) = G_{PI\_ig}(s)G_{PWM}(s)G_{pl\_ig}(s) \tag{5.32}$$

Although the controller design can be based on the zero-pole displacement, or the analysis of the root locus [12–15], the Bode plot is in focus by using the criteria of the crossover frequency and its phase margin. Although the bandwidth is defined as the frequency that rides through −3 dB in the closed-loop amplitude curve, it can be roughly considered the same with the crossover frequency, which crosses over 0 dB in the open-loop amplitude curve [2]. With the parameters of the grid-tied converter of 7.5 kW systems listed in Table 5.1, the Bode plots of open-loop grid current can be expected in terms of the amplitude-frequency and the phase-frequency characteristics. It is worth

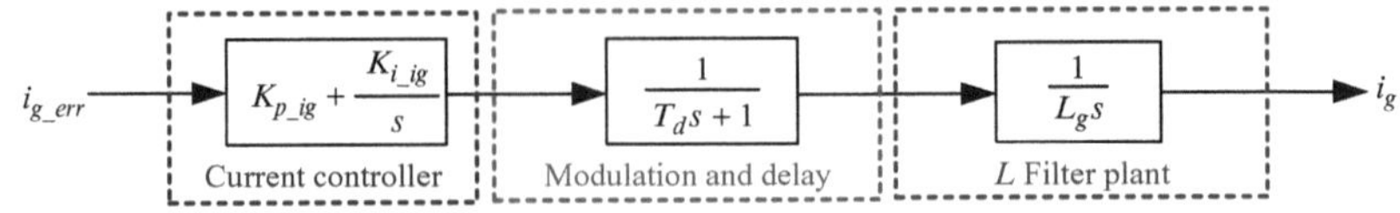

FIG. 5.18 Transfer function of open-loop gird current neglecting disturbances.

TABLE 5.1 Parameters of Grid Current Loop in 7.5 kW Grid-Tied Converter

	7.5 kW
Grid peak phase voltage U_g (V)	311
DC-link voltage V_{dc} (V)	650
Boost inductor L_g (mH)	18
Sampling frequency f_{sa} (kHz)	10
Switching frequency f_{sw} (kHz)	5
Delay time in PWM T_d (μs)	250
Proportional coefficient of PI controller K_{p_ig}	40
Integral coefficient of PI controller K_{i_ig}	120

mentioning that the doubly sample technique is applied. To achieve a fast response during the disturbance, the design target of the PI controller is to regulate the bandwidth of the current loop between 1/20 and 1/10 of the switching frequency. Meanwhile, viewed from the mitigation the harmonic around the switching frequency, the amplitude curve of the open-loop transfer function is required to cross over 0 dB with the slope rate −20 dB/dec in order to guarantee the sufficient phase margin (normally higher than 45°).

With the proportional and integral coefficients of the current controller, the boost inductance as well as the switching frequency listed in Table 5.1, the transfer function of the control plant, the current controller, and the open-loop grid current can be obtained. As shown in Fig. 5.19, it can be seen that the control plant keeps the slope −20 dB/dec because of its intrinsic zero pole. Meanwhile, due to the introduction of the pole around 1/10 switching frequency in the delay unit, another −20 dB/dec is superposed in the control plant, which results in a decrease of the phase margin. However, with the help of the current controller, which contains a zero pole and a desired zero, the bandwidth of the current loop is able to be set at 300 Hz. Meanwhile, the turning frequency of the PI controller is set at 0.5 Hz. Finally, a phase margin of 60.4° can be observed, which ensure the stability of the grid current loop.

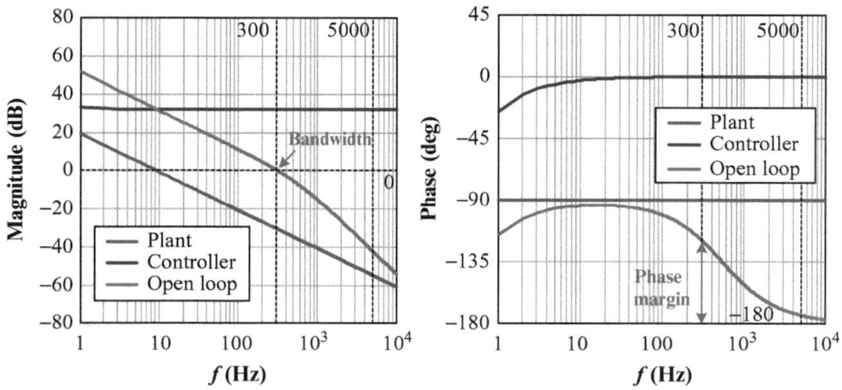

FIG. 5.19 Bode plots of open-loop current transfer function in terms of amplitude and phase characteristics.

5.3.2 Controller Design of Outer-Voltage Loop

Considering the controller design of the outer-voltage loop, the inner-current loop can be regarded as a closed-loop unit $G_{cl\_ig}(s)$. Together with the transfer function of the DC capacitor $G_{pl\_V}(s)$ as calculated in Eq. (5.28), and the DC voltage controller $G_{PI\_V}(s)$, the overall block diagram of the DC voltage loop control can be obtained as shown in Fig. 5.20. It is noted a filter unit $G_f(s)$ is also employed in the feedback loop in order to remove the high-frequency noise.

Neglecting the DC component, with the transfer function of the DC voltage controller, the grid current closed-loop, the plant of the DC capacitor, as well as the low-pass filter,

$$G_{PI\_V}(s) = K_{p\_V} + \frac{K_{i\_V}}{s} \tag{5.33}$$

$$G_{cl\_ig}(s) = \frac{G_{ol\_ig}(s)}{1 + G_{ol\_ig}(s)} \tag{5.34}$$

$$G_{pl\_V}(s) = \frac{3U_g}{2V_{dc}Cs} \tag{5.35}$$

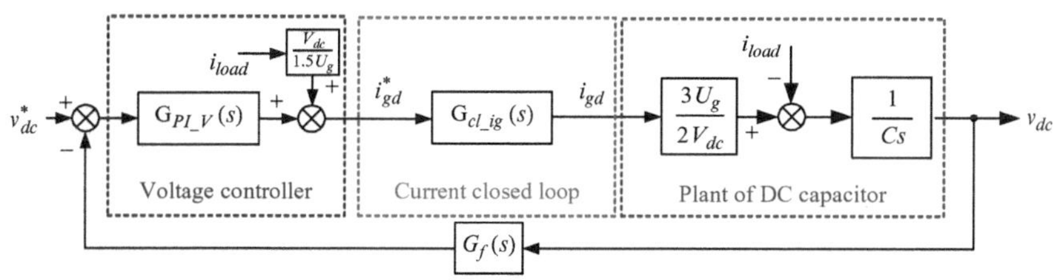

FIG. 5.20 Overall block diagram of the DC voltage loop control.

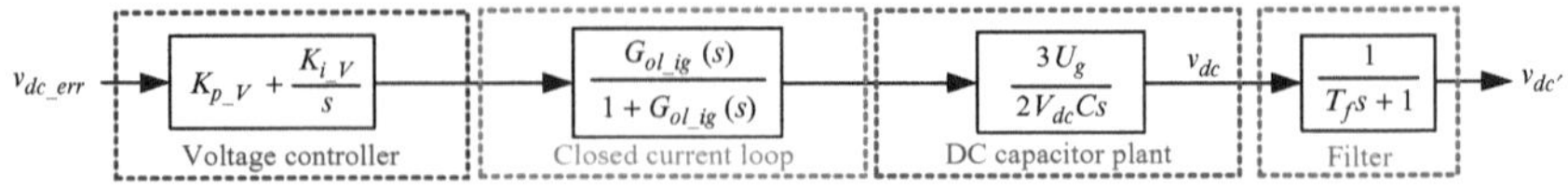

FIG. 5.21 Transfer function of open-loop DC voltage.

$$G_f(s) = \frac{1}{T_f s + 1} \tag{5.36}$$

the open-loop DC voltage can be depicted in Fig. 5.21, and it can be found as,

$$G_{ol\_V}(s) = G_{PI\_V}(s)G_{cl\_ig}(s)G_{pl\_V}(s)G_f(s) \tag{5.37}$$

In order to separate the inner loop and the outer loop for the independent design of their controllers, the bandwidth of the outer loop is limited between 1/50 and 1/10 of the inner loop. By this approach, during the design of the inner loop, the inner reference can be regarded as constant because of the much slower response of the outer loop. On the other hand, during the design of the outer loop, the inner loop can be considered as a closed-loop. As the control objective can fully track the reference within its bandwidth due to its high open-loop gain, the closed inner loop seldom affects the design of the outer loop.

With the DC-link capacitance, the time constant of the first-order filter as well as the proportional and integral coefficients of the voltage PI controller listed in Table 5.2, the transfer function of the control plant, the voltage controller and the open-loop DC voltage can be obtained. As shown in Fig. 5.22, it can be seen that, the control plant keeps −20dB/dec due to its zero pole. Moreover, the amplitude characteristic of the closed-loop grid current maintains 0dB if the frequency is lower than the crossover frequency, while it becomes −40dB/dec if the frequency is higher than the crossover frequency. If the frequency is higher than crossover frequency of the low-pass filter, a slope −20dB/dec is superposed. The amplitude curve finally behaves a slope −80dB/dec when the frequency is higher than the bandwidth of the current loop. Finally, with

TABLE 5.2 Parameters for Voltage Loop in 7.5kW Grid-Tied Power Converter

	7.5kW
DC-link capacitance C (μF)	600
Filter time constant T_f (ms)	10
Proportional coefficient of PI controller K_{p_V}	0.1
Integral coefficient of PI controller K_{i_V}	0.5

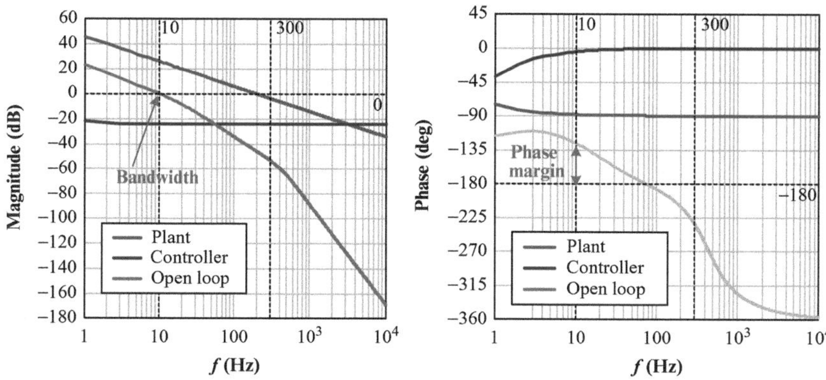

FIG. 5.22 Bode plots of open-loop voltage transfer function in terms of amplitude and phase characteristics.

the help of the PI controller, the bandwidth of the DC voltage loop is set at 10 Hz with the phase margin of 53.2°.

5.3.3 Simulation Validation

The aforementioned bandwidth design is achieved in the frequency domain, which is not easy to be validated in the simulation due to its time domain analysis. In order to create the connection between the frequency domain and the time domain, the relationship of concepts used in the frequency domain (like the bandwidth) and time domain (e.g., delay time, rise time, peak time) is a necessity. In low-order systems (first-order and second-order), it is proven that there is the strict relationship between the bandwidth and the rise time [2,16]. Furthermore, even in high-order systems, it can be equivalent to the first-order or second-order system by ignoring the minor poles but keeping the dominant poles [16].

In the loop design of the grid-tied power converter, in respect to the inner loop, the control plant can be considered as an inertial unit. Due to the fact that the PWM inertia unit can be omitted due to the much higher pole compared with the designed bandwidth, the open-loop transfer function can be considered as an integral unit, because the designed zero point of the PI controller is able to counteract the pole of the control plant. For the DC voltage outer loop, the control plant is theoretically an integral unit, but it can be regarded as the inertia unit if the equivalent loading resistance is taken into account. As the closed-loop inner current almost performs as a unit transfer function within its bandwidth, together with the PI controller, the open-loop DC voltage can be considered as an integral unit as well. As a consequence, the open-loop transfer function of the inner-current and outer DC voltage is the integral unit, which becomes a first-order system in the case of the closed-loop.

TABLE 5.3 Bandwidth Design and Expected Rise Time of 7.5 kW Power Converter

	Switching Frequency		5 kHz
Grid-tied converter	Inner loop	Bandwidth	300 Hz
		Expected rise time	1.17 ms
	Outer loop	Bandwidth	10 Hz
		Expected rise time	35 ms

The relationship between the bandwidth and the rise time during a step response can be found [16],

$$f_{BW} \cdot t_r = 0.35 \tag{5.38}$$

where f_{BW} denotes the bandwidth, and t_r denotes the rise time during the step response. And the expected rise time of the inner loop and outer loop are listed in Table 5.3.

In the following sections, the aforementioned rise time to the step response can be validated by using the simulation, where the conditions exactly follow the aforementioned parameters, and their pu values are compared in Table 5.4 as well as the conditions at rated power. Then, the step changes can be performed at the various loops in order to verify the bandwidth design.

As shown in Fig. 5.23, a 0.3 pu underexcited reactive current is injected at the time instant of 0.6 s. It can be seen that the rise time is around 1.0 ms, which proves that the bandwidth of the current loop is 300 Hz. Besides, a step change of the DC-link from 650 to 600 V is performed, the rise time of the DC-link voltage is around 35 ms, which proves that the crossover frequency of the outer-voltage loop is 10 Hz. Both of them agree with expected rise time as analyzed in Table 5.3.

TABLE 5.4 Specifications of 7.5 kW Grid-Tied Power Converter

Grid peak phase voltage	311 V	1.0 pu
Rated grid current	3.2 A	0.20 pu
Boost inductance	18 mH	0.29 pu
DC capacitance	600 μF	3.64 pu

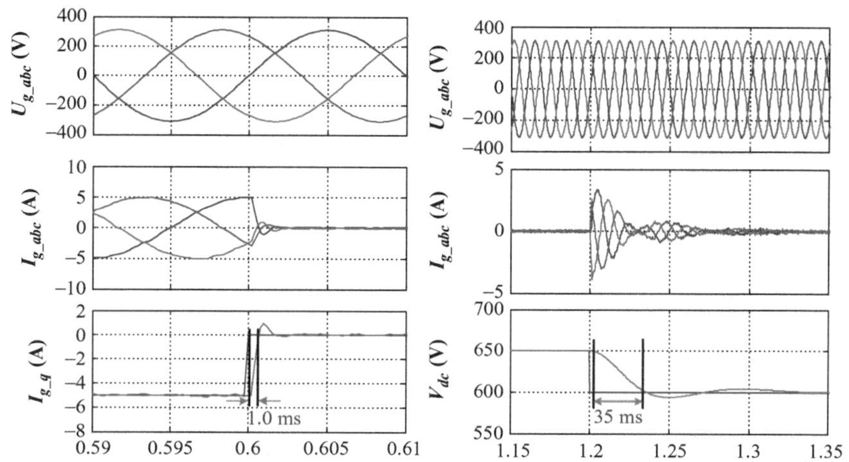

FIG. 5.23 Step response simulation of the inner loop and outer loop in 7.5 kW grid-side converter.

5.4 GRID CURRENT CONTROL BASED ON PR CONTROLLER

5.4.1 Introduction of PR Controller

The transfer function of proportional and resonant (PR) controller can be obtained by shifting the integrator part of the conventional PI controller to both positive and negative fundamental frequency,

$$C_{PR\_ideal}(s) = k_p + \frac{k_i}{s - j\omega} + \frac{k_i}{s + j\omega} = k_p + \frac{2k_i s}{s^2 + \omega^2} \tag{5.39}$$

where, k_p and k_i are the proportional and integral parameter of the PI part, and ω is the tuned resonant frequency.

However, it should be pointed out that the PR controller expression in Eq. (5.39) is an ideal one, which is quite sensitive to the grid frequency variation. Thus, the practical PR controller is adopted as following.

$$C_{PR}(s) = k_p + \frac{k_r \omega_c s}{s^2 + \omega_c s + \omega_0^2} \tag{5.40}$$

where k_r is the resonant gain, ω_c is the resonant bandwidth introduced to increase the control bandwidth and improve the robustness against the grid frequency variation, and ω_0 is the resonant frequency of 50 Hz.

The following part will discuss the parameter design of the PR controller, i.e., the proportional parameter k_p and resonant gain parameter k_r. The resonant bandwidth ω_c is adjusted between 2 and 10 rad/s to deal with different cases of the grid frequency variation, note the ω_c is particularly chosen at 5 rad/s in the following discussion.

5.4.2 Open-Loop Control Transfer Function Using PR Controller

The control block diagram of the grid-tied converter with single inductor in the stationary frame using PR controller is shown in Fig. 5.24.

The grid-side current control error can be obtained by comparing the reference signal $\boldsymbol{I}^*_{g\alpha\beta}$ and feedback signal $\boldsymbol{I}_{g\alpha\beta}$, and then the control error can be regulated by the PR controller.

The digital control delay $G_d(s)$, which is typically around 1.5 sampling period, is caused by the AD sample and PWM update, and needs to be taken into consideration during the parameter design, and it can be presented as,

$$G_d(s) = e^{-sT_d} \tag{5.41}$$

where T_d is the digital control delay of 1.5 sampling period T_s.

The control plant transfer function $G_p(s)$ is expressed as

$$G_p(s) = 1/sL_g \tag{5.42}$$

Then, based on Eqs. (5.40)–(5.42), the open-loop transfer function can be obtained as,

$$G_{ol}(s) = C_{PR}(s)G_d(s)G_p(s) \tag{5.43}$$

5.4.3 Designing of k_p and k_r

The following design regarding PR controller parameter will be conducted on the basis of the open-loop transfer function shown in Eq. (5.43). In order to ensure stable, fast, and precise tracking of the fundamental AC signal error, the two design guidelines need to be followed.

(1) Normally, the control bandwidth of the PR controller should be as high as 1/20–1/10 of the switching frequency for the sake of good dynamic response, that is, 500 Hz–1 kHz for the case of switching frequency f_{sw} = 10 kHz. In addition, the resonant part of the PR controller is tuned at fundamental frequency 50 Hz, thus it only shows AC signal tracking capability around the fundamental frequency, but no obvious influence on the rest frequency range due to magnitude dropping.

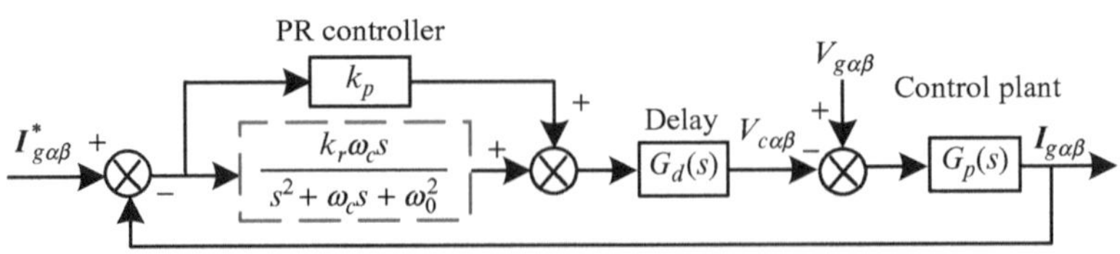

FIG. 5.24 Control block diagram of the grid-tied converter in the stationary frame using PR controller.

Based on this illustration, it can be found that the k_p can be designed on the basis of acceptable phase margin PM_P = 60° when the control bandwidth is as large as 500 Hz–1 kHz.

(2) Furthermore, a significant phase response drop can be seen at the tuned fundamental frequency due to the resonant part of the PR controller. Thus, it is essential to ensure that its lowest phase response around the fundamental frequency is still within the acceptable phase margin, therefore the parameter k_r should be designed according to this requirement.

5.4.3.1 Design Procedure of $\mathrm{k_p}$

As shown in Eq. (5.42), the control plant inherently introduces the phase delay of 90°, this means the delay caused by the digital control should be 90°—PM_P in order to achieve the expected phase margin PM_P, thus the following equation can be obtained regarding the delay caused by digital control at the 0 dB magnitude crossover frequency f_x (where the phase margin PM_P is defined),

$$T_d f_x = \left(180^\circ - 90^\circ - \mathrm{PM\_P}\right)/360^\circ \tag{5.44}$$

By substituting the $T_d = 1.5T_s = 150\,\mu s$ (due to sampling frequency = 10 kHz) and PM_P = 60°, the crossover frequency can be calculated as $f_x = 555\,\mathrm{Hz}$ according to Eq. (5.44).

Thereafter, in order to ensure the 0 dB magnitude response at the crossover frequency $f_x = 555\,\mathrm{Hz}$, the k_p parameter design can be achieved based on Eqs. (5.43) and (5.44) as,

$$k_p = 2\pi f_x L_g = 2\pi L_g \frac{\phi_P/360^\circ}{T_d}; \quad \text{where } \phi_P = 180^\circ - 90^\circ - \mathrm{PM\_P} \tag{5.45}$$

Based on Eq. (5.45), it can be found that the k_p is determined by the expected PM_P at the 0 dB crossover frequency, the digital control delay T_d, and the inductance L_g. Any variations on these parameters will result in the variation of the designed k_p.

5.4.3.2 Designing Procedure of $\mathrm{k_r}$

Once the proportional parameter k_p is designed, the k_r can be designed on the basis of the acceptable phase margin PM_R around the tuned fundamental frequency.

As shown in Fig. 5.27, the lowest phase response of the resonant part in the PR controller is achieved at the frequency of $\omega_x = \omega_0 + \omega_c$. As a consequence, the k_r should be designed considering the phase margin at the frequency of $\omega_x = \omega_0 + \omega_c$.

By substituting $s=j\omega_x=j(\omega_0+\omega_c)$ into Eq. (5.40), the following equation can be obtained,

$$C_{PR}(j\omega_x)=k_p+\frac{j\omega_x k_r\omega_c}{-\omega_x{}^2+j\omega_x\omega_c+\omega_0^2} \tag{5.46}$$

After certain mathematical deduction, the following equation of PR controller at the frequency of $\omega_x=\omega_0+\omega_c$ can be deduced as,

$$C_{\text{PR}}(j\omega_x)=\frac{k_p\left(\left(\omega_0^2-\omega_x^2\right)^2+(\omega_x\omega_c)^2\right)+k_r(\omega_x\omega_c)^2+jk_r\omega_x\omega_c\left(\omega_0^2-\omega_x^2\right)}{\left(\omega_0^2-\omega_x^2\right)^2+(\omega_x\omega_c)^2} \tag{5.47}$$

Then, based on Eq. (5.47), the phase response of PR controller at the frequency of $\omega_x=\omega_0+\omega_c$ can be expressed as,

$$\angle C_{PR}(j\omega_x)=a\tan\frac{k_r\omega_x\omega_c\left(\omega_0^2-\omega_x^2\right)}{k_p\left(\left(\omega_0^2-\omega_x^2\right)^2+(\omega_x\omega_c)^2\right)+k_r(\omega_x\omega_c)^2} \tag{5.48}$$

Similar as the case of the k_p design, the phase response of the PR controller at the frequency of $\omega_x=\omega_0+\omega_c$ should be around −30° due to the inherent phase delay of 90° in the control plant and the phase margin PM_R = 60°.

Based on the explanation and Eq. (5.48), the following equation regarding the k_r design can be deduced as,

$$k_r=\frac{k_p\tan(\phi_R)\left(\left(\omega_0^2-\omega_x^2\right)^2+(\omega_x\omega_c)^2\right)}{\omega_x\omega_c\left(\omega_0^2-\omega_x^2\right)-\tan(\phi_R)(\omega_x\omega_c)^2};\ \phi_R=-\left(180^\circ-90^\circ-\text{PM\_R}\right) \tag{5.49}$$

It can be observed from Eq. (5.49) that the k_r is determined by the k_p, the phase margin PM_R and the resonant bandwidth ω_c.

5.4.4 Theoretical Verification of the Designing of k_p and k_r

So far, the design guidelines for k_p and k_r have been introduced. In order to theoretically verify its effectiveness, the following shows the Bode diagram of the open-loop transfer function using the PR controller adopting the designed k_p and k_r.

Fig. 5.25 shows the Bode diagram of the open-loop transfer function using the PR controller with different PM_P = 45° and 60° while constant PM_R = 45°, $\omega_c=5$ rad/s. According to Eqs. (5.45) and (5.49), the k_p and k_r for the two cases can be calculated as, Case 1: PM_P = 45°, $k_p=52.3$, $k_r=262$; Case 2: PM_P = 60°, $k_p=34.9$, $k_r=175$.

As it can be seen, the phase response of the two cases is actually the same, however the different crossover frequencies of 550 and 820 Hz determine their different phase margin of 45° and 60°. Note these two crossover frequencies fit

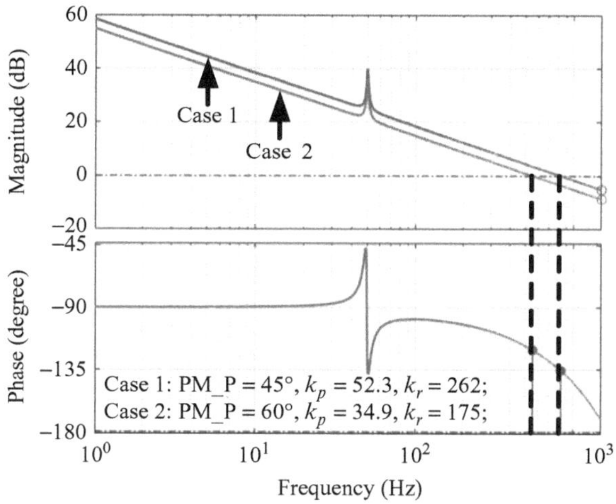

FIG. 5.25 Bode diagram of the open-loop transfer function using the PR controller with different PM_P=45° and 60°, but constant PM_R=45°.

into the control bandwidth request of 500 Hz–1 kHz (1/20–1/10 of the switching frequency = 10 kHz).

On the other hand, for the resonant part at 50 Hz, it can be observed that the phase response at 50 Hz ensures the phase margin PM_R=45°. At the same time, the magnitude response at 50 Hz is around 37 and 40 dB, respectively, which is sufficiently large to ensure the accurate tracking of 50 Hz AC signals.

Similarly, Fig. 5.26 shows the Bode diagram of the open-loop transfer function using the PR controller with constant PM_P=45° but different

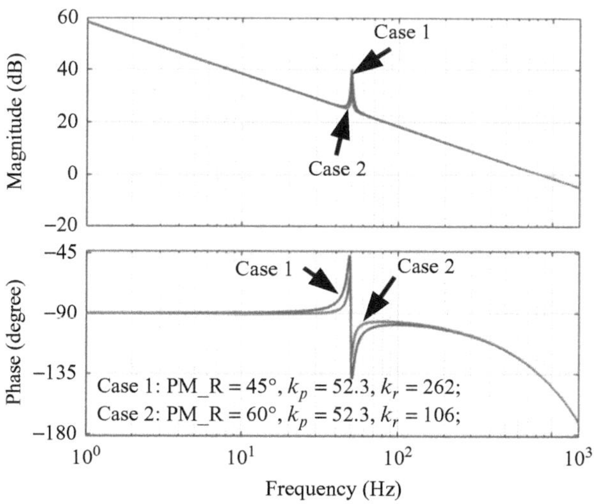

FIG. 5.26 Bode diagram of the open-loop transfer function using the PR controller with constant PM_P=45°, but different PM_R=45° and 60°.

PM_R = 45° and 60°, ω_c = 5 rad/s. According to Eqs. (5.45) and (5.49), the k_p and k_r for the two cases can be calculated as, Case 1: PM_R = 45°, k_p = 52.3, k_r = 262; Case 2: PM_R = 60°, k_p = 52.3, k_r = 106. It can be seen that the same phase margin PM_P = 45° can be achieved at 820 Hz.

On the other hand, the different phase margin at the resonant frequency 50 Hz PM_R = 45° and 60° can be achieved, and the different magnitude responses are 40 and 35 dB, respectively. It can be found that the smaller PM_R results in the higher magnitude response, which helps to better suppress the AC signal tracking error, but simultaneously results in the higher possibility of closed-loop operation instability due to the smaller phase margin. Therefore, the appropriate compromise between the resonant part magnitude response and the phase margin is critical to the satisfactory operation of the PR controller.

5.4.5 Simulation Validation of the Designing of k_p and k_r

5.4.5.1 Case 1: PM_P = PM_R = 45°, k_p = 52.3, k_r = 262

In order to validate the designed k_p and k_r, the simulation model based on MATLAB/Simulink is established. The simulation is conducted under the condition that VSC output active power P_g = 10 kW, reactive power Q_g = 0 Var. The output inductance filter L_g = 10 mH, the sampling frequency f_s, and switching frequency f_{sw} are both 10 kHz. The phase margins of PM_P and PM_R are both chosen as 45°, and then k_p = 52.3, k_r = 262 can be calculated.

Fig. 5.27 shows the simulation waveforms of the grid voltage V_{gabc} and the VSC output current I_{gabc}, as well as the FFT analysis of the output current. As it can be observed from Fig. 5.27A, the fundamental 50 Hz AC signal can be well tracked both at the output power P_g = 5 kW and 10 kW due to the satisfactory steady-state operation of the designed PR controller. Moreover the transient response at the instance of output power stepping at 0.1 s is also fast with negligible current overshoot, this result helps to verify the good dynamic tracking capability of the designed PR controller.

Furthermore, Fig. 5.27B shows the FFT analysis of the output current under the condition of P_g = 10 kW. It can be found that the most prominent harmonic components in the output current are 5th sequence with the amplitude of 1.65%, while the other harmonic sequences are below 1.0%. This output distortion behavior is acceptable in practical with single inductor filter.

5.4.5.2 Case 2: PM_P = PM_R = 60°, k_p = 34.9, k_r = 70.7

For the purpose of better comparison, the other simulation with PM_P = PM_R = 60°, and k_p = 34.9, k_r = 70.7 is conducted, while the other parameters are kept the same as Fig. 5.29. The simulation waveforms are shown in Fig. 5.28.

It can be observed that due to the requested higher phase margin, both k_p and k_r become smaller. As a consequence, the fundamental 50 Hz AC signal

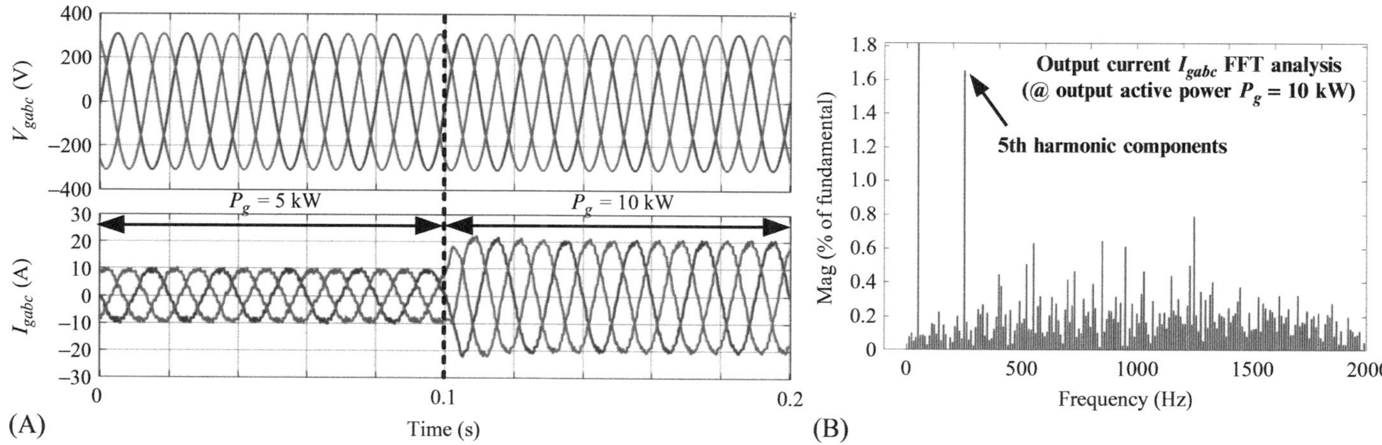

FIG. 5.27 (A) Simulation waveforms of the grid voltage V_{gabc} and the VSC output current I_{gabc}, and the (B) FFT analysis of the output current. $L_g = 10\,\text{mH}$, PM_P = PM_R = 45°, $k_p = 52.3$, $k_r = 262$, $\omega_c = 5\,\text{rad/s}$, $f_s = f_{sw} = 10\,\text{kHz}$.

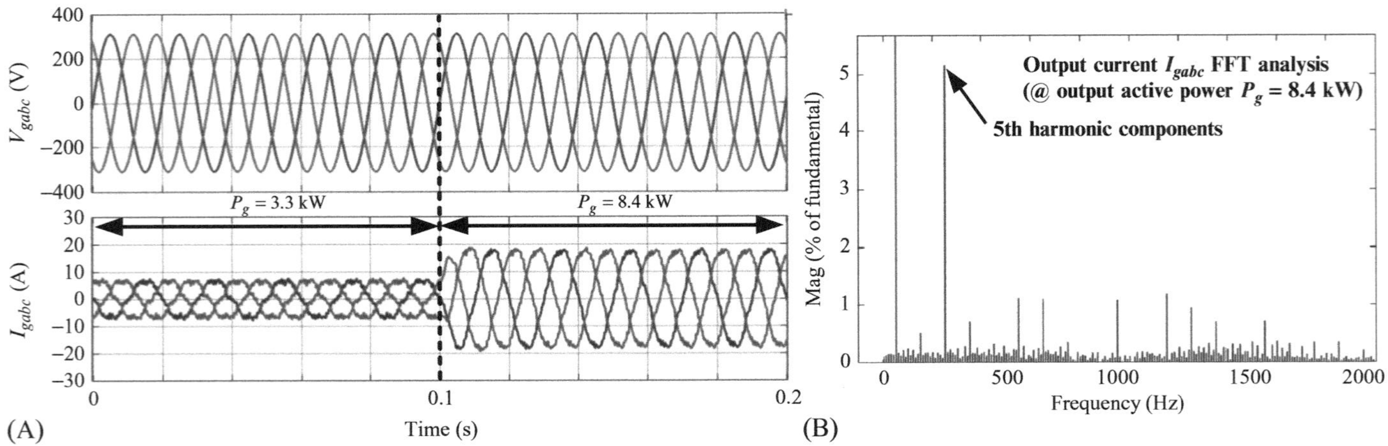

FIG. 5.28 (A) Simulation waveforms of the grid voltage V_{gabc} and the VSC output current I_{gabc}, and (B) the FFT analysis of the output current. $L_g = 10\,\text{mH}$, PM_P=PM_R=60°, $k_p = 34.9$, $k_r = 70.7$, $\omega_c = 5\,\text{rad/s}$, $f_s = f_{sw} = 10\,\text{kHz}$.

tracking has a larger control error, which can be verified by the smaller output power as shown in Fig. 5.28, i.e., P_g=3.3 and 8.4kW, respectively.

Moreover, due to the decreasing of k_p, the PR controller has weaker control capability with smaller bandwidth. As a result, the harmonic distortion components of the output current become higher, i.e., the 5th sequence becomes 5%, and the other harmonic sequences are close to 1.0%.

5.4.5.3 Case 3: PM_P=PM_R=30°, k_p=69.8, k_r=2366

Furthermore, it is also necessary to discuss the PR controller performance if too small phase margins are assigned as PM_P=PM_R=30°, then k_p=69.8, k_r=2366, the other parameters are kept the same as the earlier two cases. The simulation waveforms are shown in Fig. 5.29.

As it can be seen, the output active power is precisely tracked due to the high k_p and k_r. Nevertheless, the transient response at the active power stepping instant sees obvious current overshoot, then decreases, and finally increases to the steady state. This transient takes the period around 40ms, which is much longer than that of Case 1 as 10ms. Clearly, this unfavorable transient response is due to the too high k_p.

On the other hand, according to the FFT results in Fig. 5.29B, it can be observed that although the 5th harmonic sequence can be better suppressed due to the high k_p, it also makes the system more sensitive and vulnerable to the harmonic spectrum, thus more abundant harmonic sequences can be a consequence.

By comparing these three simulated cases, the following conclusions can be drawn,

(1) From the perspective of stable, fast, and precise tracking of 50Hz AC signal using the PR controller, it is important to adopt the appropriate phase margin as PM_P=PM_R=45°, then the k_p and k_r can be calculated according to Eqs. (5.45) and (5.49).

(2) Too high phase margin results in smaller k_p and k_r, which in turn impairs the PR controller precision with higher error. On the contrary, too small phase margin results in higher k_p and k_r, which causes overshoot in the current transient response and also makes the system more sensitive to harmonic sequences.

5.5 SUMMARY

By using the different reference frames, this chapter begins with the modeling of the three-phase grid-tied converter. In order to synchronize with the power grid, the phasor locked loop with the structures of the synchronous reference frame and dual second-order generalized integrator are described and compared at the normal and distorted grid conditions. With the help of the Bode plot, the grid current with the PI (Proportion + Integral) and PR (Proportion + Resonant)

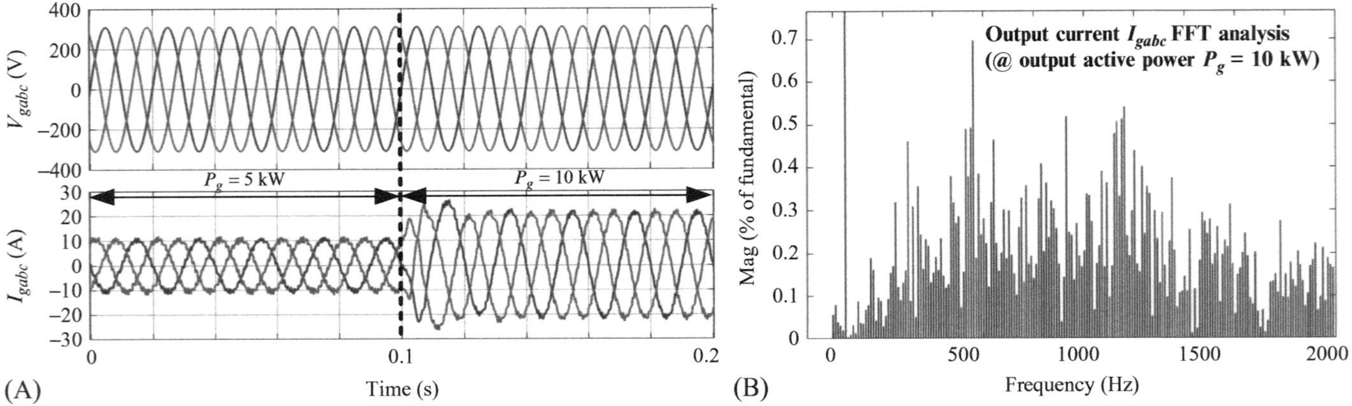

FIG. 5.29 (A) Simulation waveforms of the grid voltage V_{gabc} and the VSC output current I_{gabc}, and (B) the FFT analysis of the output current. $L_g = 10\,\text{mH}$, PM_P=PM_R=30°, $k_p = 69.8$, $k_r = 2366$, $\omega_c = 5\,\text{rad/s}$, $f_s = f_{sw} = 10\,\text{kHz}$.

controller are designed under the synchronous reference frame and the stationary reference frame, respectively. It is proven that the step response of the designed PI and PR controller effectively agree with the desired bandwidth and its corresponding phase margin.

REFERENCES

[1] M.P. Kazmierkowski, R. Krishnan, F. Blaabjerg, et al., Control in Power Electronics: Selected Problems, Academic Press, 2002.
[2] R.W. Erickson, M. Dragan, Fundamentals of Power Electronics, Springer Science & Business Media, 2007.
[3] M. Liserre, F. Blaabjerg, S. Hansen, Design and control of an LCL-filter-based three-phase active rectifier, IEEE Trans. Ind. Appl. 41 (5) (September 2005) 1281–1291.
[4] S. Golestan, J.M. Guerrero, J.C. Vasquez, Three-phase PLLs: a review of recent advances, IEEE Trans. Power Electron. 32 (3) (March 2017) 1894–1907.
[5] S. Golestan, F.D. Freijedo, A. Vidal, A.G. Yepes, J.M. Guerrero, J. Doval-Gandoy, An efficient implementation of generalized delayed signal cancellation PLL, IEEE Trans. Power Electron. 31 (2) (February 2016) 1085–1094.
[6] L. Tong, X. Zou, S. Feng, Y. Chen, Y. Kang, Q. Huang, Y. Huang, An SRF-PLL-based sensorless vector control using the predictive deadbeat algorithm for the direct-driven permanent magnet synchronous generator, IEEE Trans. Power Electron. 29 (6) (June 2014) 2837–2849.
[7] S. Golestan, J.M. Guerrero, Conventional synchronous reference frame phase-locked loop is an adaptive complex filter, IEEE Trans. Power Electron. 62 (3) (March 2015) 1679–1682.
[8] H. Yi, X. Wang, F. Blaabjerg, F. Zhuo, Impedance analysis of SOGI-FLL-based grid synchronization, IEEE Trans. Power Electron. 32 (10) (October 2017) 7409–7413.
[9] S. Golestan, S.Y. Mousazadeh, J.M. Guerrero, J.C. Vasquez, A critical examination of frequency-fixed second-order generalized integrator-based phase-locked loops, IEEE Trans. Power Electron. 32 (9) (September 2017) 6666–6672.
[10] D.G. Holmes, T.A. Lipo, Pulse Width Modulation for Power Converters: Principles and Practice, John Wiley & Sons, 2003.
[11] D.G. Holmes, T.A. Lipo, B.P. McGrath, W.Y. Kong, Optimized design of stationary frame three phase AC current regulators, IEEE Trans. Power Electron. 24 (11) (November 2009) 2417–2426.
[12] B. Wen, D. Boroyevich, R. Burgos, P. Mattavelli, Z. Shen, Small-signal stability analysis of three-phase AC systems in the presence of constant power loads based on measured d-q frame impedances, IEEE Trans. Power Electron. 30 (10) (October 2015) 5952–5963.
[13] H. Mao, D. Boroyevich, F.C.Y. Lee, Novel reduced-order small-signal model of a three-phase PWM rectifier and its application in control design and system analysis, IEEE Trans. Power Electron. 13 (3) (May 1998) 511–521.
[14] R. Teodorescu, F. Blaabjerg, Flexible control of small wind turbines with grid failure detection operating in stand-alone and grid-connected mode, IEEE Trans. Power Electron. 19 (5) (September 2004) 1323–1332.
[15] T. Nussbaumer, G. Gong, M.L. Heldwein, J.W. Kolar, Modeling and robust control of a three-phase Buck+Boost PWM rectifier (VRX-4), IEEE Trans. Ind. Appl. 44 (2) (March 2008) 650–662.
[16] A.E. Bryson, Applied Optimal Control: Optimization, Estimation and Control, CRC Press, 1975.

Chapter 6

Control of Single-Phase and Three-Phase DC/AC Converters

Ariya Sangwongwanich*, Ahmed Abdelhakim†, Yongheng Yang* and Keliang Zhou‡
**Aalborg University, Aalborg, Denmark, †University of Padova, Padua, Italy, ‡University of Glasgow, Glasgow, United Kingdom*

6.1 INTRODUCTION

Power electronic converters have been widely used in industrial and residential applications to enable proper, reliable, and efficient power control between the source and the load. There are several types of power conversion systems categorized by the combination of sources and load types (DC or AC), as also being discussed in previous chapters of this book. Among other power conversion systems, the DC/AC converters are one of the mainstream power conditioning systems, which cover a wide range of applications, since most of the industrial loads are AC (e.g., motor drives and the utility grid). For instance, the DC/AC converters are commonly employed in motor drive systems to improve the conversion efficiency and to enable a variable speed operation [1–3]. In that case, the control target of the DC/AC converters is to regulate the motor torque and speed, and the frequency of the AC output voltage should be fully controllable.

In the last decade, the DC/AC power converters have also been widely adopted in renewable energy applications as an adaptor to the grid. For instance, in wind turbine (WT) and photovoltaic (PV) systems, the DC/AC converters are employed as a grid-side converter to transfer the extracted power from renewable sources to the AC grid [4–7]. The main objective of the grid-side DC/AC converter is to deliver the active power to the grid with a satisfied power quality, e.g., total harmonic distortion level according to the grid requirements [8]. Additionally, grid support functionalities such as reactive power injection or active power control capabilities have also been imposed in grid regulations along with the continuously increasing penetration of grid-connected WT and PV systems [9–13]. These stringent requirements are demanded in order to avoid the adverse impacts from the intermittent power injection from the

Control of Power Electronic Converters and Systems. https://doi.org/10.1016/B978-0-12-805245-7.00006-8

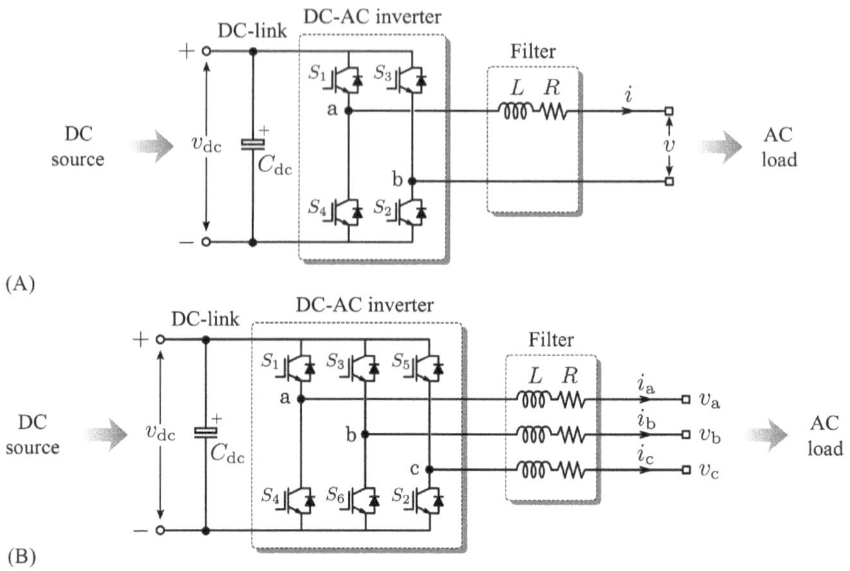

FIG. 6.1 Circuit diagram of DC/AC converters with an L filter to smoothen the current from the switching (R is its internal resistance): (A) single-phase and (B) three-phase, where C_{dc} is the DC-link capacitor.

renewables. Therefore, the control of the DC/AC converter plays an important role in achieving the above requirements and ensuring a high-performance power conversion from DC to AC.

In this chapter, the control of DC/AC converters will be discussed in details for both single- and three-phase systems. Here, the most common single- and three-phase DC/AC converters shown in Fig. 6.1 are considered. The model of the DC/AC power converter will be introduced to represent the system control dynamics. Then, the control of the three-phase DC/AC converter in the *dq*-reference frame with proportional integral (PI) controllers will be discussed, including the current controller, DC-link voltage controller, and active power and reactive power controller (also referred to as *PQ* controller). The design of each controller will be demonstrated and verified with case studies.

6.2 MODELING OF DC/AC CONVERTERS

In this part, the modeling of DC/AC converters will be provided in detail. The reference frame transformation will be first discussed, where the AC control variables (e.g., voltages and currents) will be transformed into two DC quantities through the Clarke and Park transformations. This is applied to both single- and three-phase systems. Afterward, system modeling of DC/AC converters with an L filter, including the current controller, the *PQ* controller, and the DC-link voltage controller will be elaborated.

6.2.1 Reference Frame Transformation

6.2.1.1 *Three-Phase to the Stationary Reference Frame (*abc→αβ*)*

Fig. 6.1B shows the circuit diagram of a three-phase, two-level DC/AC converter. As seen, there are two pairs of AC variables (voltages: v_a, v_b, v_c and currents: i_a, i_b, i_c), which are time varying according to the dynamics of each individual phase. From the modeling perspective, the mathematical model of such a system can be complicated. However, considering a balanced three-phase system, it can be represented by a two-phase system on the stationary reference frame (i.e., the $\alpha\beta$-reference frame) as

$$\begin{bmatrix} x_\alpha \\ x_\beta \end{bmatrix} = \frac{2}{3}\begin{bmatrix} 1 & -\dfrac{1}{2} & -\dfrac{1}{2} \\ 0 & \dfrac{\sqrt{3}}{2} & -\dfrac{\sqrt{3}}{2} \end{bmatrix}\begin{bmatrix} x_a \\ x_b \\ x_c \end{bmatrix} \tag{6.1}$$

where x_a, x_b, and x_c represent the voltages or currents of the system and x_α, x_β are variables on the $\alpha\beta$-reference frame. In practice, the transformation from the *abc*-reference frame to the $\alpha\beta$-reference frame is also called the Clarke transformation.

6.2.1.2 *Stationary Reference Frame to the Synchronous Reference Frame (*αβ→dq*)*

By applying the Clarke transformation in Eq. (6.1), the control variables (x_a, x_b, and x_c) in any balanced three-phase system are reduced to two components (x_α, x_β). Nevertheless, the $\alpha\beta$ components are still AC variables rotating at the same speed (denoted as ω) of the original three-phase AC variables. In that case, the control design is not straightforward, and the well-known proportional integral (PI) controllers cannot achieve zero-error tracking of the $\alpha\beta$ components [14]. Thus, the Park transformation ($\alpha\beta \rightarrow dq$) is applied to the $\alpha\beta$ variables, and also it gives two variables on the synchronous reference frame (i.e., the *dq*-reference frame) as

$$\begin{bmatrix} x_d \\ x_q \end{bmatrix} = \begin{bmatrix} \cos\theta & \sin\theta \\ -\sin\theta & \cos\theta \end{bmatrix}\begin{bmatrix} x_\alpha \\ x_\beta \end{bmatrix} \tag{6.2}$$

in which x_d, x_q are the components on the *dq*-reference frame and $\theta = \omega t$ is the angular position or phase with ω being the frequency. When the Park transformation in Eq. (6.2) is adopted, the two orthogonal variables on the $\alpha\beta$-reference frame will become two DC quantities on the *dq*-reference frame. This is usually preferable in the control design, e.g., of PI controllers.

Notably, the Park transformation can also be employed in single-phase systems. In that case, an orthogonal signal generator (OSG) is required to create a fictitious component (i.e., x_β) in-quadrature with the original variable (i.e., x_α). There are several OSG systems, which has been discussed in Ref. [5]. Once the orthogonal variables on the $\alpha\beta$-reference frame, i.e., x_α and x_β, are obtained, the transformation to the dq-reference frame in Eq. (6.2) can be applied, similar to that for three-phase systems.

6.2.2 System Modeling

A system diagram of the two-level three-phase DC/AC converter is illustrated in Fig. 6.2, where a cascaded-control loop is employed. At the outer control loop, the active and reactive power is controlled either directly by a *PQ* controller (as shown in Fig. 6.2) or indirectly through the DC-link voltage regulation [15]. This is mainly determined by applications. For instance, for grid-connected photovoltaic systems, where a DC-DC stage is usually employed to optimize the power extraction, the DC-link voltage control is usually taken as the outer loop. Nevertheless, the inner control loop is generally to control the current injected to the AC load. Herein, the objective is to regulate the output current according to the reference set by the outer control loop, as demonstrated in Fig. 6.2. An output voltage reference (i.e., v_{inv}^*) of the DC/AC converter is obtained from the current controller with an inverse Park and Clark transformation. The voltage reference

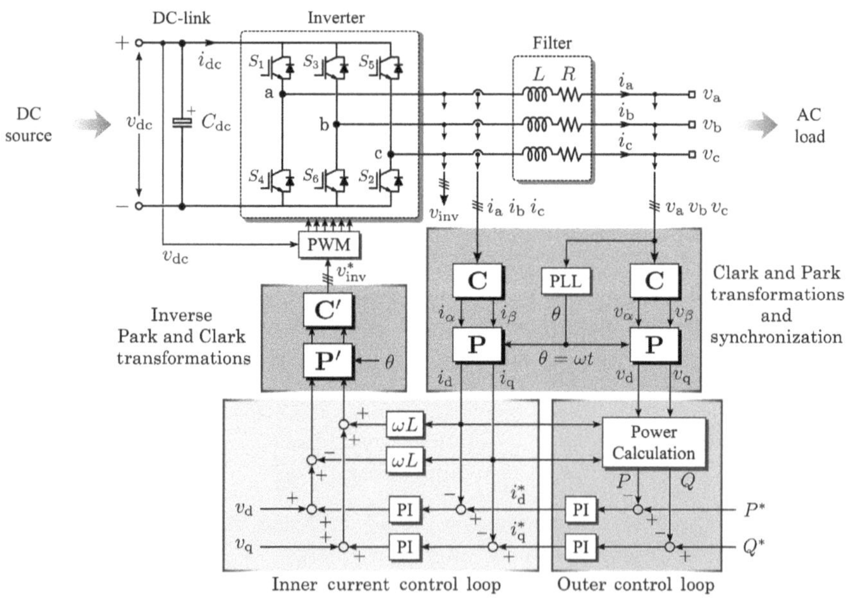

FIG. 6.2 System diagram of the DC/AC two-level converter with a cascaded-control loop in the synchronous (dq) reference frame (PLL—phase locked loop; PWM—pulse width modulation), where **C**,**P** represents the Clark and Park transformation with **C**′,**P**′ being the corresponding inverse transformation, respectively, and $\theta = \omega t$ is the phase of the output voltage with ω being the frequency. The outer control loop is a power control loop, which is able to control both active power P and reactive power Q. The subscript represents the variable on the corresponding reference frame. In certain applications, the frequency ω is assigned arbitrarily, where the PLL unit may not be necessary.

is then used for the pulse-width modulation (PWM) in order to synthesize the corresponding output voltage $v_{\rm inv}$ by switching the DC/AC converter. This closes the entire DC to AC power processing. Notably, a phase-locked loop (PLL) as the synchronization unit is required for the reference frame transformation to keep synchronizing with the grid. However, in certain applications, the frequency ω can be assigned arbitrarily, where the PLL may be absent. Nevertheless, the entire system should be modeled in order to design proper controllers for the DC/AC power conversion as shown in Fig. 6.2. Hereafter, the mathematical model of the entire system is derived in terms of the modeling for the current control loop, the power loop, and the DC-link.

6.2.2.1 *Modeling of the Current Control Loop*

The main aim of the current controller is to regulate the converter output current by generating a proper output voltage reference. Following the circuit diagram in Fig. 6.1, according to the Kirchhoff's law, the DC/AC converter output current (i.e., the current injected to the load) can be expressed in the single-phase reference frame as

$$L\frac{di}{dt}+Ri=v_1-v \tag{6.3}$$

in which L and R are the inductance and the resistance of the filter, v_1, i, and v are the converter output voltage, the current injected to the load, and the load voltage.

Similarly, according to Figs. 6.1 and 6.2, the DC/AC converter output currents for the three-phase system can be expressed as

$$\begin{cases} L\dfrac{di_{\rm a}}{dt}+Ri_{\rm a}=v_{\rm a1}-v_{\rm a} \\ L\dfrac{di_{\rm b}}{dt}+Ri_{\rm b}=v_{\rm b1}-v_{\rm b} \\ L\dfrac{di_{\rm c}}{dt}+Ri_{\rm c}=v_{\rm c1}-v_{\rm c} \end{cases} \tag{6.4}$$

in which v_{p1}, i_p, and v_p are the converter output voltage, the current injected to the load, and the load voltage, with $p=a,b,c$ being the phase index. Applying the reference frame transformations in Eqs. (6.1), (6.2) to the three-phase model results in

$$\begin{cases} L\dfrac{di_d}{dt}+Ri_d-\omega Li_q=v_{d1}-v_d \\ L\dfrac{di_q}{dt}+Ri_q+\omega Li_d=v_{q1}-v_q \end{cases} \tag{6.5}$$

where i_d and i_q are the currents injected to the load on the dq-reference frame, v_{d1} and v_{q1} are the converter output voltages on the d- and q-axis, v_d and v_q are the load voltages on the d- and q-axis, respectively, and ω is the frequency of the system.

As discussed previously in Section 6.2.1, in the case of single-phase systems, a fictitious component (i.e., x_β) in-quadrature with the original variable (i.e., x_α) should be generated from the OSG system before applying the reference frame transformation. Afterward, the transformation to the dq-reference frame in Eq. (6.2) can be applied to Eq. (6.3) and the fictitious system, resulting in a similar system model as in Eq. (6.5). Seen from this viewpoint, the system model in the dq-reference frame is similar for both single- and three-phase systems.

It is clearly seen in Eq. (6.5) that the injected AC current can be controlled by regulating the output voltage of the DC/AC converter. However, the d- and q-axis output currents are coupled to each other, leading to a slightly complicated system from the control perspective. In addition, the load voltage also has an influence on the control dynamics. Therefore, the output voltage references (i.e., v_{d1}^* and v_{q1}^*) are modified by means of adding one decoupling term and one feed-forward voltage as

$$\begin{cases} v_{d1}^* = v_{d1} + \omega L i_q - v_d \\ v_{q1}^* = v_{q1} - \omega L i_d - v_q \end{cases} \tag{6.6}$$

The current control loop model after the modification can then be rewritten as

$$\begin{cases} L\dfrac{di_d}{dt} + Ri_d = v_{d1}^* \\ L\dfrac{di_q}{dt} + Ri_q = v_{q1}^* \end{cases} \tag{6.7}$$

in which the d- and q-axis output currents (i.e., the current injected to the load) are decoupled, as shown in Fig. 6.3. Moreover, as it is shown in Eq. (6.7), the d- and q-axis output current expressions are identical. Thus, the analysis on one axis is sufficient, as they share the same dynamics. Accordingly, the plant transfer functions from the converter output voltage references to the output current currents can be obtained as [16]

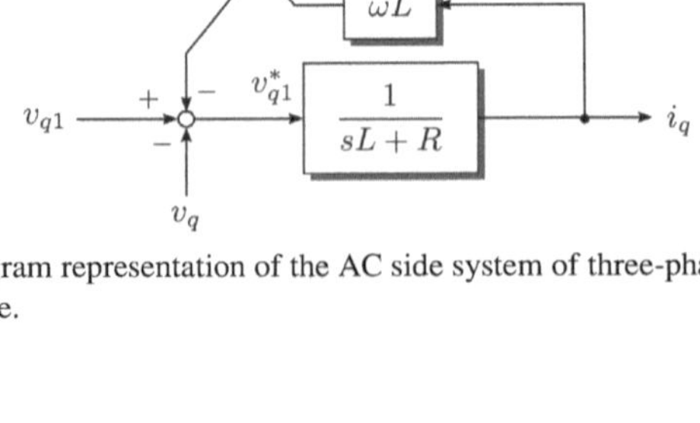

FIG. 6.3 Block diagram representation of the AC side system of three-phase DC/AC inverters in the dq-reference frame.

$$\frac{i_d(s)}{v_{d1}^*(s)} = \frac{i_q(s)}{v_{q1}^*(s)} = \frac{1}{Ls+R} \tag{6.8}$$

also indicating that the d- and q-axis output currents have the same dynamics.

6.2.2.2 *Modeling of the* PQ *Control Loop*

The concept of the *PQ* control (i.e., the outer control loop in Fig. 6.2) is based on the *instantaneous power theory* proposed by Akagi [17]. In the synchronous *dq*-reference frame, the instantaneous active power P and reactive power Q can be calculated as

$$\begin{cases} P = \frac{3}{2}\left(v_d i_d + v_q i_q\right) \\ Q = \frac{3}{2}\left(v_q i_d - v_d i_q\right) \end{cases} \tag{6.9}$$

Assuming that the PLL is aligned with the load voltage vector to the d-axis of the dq-reference frame (i.e., $v_q = 0$), the transfer functions from the d- and q-axis output currents to the active and reactive power can be calculated as

$$\frac{P(s)}{i_d(s)} = \frac{3}{2} v_d(s) = \frac{3}{2} V_m \tag{6.10}$$

$$\frac{Q(s)}{i_q(s)} = -\frac{3}{2} v_d(s) = -\frac{3}{2} V_m \tag{6.11}$$

which are simply proportional gains with V_m being the amplitude of the output load voltage. It can be noticed from Eqs. (6.10), (6.11) that, in theory, an open-loop control is sufficient to regulate the active and reactive power (assuming an ideal converter). However, a closed-loop control is practically employed to improve the control performance due to uncertainties in the system (e.g., load voltage variations and power losses), as demonstrated in Fig. 6.2.

6.2.2.3 *Modeling of the DC-Link Control Loop*

One alternative to control the output power of the DC/AC power converter is through the regulation of the DC-link voltage [15]. Maintaining the DC-link voltage constant (at the nominal value) ensures that all the extracted power from the DC source can be delivered to the AC load. This control scheme is widely used in photovoltaic (PV) systems, where the input power is continuously fed by the PV arrays. In addition, in the case of back-to-back power conversion systems, the (average) DC-link voltage is also commonly controlled at a desired level. This is also to ensure a proper power injection to the load, e.g., the DC-link voltage has to be at least higher than the peak grid voltage in the case of grid-connected applications.

Assuming a loss-less DC/AC power converter, the input DC power and the output active power should be balanced according to the instantaneous power theory. In that case, it gives

$$\overbrace{v_{\mathrm{dc}} C_{\mathrm{dc}} \frac{dv_{\mathrm{dc}}}{dt}}^{\text{DC input power}} = \overbrace{\frac{3}{2}\left(v_d i_d + v_q i_q\right)}^{\text{Output active power}} \tag{6.12}$$

where C_{dc} is the DC-link capacitance as shown in Fig. 6.2 and i_{dc} is the output current of the DC source. If the d-axis of the dq-reference frame is aligned with the load voltage vector (by the PLL), the q-axis component of the grid voltage becomes zero (i.e., $v_q = 0$). Applying the small-signal analysis to Eq. (6.12) can linearize the model. By doing so, the transfer function of the DC-link from the output d-axis current to the DC-link voltage can be simplified as

$$\frac{v_{\mathrm{dc}}(s)}{i_d(s)} = \frac{3}{2} \frac{V_m}{V_{\mathrm{dc}} C_{\mathrm{dc}} s} \tag{6.13}$$

with V_{dc} and I_{dc} being the average voltage and current of the DC source.

It should be noted that, although the model in Eq. (6.13) shows the control of the d-axis current through the regulation of the DC-link voltage, a reactive power injection is still possible if required. In that case, the reactive power can separately be controlled following Eq. (6.11) to generate the q-axis current reference. The decoupling of the two currents on the dq-reference frame also ensures the individual control.

6.3 CONTROLLER DESIGN

As previously mentioned, classical PI controllers are commonly used in current-controlled DC/AC converters [18,19]. This is also depicted in Fig. 6.2, in which the three-phase DC/AC converter is controlled as a current source in order to track a certain current reference in the synchronous dq-reference frame. The d- and q-axis current references (i_d^* and i_q^*) can be injected directly or calculated from the desired active and reactive power. In this case, the d-axis current i_d^* is determined by the desired active power P as shown in Eq. (6.10), while the q-axis current i_q^* is obtained from the set-point of the reactive power Q according to Eq. (6.11). Notably, an outer power loop can be employed in order to regulate the active and reactive power, as exemplified in Fig. 6.2. It is also worth mentioning that the open-loop case is sufficient in most applications.

On the other hand, the outer control loop can also be realized by the DC-link voltage regulation, which generates the d-axis reference current i_d^* [19]. This control scheme is widely used in grid-connected application, e.g., PV applications. In some cases, e.g., single-phase system, current controllers in the stationary reference frame (or corresponding to the abc-reference frame) become more attractive with a simpler control structure due to the absence of reference frame transformations. This will be further discussed in the following section.

Because of the cascaded control structure shown in Fig. 6.2, the controller design can be challenging due to possible dynamic interactions between different controllers. As a result, it is of importance to properly design the current and the DC-link voltage controllers in order to ensure a high-control performance and proper operation of the DC/AC converters. In practice, the response of the inner control loop, i.e., the current controller, should be designed to be much faster than the response of the outer control loop, i.e., the DC-link voltage controller [19]. Notably, the analysis is based on the previously discussed models in the s-domain with an assumption that the controller sampling rate is fast enough and the impact from discretization can be neglected. Accordingly, this section focuses on the design procedure of the current-controlled DC/AC converter, where the next subsections will cover the design steps of the current and DC-link voltage controllers.

6.3.1 Current Controller

According to the prior system modeling, the inner current control loops for the d- and q-axis components have the same dynamics, as implied in Eq. (6.8). Hence, the design procedure can be demonstrated only on the d-axis (i.e., for the d-axis current i_d) as shown in Fig. 6.4, in which $G_{\mathrm{PI}}^{d}(s), G_{\mathrm{delay}}(s)$, and $G_f(s)$ represent the d-component PI current controller, the elapsed delay due to the PWM and computations in the control system, and the filter (plant) transfer function.

The controller design for the q-axis current can be achieved following the same method. Nevertheless, these transfer functions in Fig. 6.4 can be given as

$$G_{\mathrm{PI}}^{d}(s) = k_{pd} + \frac{k_{id}}{s} = \frac{k_{pd}(1+T_{id}s)}{T_{id}s} \tag{6.14}$$

$$G_{\mathrm{delay}}(s) = \frac{1}{1+1.5T_s s} \tag{6.15}$$

$$G_f(s) = \frac{1}{R+Ls} = \frac{T_f}{L(1+T_f s)} \tag{6.16}$$

in which k_{pd} and k_{id} are the proportional and the integral gains of the PI current controller, $T_{id}=k_{pd}/k_{id}$ is the integrator time constant, T_s is the sampling time,

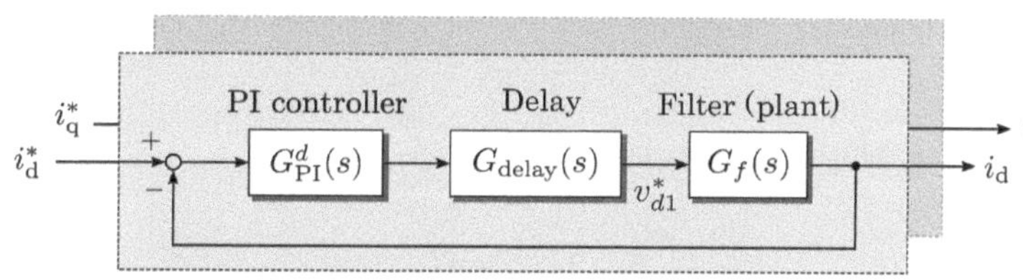

FIG. 6.4 Current control loops in the synchronous dq-reference frame, where the cross-coupling and the feed-forward terms are neglected.

L is the inductance of the employed filter, R is the equivalent series resistance of the inductor L, and $T_f = L/R$ is the filter time constant.

As shown in Fig. 6.4, the cross-coupling ($\omega L i_q$) and the voltage feed-forward (v_d) terms for the d-axis current component control loop are neglected. In that case, the two terms are considered as disturbances in the system. Accordingly, the open-loop transfer function can be expressed as

$$G_{ol}^{d}(s) = G_{\mathrm{PI}}^{d}(s) \cdot G_{\mathrm{delay}}(s) \cdot G_f(s) = \frac{k_{pd} T_f (1 + T_{id} s)}{T_{id} L s (1 + 1.5 T_s s)(1 + T_f s)} \tag{6.17}$$

Hence, the closed-loop transfer function is obtained as

$$G_{cl}^{d}(s) = \frac{G_{ol}^{d}(s)}{1 + G_{ol}^{d}(s)} = \frac{k_{pd} T_f (1 + T_{id} s)}{T_{id} L s (1 + 1.5 T_s s)(1 + T_f s) + k_{pd} T_f (1 + T_{id} s)} \tag{6.18}$$

which is also applicable for the q-axis current.

In order to analytically design the parameters of the PI current controller, the closed-loop transfer function shown in Eq. (6.18) can be simplified by choosing the integrator time constant T_{id} to be equal to the filter time constant T_f. By doing so, the closed-loop transfer function can be expressed as

$$G_{cl}^{d}(s) = \frac{k_{pd}}{L s (1 + 1.5 T_s s) + k_{pd}} = \frac{\frac{2 k_{pd}}{3 T_s L}}{s^2 + \frac{2}{3 T_s} s + \frac{2 k_{pd}}{3 T_s L}} \tag{6.19}$$

being a typical second-order system with

$$\omega_n^2 = \frac{2 k_{pd}}{3 T_s L} \quad \text{and} \quad 2 \zeta \omega_n = \frac{2}{3 T_s} \tag{6.20}$$

where ω_n is the natural frequency and ζ is the damping ratio.

In practice, $\zeta = 1/\sqrt{2}$ is designed for an optimally damped system, which will result in an overshoot of 5% for a step response [5]. As a consequence, the proportional and integral gains can be obtained as

$$k_{pd} = \frac{L}{3 T_s} \quad \text{and} \quad k_{id} = \frac{L}{3 T_s T_f} \tag{6.21}$$

Finally, with the assumption that the current control loop is optimally designed, the closed-loop transfer function can then be approximated by as

$$G_{cl}^{d}(s) \approx \frac{1}{1 + 3 T_s s} = \frac{1}{1 + \tau s} \tag{6.22}$$

from which the bandwidth can be estimated as

$$f_{bw}^{d} \approx \frac{1}{2 \pi \tau} = \frac{1}{6 \pi T_s}. \tag{6.23}$$

On the other hand, the PI controller can be tuned using MATLAB, where an interesting tool called "*pidTuner*" can be used. Using this tool, the parameters of the PI controller can be selected by importing the plant and the delay transfer functions, i.e., $G_f(s)$ and $G_{\text{delay}}(s)$, and then tuning the PI controller according to the desired bandwidth or the response time and the phase margin or the transient behavior (e.g., the rise time).

In some applications, e.g., single-phase PV systems, the DC/AC converter is normally required to operate with a unity power factor. In that case, the reference output current of the DC/AC converter is one single sinusoidal time-varying signal following the phase of the grid voltage. As discussed previously, PI controllers can be employed in the dq-reference frame. However, this can increase control complexity in terms of reference frame transformations. An alternative solution to regulate the AC signal is to use a proportional-resonant (PR) controller in the $\alpha\beta$-reference frame [20], which has a controllability of the AC signal at a specific frequency ω, e.g., the grid fundamental frequency. The transfer function of the PR controller is given as

$$G_{\text{PR}}(s) = k_{pr} + \frac{k_{ir}s}{s^2 + \omega^2} \tag{6.24}$$

in which k_{pr} and k_{ir} are the proportional and the integral gains, respectively.

From the transfer function in Eq. (6.24), the PR controller has an infinite gain at a particularly tuned frequency ω, making it capable of tracking an AC signal. The proportional gain k_{pr} can be selected in the similar way as in Eq. (6.21), while the integral gain k_{ir} is recommended as

$$k_{ir} = 2\alpha_h k_{pr} \tag{6.25}$$

where α_h is the resonant bandwidth, which should be much lower than the current controller bandwidth (e.g., $\alpha_h \ll 2\pi f_{bw}^d$).

In some cases, the PR controller has its implementation advantage due to the absence of the reference frame transformation. However, the ideal PR controller in Eq. (6.24) is sensitive to even a small frequency variation (which is typical grid-connected applications), leading to stability issues [20]. To address those, the nonideal PR controller can be employed in practice, whose transfer function is given as

$$G_{\text{PR}}(s) = k_{pr} + \frac{k_{ir}\omega_c s}{s^2 + 2\omega_c s + \omega^2} \tag{6.26}$$

where ω_c is the cut-off frequency that can be used to adjust the selectivity of the tracking frequency with the trade-off of the controller gain. It is recommended in Ref. [21] that the cut-off frequency ω_c should be selected in the range of 5–15 rad/s. More detailed analysis and design procedure can be found in Refs. [20–24].

6.3.2 DC-Link Voltage Controller

It has been illustrated in previous sections that the DC-link voltage can be controlled for the inner d-axis current control loop. That is, the DC-link voltage controller represents an outer loop to generate the reference for the d-axis current. This is depicted in Fig. 6.5, in which $G_{\mathrm{PI}}^{v}(s), G_v(s)$ represents the DC-link voltage PI controller and the plant (i.e., the DC-link voltage model), respectively. The transfer functions can be given as

$$G_{\mathrm{PI}}^{v}(s) = k_{pv} + \frac{k_{iv}}{s} = \frac{k_{pv}(1 + T_{iv}s)}{T_{iv}s} \tag{6.27}$$

$$G_v(s) = \frac{3}{2} \frac{V_m}{V_{\mathrm{dc}} C_{\mathrm{dc}} s} \tag{6.28}$$

where k_{pv} and k_{iv} are the proportional and the integral terms of the DC-link voltage controller, respectively, and T_{iv} is the integrator time constant. In the case of grid-connected applications, the DC-link voltage V_{dc} should be higher than the minimum required DC-link voltage, in order to allow the current controllability and avoid the over modulation, which can be determined as

- $V_{\mathrm{dc}} \geq V_m$ for single-phase systems
- $V_{\mathrm{dc}} \geq \sqrt{3}V_m$ for three-phase systems with space vector modulation (SVM) scheme
- $V_{\mathrm{dc}} \geq 2V_m$ for three-phase systems with a sinusoidal PWM scheme (SPWM)

On the other hand, the average DC-link voltage V_{dc} should not be much higher than $\sqrt{3}V_m$ when using the SVM scheme in order to avoid to high power losses.

From Fig. 6.5, the open loop transfer function of the DC-link voltage control loop can be expressed by

$$G_{ol}^{v}(s) = G_{\mathrm{PI}}^{v}(s) \cdot G_{cl}^{d}(s) \cdot G_v(s) \tag{6.29}$$

With the simplified current control loop in Eq. (6.22), the transfer function in Eq. (6.29) is given as

$$G_{ol}^{v}(s) \approx \frac{3V_m k_{pv}(1 + T_{iv}s)}{2T_{iv}V_{\mathrm{dc}}C_{\mathrm{dc}}s^2(1 + 3T_s s)} \tag{6.30}$$

Then, the phase crossover frequency ω_c can be expressed as

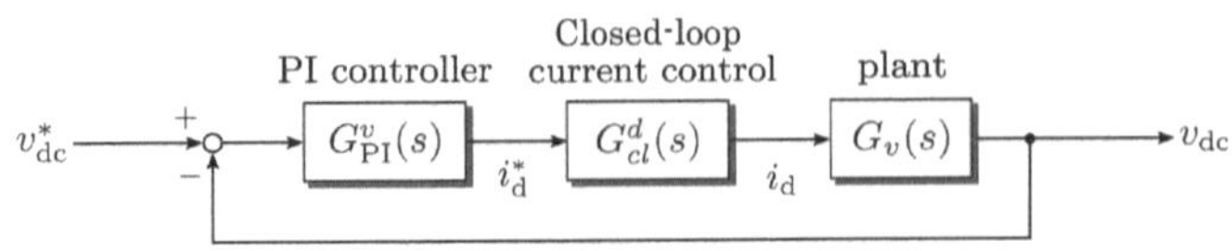

FIG. 6.5 Block diagram of the DC-link voltage control loop.

$$\omega_c = \frac{1}{\sqrt{3T_s T_{iv}}} \tag{6.31}$$

Hence, the control parameters k_{pv} and k_{iv} at the phase crossover frequency ($\omega_{\rm pc}$) are given by

$$k_{pv} = \frac{C_{\rm dc}}{2\sqrt{T_s T_{iv}}} \text{ and } k_{iv} = \frac{C_{\rm dc}}{2\sqrt{T_s T_{iv}^3}} \tag{6.32}$$

Similarly, the PI controller can be tuned using the "*pidTuner*" in MATLAB, where the plant transfer function and the exact closed-loop transfer function of the d-axis current (i.e., $G_{cl}^d(s) \cdot G_v(s)$) can be imported. The parameters can be obtained according to the desired bandwidth or the response time, the phase margin, or the transient behavior of the system.

6.3.3 Case Study

In order to show a numerical example for the previously discussed design procedure, the controllers of a 10-kW grid-connected boost converter-fed three-phase DC/AC converter are designed in this subsection. The schematic and the entire control are shown in Fig. 6.6, and the system parameters of the double-stage DC/AC system are shown in Table 6.1. Notably, the case study

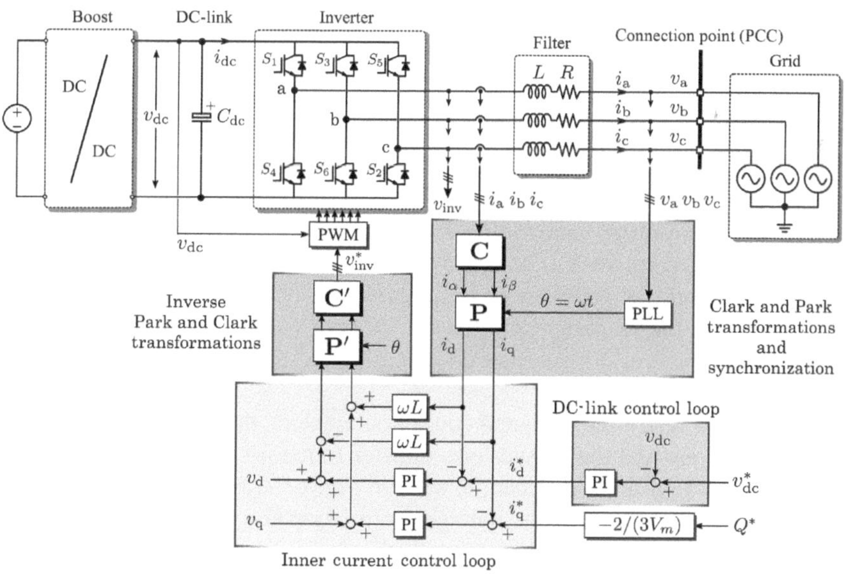

FIG. 6.6 Control structure of the DC/AC stage in a double-stage three-phase grid-connected system in the synchronous reference frame, where the outer control loop is a voltage control and the reactive power can be directly controlled and V_m is the grid phase voltage amplitude. The PLL is used for reference transformations.

TABLE 6.1 System Parameters of the 10-kW Grid-Connected Three-Phase DC/AC Converter

Parameter	Symbol	Value
DC-link voltage reference	V_{dc}^*	800 V
DC-link capacitor	C_{dc}	500 μF
Grid phase voltage amplitude	V_m	311 V
Filter inductance	L	5 mH
Filter resistance	R	0.1 Ω
Switching frequency	f_{sw}	20 kHz
Sampling frequency	$f_s = 1/T_s$	20 kHz

is to demonstrate the controller parameter design for the DC/AC power converter (i.e., the inverter), and thus it is assumed that the control of the boost converter is robust. That is, a constant power is fed to the DC-link.

In order to design the current controller analytically, Eq. (6.21) can be used with the parameters specified in Table 6.1. Accordingly, the PI control parameters for the d-axis current loop are obtained as

$$k_{pd} = 33.3 \quad \text{and} \quad k_{id} = 666.7 \tag{6.33}$$

where the bandwidth can be approximated using Eq. (6.23) as $f^d_{bw} = 1\,\text{kHz}$. Then, the DC-link voltage controller can be analytically designed on a basis of Eq. (6.32). In practice, the outer loop should be much slower than the inner current control loop in order to separate the cascaded control system [25]. With this consideration, the bandwidth of the outer DC-link voltage control loop should be limited between 1/50 and 1/10 of the current control loop. In this case study, a bandwidth of 100 Hz (i.e., 10 times slower than the inner current control loop) and then the PI parameters for the DC-link voltage are determined as

$$k_{pv} = 0.27 \quad \text{and} \quad k_{iv} = 16.11 \tag{6.34}$$

In order to verify the designed control parameters, the frequency responses of the open-loop and the closed-loop transfer functions of these control loops are plotted in Fig. 6.7. As it can be seen in Fig. 6.7A that a phase margin of 78.8° and 65.5° for the DC-link voltage control loop and the d-axis current control loop, respectively, is achieved with the designed parameters. The phase margins are sufficient to ensure the stability of the closed-loop control systems, as verified by Fig. 6.7B. However, as it is also shown in Fig. 6.7B, the bandwidth of the closed-loop d-axis current control (1.45 kHz) is higher than the approximated one (1 kHz) according to Eq. (6.23). Nevertheless, the outer

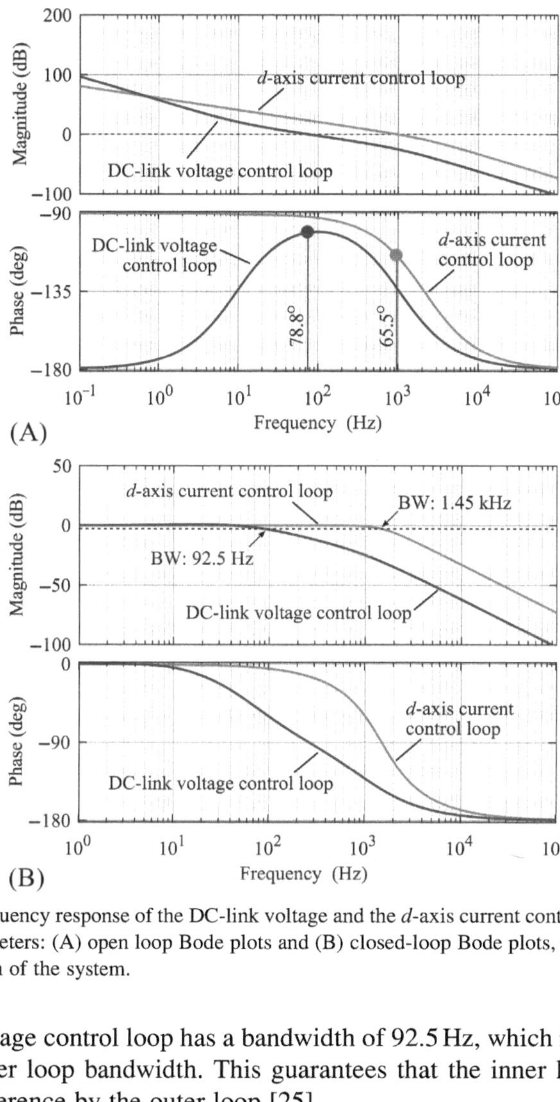

FIG. 6.7 Frequency response of the DC-link voltage and the *d*-axis current control loops with the designed parameters: (A) open loop Bode plots and (B) closed-loop Bode plots, where BW represents bandwidth of the system.

DC-link voltage control loop has a bandwidth of 92.5 Hz, which is much lower than the inner loop bandwidth. This guarantees that the inner loop can fully track the reference by the outer loop [25].

Furthermore, Fig. 6.8 shows the step response of these control loops. Clearly, as shown in Fig. 6.8 that the designed parameters for the DC/AC converter system ensure a much faster inner control loop. Therefore, the design is proper and the parameters can be used to control the double-stage DC/AC grid-connected converter system. This will be verified by the following simulation case in MATLAB.

Referring to Fig. 6.6, the designed controller parameters (i.e., Eqs. (6.33), (6.34)) are thus applied to a three-phase, grid-connected DC/AC power conversion system, and the model is built up in MATLAB. Simulation results are

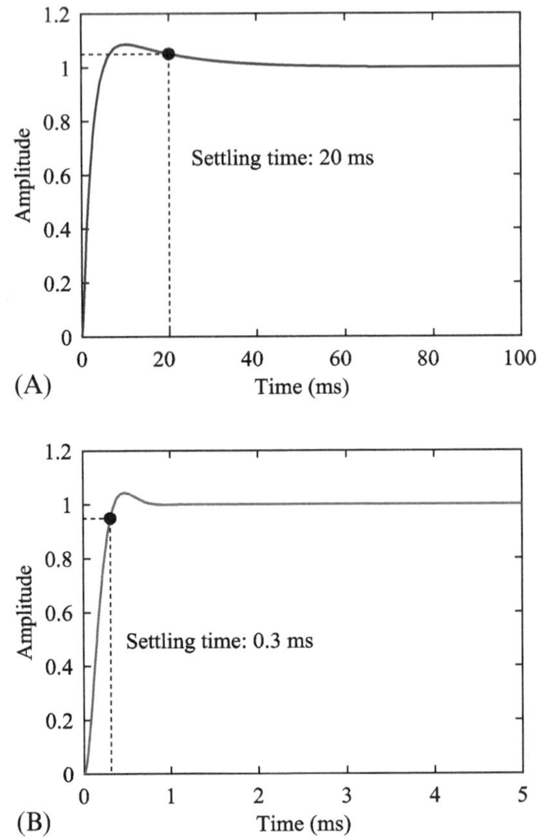

FIG. 6.8 Step responses of the closed-loop control systems for the DC/AC converter with the designed parameters: (A) the DC-link voltage control loop and (B) the *d*-axis current control loop.

reported in Fig. 6.9, in which the grid line-to-line voltages, grid line currents, DC-link voltage, and *dq*-current components are shown. In this case study, two transients have been employed, where the first transient is a step change in the active power from 0 to 8 kW, which is achieved through the boost converter control at $t = 0.2$ s; while the second one is a step change of the reactive power from 0 to 6 kVAR, which has been achieved through the *q*-current component at $t = 0.4$ s. As observed in Fig. 6.9, the designed controllers can follow the commands or references with high accuracy in the steady state. In the case of dynamic changes (i.e., at $t = 0.2$ s and $t = 0.4$ s), the control in the *dq*-reference frame enables the DC/AC system quickly to response to the disturbances. More specifically, when the input power is changed, the *d*-axis current will follow and react this change. This is also reflected at the DC-link, where the voltage presents an overshoot of about 7.5%, but it comes to steady state in one cycle. In contrast, the step change in the *q*-axis current is achieved by directly setting the

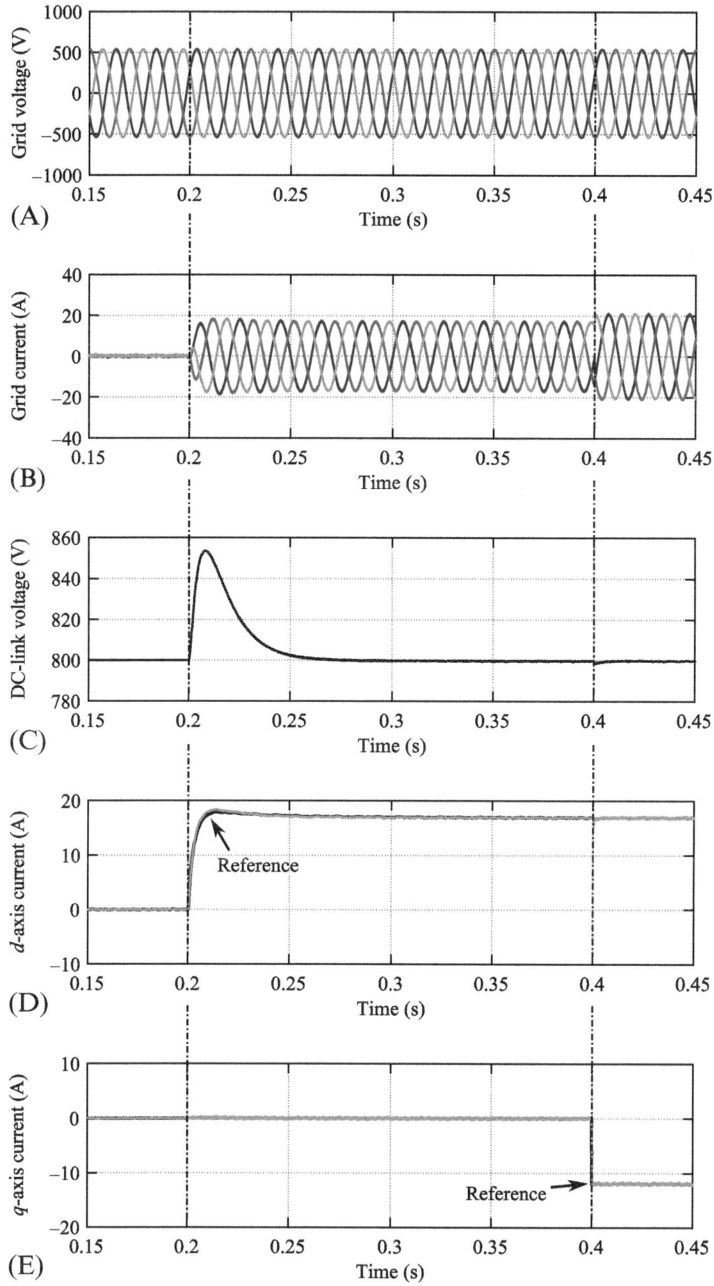

FIG. 6.9 Simulation results for the grid-connected three-phase AC/DC converter system controlled in the synchronous reference frame with the designed parameters: (A) grid line-to-line voltages, (B) grid currents, (C) DC-link voltage, (D) d-axis current component, and (E) q-axis current.

reactive power reference according to Fig. 6.6. Hence, there is a step-change in the q-axis current reference, which can be considered as a harsh operating condition in practice. Nevertheless, the designed PI controller for the q-axis current control loop can still quickly regulate the q-axis current with minor disturbances in the d-axis current, as shown in Fig. 6.9. In all, the designed controllers can properly control three-phase DC/AC converter with desired performances in terms of dynamics and control accuracy. Additionally, practical DC/AC converters may be connected to a weak grid with many low-order background harmonics and also to nonlinear loads (e.g., diode rectifiers). In that case, the currents injected to the load may be highly distorted with high harmonics. To improve the current quality, a harmonic compensation control should be implemented. This can also be done in either the dq-reference frame or the $\alpha\beta$-reference frame [15]. The audience of this book are advised to study this as an exercise to better understand the control of DC/AC converters.

The above has demonstrated the control of a three-phase DC/AC power converter, when connected to a grid. The design procedures presented in Sections 6.3.1 and 6.3.2 are validated by the case study. However, in practice, there are many applications employing single-phase full-bridge DC/AC power converters. As discussed previously, a fictitious system should be created in order to perform the Park transformations and then enable the application of PI controllers to regulate AC variables. If so, the presented controller design guidelines are still valid. Alternatively, the control can be achieved with proportional resonant (PR) controllers and repetitive controllers, which show high performance in tracking of AC variables [26,27]. In this case, the control loop can still be taken as a cascaded-control system. On the other hand, for grid-connected DC/AC converters, synchronization is mandatory, which is also a challenging issue, as discussed in Chapter 4. Typically, an orthogonal signal generator can be adopted for synchronization, which thus is beneficial to the control of single-phase DC/AC converters. That is, the design can be applied in single-phase grid-connected DC/AC converters without further efforts for creating the fictitious system that is in-quadrature with the original system.

In order to demonstrate this, an example of the current controller implemented with the PR controller in single-phase DC/AC converter will be shown in the following. The system and controller parameters are given in Table 6.2, while the controller parameters are designed following the discussion in Section 6.3, which gives

$$k_{pr} = 33.3 \quad \text{and} \quad k_{ir} = 13{,}334$$

The simulation results are shown in Fig. 6.10, where the reference output current experienced a step-change from 5 to 10 A at $t = 0.205$ s. The current responses of the single-phase DC/AC converter are transformed into the responses in the dq-reference frame, as shown in Fig. 6.10B. It can be seen from the results in Fig. 6.10 that the output current can follow the reference value,

TABLE 6.2 System Parameters of the 3.5-kW Grid-Connected Single-Phase DC/AC Converter

Parameter	Symbol	Value
DC-link voltage reference	V_{dc}^*	400 V
DC-link capacitor	C_{dc}	1000 μF
Grid phase voltage amplitude	V_m	230 V
Filter inductance	L	7.6 mH
Filter resistance	R	0.08 Ω
Switching frequency	f_{sw}	10 kHz
Sampling frequency	$f_s = 1/T_s$	10 kHz

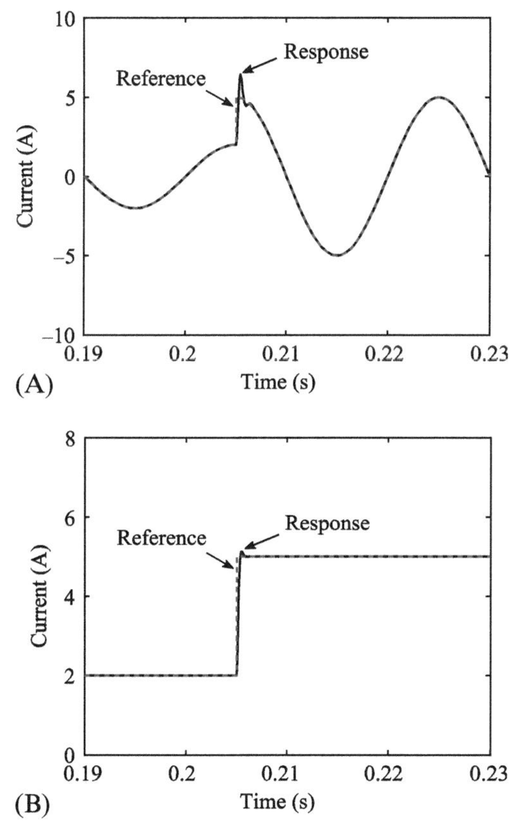

FIG. 6.10 Current step responses in single-phase DC/AC converters with the PR controller: (A) AC current (i.e., the α-axis current) and (B) the d-axis current.

ensuring the controllability of the current controller. However, as observed, the dynamics are slightly different in terms of overshoots and settling time. It is probably related to the coupling effects and the reference frame transformations in the control system. Nevertheless, this case study indicates that the control of single-phase DC/AC converters can be implemented in both the $\alpha\beta$-reference frame and the dq-reference frame.

6.4 SUMMARY

This chapter has presented the basic control of a three-phase two-level DC/AC power converters, where the detailed design of the PI controllers in the synchronous dq-reference frame was illustrated. First, the modeling of the three-phase DC/AC power converters was presented in terms of the current control loop in the dq-reference frame, the DC-link voltage, and the power control loop. The models enable the use of well-developed PI controllers to regulate the DC-voltage and the dq-axis currents for three-phase DC/AC converters, which have been demonstrated on a 10-kW three-phase grid-connected system. In respect to the control of single-phase DC/AC converters, similar modeling can be done, where a virtual system should be established in order to transform the AC control variables to DC quantities. A case study on a single-phase converter was demonstrated to show the controllability in the $\alpha\beta$-reference frame.

REFERENCES

[1] N. Mohan, T.M. Undeland, B. Robbins, Power Electronics, J. Wiley, New York, 1989.
[2] D.W. Novotny, T.A. Lipo, Vector Control and Dynamics of AC Machines, Clarendon Press, Oxford, 1996.
[3] B.K. Bose, Power electronics and motor drives recent progress and perspective, IEEE Trans. Ind. Electron. 56 (2) (2009) 581–588.
[4] F. Blaabjerg, Z. Chen, S.B. Kjaer, Power electronics as efficient interface in dispersed power generation systems, IEEE Trans. Power Electron. 19 (5) (2004) 1184–1194.
[5] R. Teodorescu, M. Liserre, P. Rodriguez, Grid Converters for Photovoltaic and Wind Power Systems, Wiley, New York, NY, USA, 2011.
[6] S.B. Kjaer, J.K. Pedersen, F. Blaabjerg, A review of single-phase grid-connected inverters for photovoltaic modules, IEEE Trans. Ind. Appl. 41 (5) (2005) 1292–1306.
[7] E. Romero-Cadaval, B. Francois, M. Malinowski, Q.C. Zhong, Grid-connected photovoltaic plants: an alternative energy source, replacing conventional sources, IEEE Ind. Electron. Mag. 9 (1) (2015) 18–32.
[8] IEEE Standard for Interconnecting Distributed Resources With Electric Power Systems, IEEE Std 1547-2003, 2003.
[9] European Network of Transmission System Operators for Electricity, Network code for requirements for grid connection applicable to all generators, Tech. Rep., 2013. Available: https://www.entsoe.eu.2013.

[10] Energinet.dk, Technical regulation 3.2.5 for wind power plants with a power output above 11 kW, Tech. Rep., 2010.
[11] Energinet.dk, Technical regulation 3.2.2 for PV power plants with a power output above 11 kW, Tech. Rep., 2015.
[12] Y. Yang, P. Enjeti, F. Blaabjerg, H. Wang, Wide-scale adoption of photovoltaic energy: grid code modifications are explored in the distribution grid, IEEE Ind. Appl. Mag. 21 (5) (2015) 21–31.
[13] Y.K. Wu, J.H. Lin, H.J. Lin, Standards and guidelines for grid-connected photovoltaic generation systems: a review and comparison, IEEE Trans. Ind. Appl. 53 (4) (2017) 3205–3216.
[14] R. Teodorescu, F. Blaabjerg, M. Liserre, P.C. Loh, Proportional-resonant controllers and filters for grid-connected voltage-source converters, IEE Proc. Electric Power Appl. 153 (5) (2006) 750–762.
[15] F. Blaabjerg, R. Teodorescu, M. Liserre, A.V. Timbus, Overview of control and grid synchronization for distributed power generation systems, IEEE Trans. Ind. Electron. 53 (5) (2006) 1398–1409.
[16] V. Blasko, V. Kaura, A new mathematical model and control of a three-phase AC-DC voltage source converter, IEEE Trans. Power Electron. 2 (1) (1997) 116–123.
[17] H. Akagi, Y. Kanazawa, A. Nabae, Instantaneous reactive power compensators comprising switching devices without energy storage components, IEEE Trans. Ind. Appl. IA-20 (3) (1984) 625–630.
[18] J. Kan, S. Xie, Y. Wu, Y. Tang, Z. Yao, R. Chen, Single-stage and boost-voltage grid-connected inverter for fuel-cell generation system, IEEE Trans. Ind. Electron. 62 (9) (2015) 5480–5490.
[19] A. Abdelhakim, P. Mattavelli, V. Boscaino, G. Lullo, Decoupled control scheme of grid-connected split-source inverters, IEEE Trans. Ind. Electron. 64 (8) (2017) 6202–6211.
[20] R. Teodorescu, F. Blaabjerg, M. Liserre, P.C. Loh, Proportional-resonant controllers and filters for grid-connected voltage-source converters, IEE Proc. Electric Power Appl. 153 (5) (2006) 750–762.
[21] P.C. Tan, P.C. Loh, D.G. Holmes, High performance harmonic extraction algorithm for a 25 kV traction power conditioner, IEE Proc. Electric Power Appl. 151 (2004) 505–512.
[22] D.G. Holmes, T.A. Lipo, B.P. McGrath, W.Y. Kong, Optimized design of stationary frame three phase AC current regulators, IEEE Trans. Power Electron. 24 (11) (2009) 2417–2426.
[23] C. Lascu, L. Asiminoaei, I. Boldea, F. Blaabjerg, High performance current controller for selective harmonic compensation in active power filters, IEEE Trans. Power Electron. 22 (5) (2007) 1826–1835.
[24] A.G. Yepes, F.D. Freijedo, Ó. Lopez, J. Doval-Gandoy, Analysis and design of resonant current controllers for voltage-source converters by means of Nyquist diagrams and sensitivity function, IEEE Trans. Ind. Electron. 58 (11) (2011) 5231–5250.
[25] D. Zhou, F. Blaabjerg, Bandwidth oriented proportional-integral controller design for back-to-back power converters in DFIG wind turbine system, IET Renew. Power Gener. 11 (7) (2017) 941–951.
[26] K. Zhou, D. Wang, Y. Yang, F. Blaabjerg, Periodic Control of Power Electronic Converters, IET Publisher, London, UK, 2016.
[27] Y. Yang, Advanced Control Strategies to Enable a More Wide-Scale Adoption of Single-Phase Photovoltaic Systems, Ph.D. Dissertation, Department of Energy Technology, Aalborg University, 2014. Available: http://vbn.aau.dk/ws/files/203989175/yongheng_yang.pdf.

Chapter 7

Modeling and Control of AC/AC Converter

Kyo-Beum Lee and Yeongsu Bak
Ajou University, Suwon, South Korea

7.1 FUNDAMENTALS OF A MATRIX CONVERTER

Recently, motor drives capable of variable speed operation (ASD: adjustable-speed drives) via an inverter are widely used in industry applications. A voltage source inverter as a power conversion device having a diode rectifier circuit on the input stage is the most widely used in motor drive systems; however, the inverter may be inefficient because of the considerable amount of harmonics in the input current. In addition, the inverter has the drawback that the energy cannot be regenerated. A voltage source inverter using a pulse width modulation (PWM) converter in place of the diode rectifier circuit on the input stage has energy regeneration and power factor control capabilities. Additionally, the input stage harmonics can be reduced through the use of a reactor. However, this type of inverter has disadvantages of higher switching loss and increased whole construction volume compared to the voltage source inverter using diode rectification. In the case of these two types of voltage source inverters, the lifetime of the overall system may also be limited owing to the use of a capacitor as an intermediate energy storage element.

A matrix converter has been developed as an alternative to the conventional indirect AC/AC power converters. It is a type of power conversion circuit that can perform direct AC/AC power conversion without the use of a DC link. Therefore, in comparison with the voltage source inverter, the matrix converter has a longer lifetime because it does not have the intermediate DC-link circuit. In addition, direct control of the input power factor and energy regeneration are possible in the case of matrix converters. Hence, the use of a matrix converter that can be an effective method of harmonic suppression of the current and also provide regeneration of the power during deceleration operation of a motor is gaining more importance.

In this chapter, an overview and a control method of the matrix converter are provided. Additionally, a system is described used for driving an induction

Control of Power Electronic Converters and Systems. https://doi.org/10.1016/B978-0-12-805245-7.00007-X

motor by using a matrix converter in detail and simulation results are also provided.

7.1.1 History of Matrix Converter

Traditionally, AC/AC converters using semiconductor devices are classified into two different groups. The first group is two-stage AC/DC and DC/AC conversion, and the second group is one-stage AC/AC cycloconverter. Cycloconverters usually operate thyristors and are used in phase-controlled applications because of their ease of phase commutation. In addition, they are used in high-power applications such as an induction motor and permanent magnet synchronous motor driving system. AC/AC matrix converter technology belongs to the class of cycloconverters. It was first announced in 1976 [1], though an US patent had been filed already in 1969 [2]. Substantial research of the matrix converter has been initiated chiefly owing to the topology and control algorithm proposed by Venturini and Alesina in the late 1970s [3–5]. They presented the power circuit configured by bidirectional power switches in the matrix form, and hence it was named the "'matrix converter." The output voltage of the matrix converter can be expressed by the product of the input voltage and the modulation function, also called the direct transfer function. The voltage transfer ratio is 87%, and it is theoretically the maximum value. In 1983, Rodriguez introduced a control scheme based on a new concept by using a fictitious DC link [6]. The then-existing PWM method was applied directly, and the switching was performed using the maximum input phase voltage and the minimum input phase voltage. The modulation function used was called the indirect transfer function. Since the 1980s, the space vector modulation (SVM) method has been applied to the control method of the matrix converter, and the matrix converter has an adjustable phase angle and sinusoidal input current. In 1985, the matrix converter was firmly implemented by Kastner and Rodriguez [7]. In 1992, vector control of the induction motor having input and output current of high quality was effectively performed by Neft and Schauder using nine switches [8]. However, the rectification method using the bidirectional switches resulted in the destruction of the power semiconductor devices due to overvoltage and overcurrent sparks. These facts limited the actual implementation, adversely affecting the implementation of the matrix converter. However, these problems have been resolved by some of the new rectification methods, which allow a safe operation of the switches. In 1989, Burany introduced the semisoft current rectification method [9].

Thereafter, the research field of the matrix converter has mostly dealt with various modulation methods and control algorithms, and several papers about the hardware design of the matrix converter were also published. More recently, papers about research being conducted to overcome the drawbacks of the matrix converter have been published, and the applications fields are also varied.

7.1.2 Research Trends of Matrix Converter

A three-phase matrix converter is configured by using only nine bidirectional switches in the power circuit. Additionally, it is composed of the inductor-capacitor filter of the input stage to satisfy the harmonic standards such as IEEE 519, and a clamp circuit for the protection of the switches from the overvoltage. The matrix converter has advantages of driver reliability, long lifetime, small size, and low weight because the DC link does not use electrolytic capacitors, which are large in size and they have also reliability issues. Additionally, the matrix converter offers advantages such as efficiency improvement and reduced maintenance. The matrix converter can be used in a variety of applications; particularly, the outstanding regeneration operation capability of the matrix converter leads to energy-saving operation of equipment such as elevator, crane, and press equipment. In addition, it can be applied in fields, where electronic devices are affected by noise like in hospitals and computer centers, and where harmonic regulation of the power supply devices has been strictly applied. Moreover, it is expected that the application of the matrix converter for smart transformers and wind power generation systems will be expanded. Research areas of the matrix converter in fields where the development is either advanced or completed are introduced here.

- Integrated driver: The miniaturization depending on the high power density properties of the matrix converter enables the integrated driver that packages the power converter in an AC motor case. Integrated driver is required in the aerospace and the industrial variable speed drive applications.
- Variable speed drive of permanent magnet synchronous motor and induction motor: This area has been researched most actively. Recently, vector control of variable speed drives has been developed and researched by Yaskawa, Rockwell, and Otis. Yaskawa has been focused on the fields such as crane and elevator as matrix converter products. Otis has developed the matrix converter capable of driving the permanent magnet synchronous motor, induction motor, and DC motor that are used in elevators and escalators.
- Military fighting vehicles: For the US Army, the matrix converter has been developed as a 10 kW power supply device of the engine-generator system for next-generation combat vehicles. It is employed owing to its fuel efficiency, variable electric characteristic ability required in various fighting vehicles, reliability, and long lifetime guarantee under severe operating conditions. In addition, various bidirectional switching elements such as MCT, MTO, and SiC have been developed and verified.
- Wind power generator: In general, a constant speed type driving technique using an induction type electric motor/generator is employed. However, Vestas in Denmark held a patent for a variable speed driving technique using the matrix converter.

7.2 TOPOLOGIES AND MODULATION TECHNIQUES

The power conversion system for AC/AC power conversion can be classified into two main types. One is a two-step indirect power conversion system of fundamental AC-DC-AC such as that shown in Fig. 7.1, and the other one is a one-step direct power conversion system of AC-AC such as that shown in Fig. 7.2.

The fundamental indirect power conversion system uses a rectifier to convert the input AC voltage into a DC voltage that contains ripple components. The ripple components are removed by DC-link energy storage elements having a large capacity such as electrolytic capacitors. The DC-link voltage without ripple components is used for an inverter and the inverter generates an output AC voltage having the desired magnitudes and frequencies [10].

On the other hand, the direct power conversion system is only composed of power semiconductor switches connected to the input stage and the output stage, without the use of energy storage elements. In this case, the input AC voltage is directly converted to the output AC voltage of desired magnitudes and frequencies through suitable switch control of the direct power conversion system that is called the matrix converter [11]. The matrix converter is classified into two types depending on their structure, namely, direct matrix converter and indirect matrix converter.

7.2.1 Circuit Configuration of Matrix Converter

The matrix converter is composed of power semiconductor switches that can establish a direct connection between the input and output stages. In addition, it has a function that generates an AC output voltage of unselected frequency and magnitude from the AC voltage source of the input stage that has unselected frequency and amplitude.

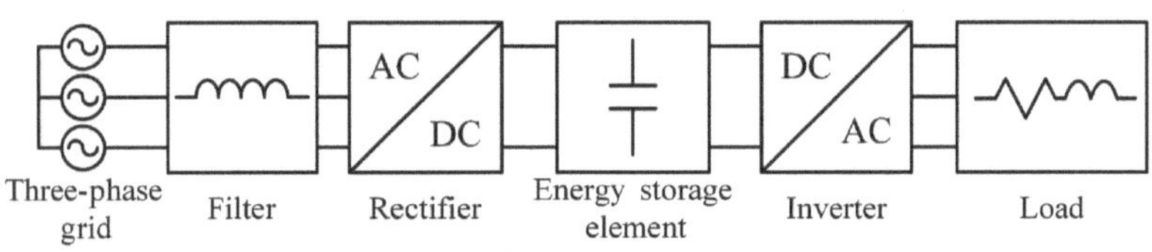

FIG. 7.1 AC-DC-AC indirect power conversion system.

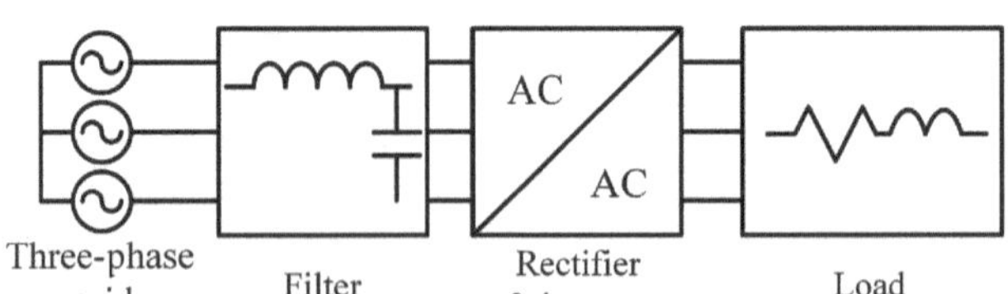

FIG. 7.2 AC-AC direct power conversion system.

7.2.1.1 Direct Matrix Converter

Fig. 7.3 shows the configuration of a three-phase direct matrix converter that consists of nine bidirectional switches. The direct matrix converter is not composed of the internal passive elements such as an inductor and a capacitor. Therefore, it has advantages such as less necessity of maintenance and longer lifetime than AC-AC power conversion systems adopting the indirect power conversion method with DC-link elements [12–14]. In Fig. 7.3, points a, b, and c indicate the phases of the input stage, whereas points A, B, and C indicate the phases of the output stage. In addition, the AC input voltages are represented by v_{sa}, v_{sb}, and v_{sc}, whereas the AC output voltages are represented by v_{oA}, v_{oB}, and v_{oC}. The output voltage of the direct matrix converter is determined by the input source voltage and switching states of the direct matrix converter [15,16].

The bidirectional switch used for the direct matrix converter can allow the flow of current in both directions during its turn-on period, and the switch has a voltage-blocking characteristic during its turn-off period. Commercialized power semiconductor switches are mostly elements that have unidirectional voltage and current features. In this case, the bidirectional switch composed of a diode and a unidirectional switch is used. Fig. 7.4 shows the methods of composition for a bidirectional switch.

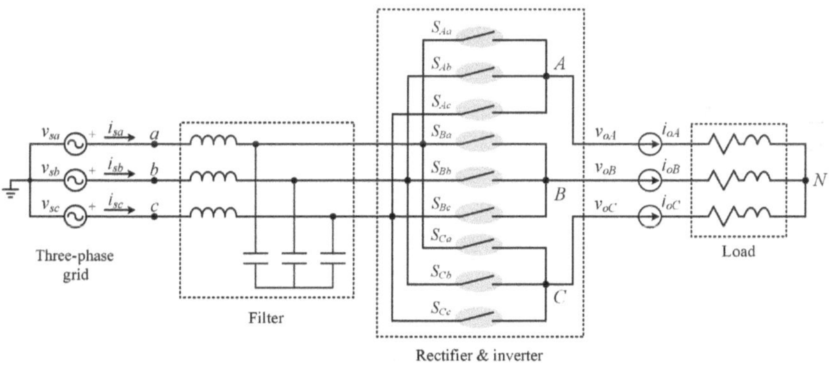

FIG. 7.3 Configuration of three-phase direct matrix converter.

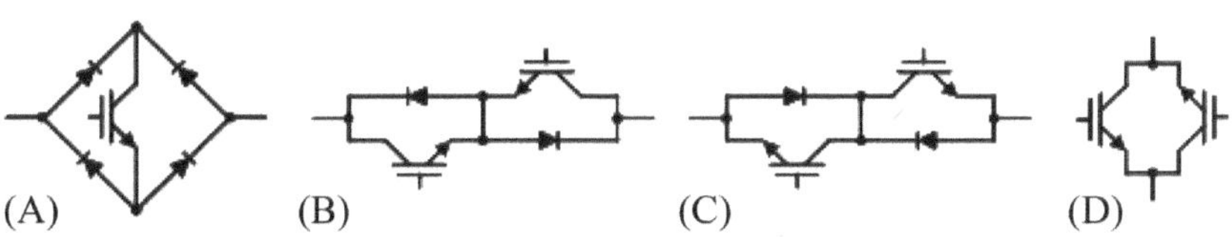

FIG. 7.4 Configuration of bidirectional switch used in the matrix converter topology in Fig. 7.3. (A) Diode embedded IGBTs, (B) common emitter IGBTs, (C) common collector IGBTs, and (D) antiparalleled reverse blocking IGBTs.

7.2.1.2 Indirect Matrix Converter

In the case of the direct matrix converter discussed in Section 2.1.1, extending the output stage is not possible because the matrix converter does not have a DC link. To improve this disadvantage of the direct matrix converter, an indirect matrix converter can be applied, and it is shown in Fig. 7.5.

The structure of the indirect matrix converter can be obtained by removing the capacitor in the DC link, the energy storage element of the DC link from the structure of the conventional AC-DC-AC indirect power conversion device. Moreover, it is possible to obtain bidirectional power flow because the rectifier stage is organized by the bidirectional switches instead of the diode bridge. This structure and performance are similar to that of the direct matrix converter. Besides, the indirect matrix converter has the advantage that the conventional widely known modulation method can be applied because the rectifier stage and inverter stage are separated physically. In addition, a number of output stages can be connected to an input stage by extending the output stage because there is the fictitious DC link.

The indirect matrix converter consists of 18 unidirectional switches similar to the direct matrix converter. However, if the zero-current switching in the rectifier stage is assured, the number of power semiconductor switches can be decreased as shown in Fig. 7.6.

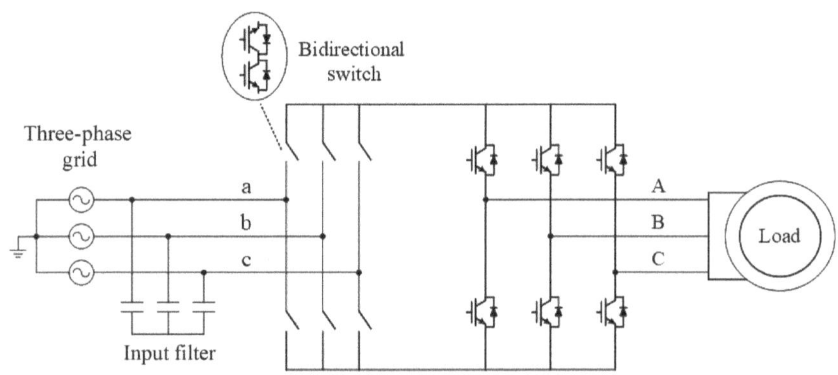

FIG. 7.5 Configuration of three-phase indirect matrix converter.

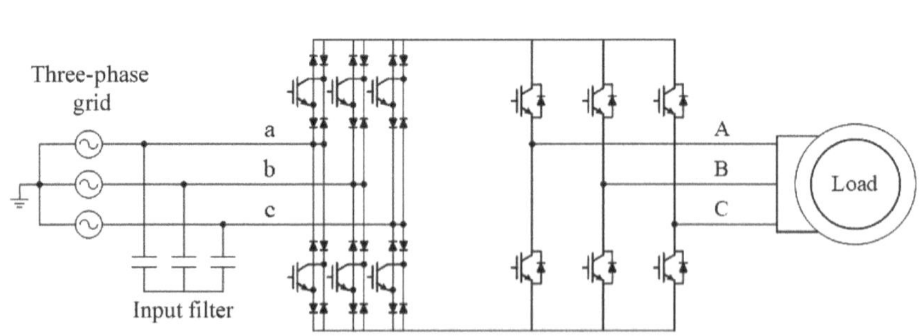

FIG. 7.6 Indirect matrix converter that reduces the number of switches at the rectifier stage.

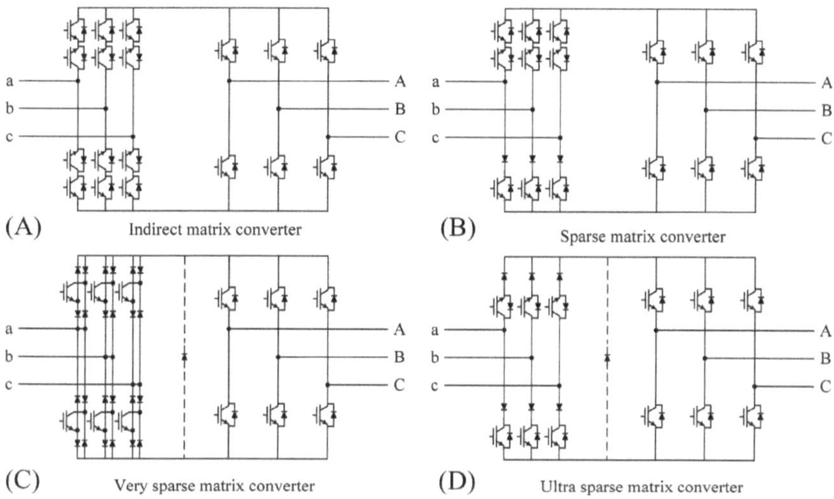

FIG. 7.7 Different topologies of indirect matrix converter.

Fig. 7.7 shows the indirect matrix converter topologies depending on the number of the switches and diodes. In these topologies, the performance is equal to the performance of the direct matrix converter although the rectifier stage and inverter stage are physically separated. However, in the case of Fig. 7.7D ultra-sparse matrix converter, unidirectional power flow can be possible. In other words, other matrix converter topologies except the ultrasparse matrix converter can be controlled to a bidirectional power flow. Therefore, the matrix converter can be controlled in variable load conditions. Another advantage is that there is no complicated modulation problem related to the rectifier stage because the zero-current switching method is performed to the rectifier stage [17,18].

7.2.2 Interaction Equation for Input-Output Voltage and Current of Matrix Converter

The three-phase matrix converter is composed of nine bidirectional switches. Its structure enables to connect the unselected phase of the input stage to the unselected phase of the output stage. Therefore, it is possible that the matrix converter has 512 ($=2^9$) switching states; however, it has only 27 switching states actually because of the limitation of the circuit structure. To describe in detail, the A-phase branch can be connected to the three-phase voltage of the input stage as shown in the dotted line box of Fig. 7.8.

- Kirchhoff's voltage law (KVL) limitation condition: Only one of the three switches connected to A-phase has to be turned on because the input stage of the matrix converter is a voltage source. If more than two of the switches are

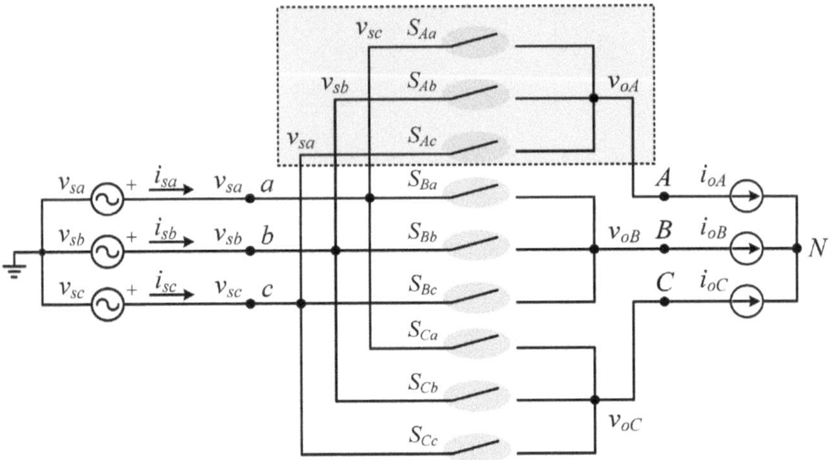

FIG. 7.8 Configuration of A-phase branch having A-phase output voltage.

turned on, a high short-circuit current will flow to the switches because of the short circuit formed. Similarly, only one of the switches has to be turned on of the three switches connected to the branches of the B-phase and C-phase.

- Kirchhoff's current law (KCL) limitation condition: The load connected to the output of the matrix converter seems to be a current source. Therefore, at least one of the three switches connected to A-phase has to be turned on. If all the switches of the branch of A-phase are turned off, an abnormal surge voltage may occur on the switches S_{Aa}, S_{Ab}, and S_{Ac} because the path of load current i_{oA} is not assured. Similarly, at least one switch has to be turned on of the three switches connected to the branches of B-phase and C-phase.

Considering simultaneously the KVL and KCL limitation conditions, in the operation condition of the three switches connected to the branch of the A-phase, only one of the switches among switches S_{Aa}, S_{Ab}, and S_{Ac} has to be turned on in the unselected instant. Therefore, the operation condition of the switches connected to the A-phase branch is expressed in Eq. (7.1).

$$s_{Aa} + s_{Ab} + s_{Ac} = 1, \tag{7.1}$$

where s_{Aa}, s_{Ab}, and s_{Ac} are represented for the presence function of S_{Aa}, S_{Ab}, and S_{Ac}, respectively.

Fig. 7.9 shows an example of the presence function when only one of the switches is turned on during the switching period T_s among the switches of the branch in the A-phase.

Similarly, the operation condition of the three switches connected to each branch of B-phase and C-phase is expressed in Eqs. (7.2), (7.3), respectively.

$$s_{Ba} + s_{Bb} + s_{Bc} = 1, \tag{7.2}$$

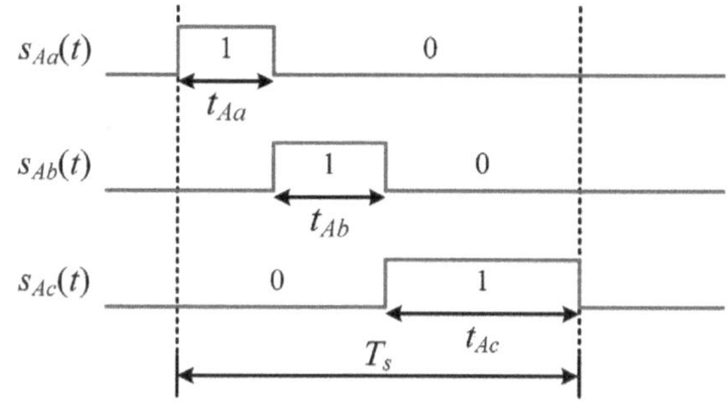

FIG. 7.9 Operation of the A-phase switches in Fig. 7.8.

$$s_{Ca} + s_{Cb} + s_{Cc} = 1, \tag{7.3}$$

where s_{jk} is the presence function of switch S_{jk} and is expressed in Eq. (7.4).

$$s_{jk} = \begin{cases} 1, & S_{jk} \text{ is on} \\ 0, & S_{jk} \text{ is off} \end{cases}, \quad j \in \{A, B, C\}, \quad k \in \{a, b, c\}. \tag{7.4}$$

In other words, depending on the condition of Eqs. (7.1)–(7.3), the three-phase matrix converter only has 27 (=3 × 3 × 3) switching states. The output voltages are defined by the instantaneously presence functions in Eq. (7.5).

$$\begin{aligned} v_{oA} &= s_{Aa}v_{sa} + s_{Ab}v_{sb} + s_{Ac}v_{sc}, \\ v_{oB} &= s_{Ba}v_{sa} + s_{Bb}v_{sb} + s_{Bc}v_{sc}, \\ v_{oC} &= s_{Ca}v_{sa} + s_{Cb}v_{sb} + s_{Cc}v_{sc}. \end{aligned} \tag{7.5}$$

If the switching period T_s is sufficiently small, the input voltages v_{sa}, v_{sb}, and v_{sc} are fixed during T_s and output voltage v_{oA} is also equal to the average value of v_{oA} during T_s as given in Eq. (7.6).

$$v_{oA} \cong < v_{oA} > = d_{Aa}v_{sa} + d_{Ab}v_{sb} + d_{Ac}v_{sc}, \tag{7.6}$$

where d_{Aj} is t_{Aj}/T_s ($j = a, b, c$) and is the duty ratio during T_s. Similarly, the output voltages of B-phase and C-phase are expressed in Eq. (7.7).

$$\begin{aligned} v_{oB} &= d_{Ba}v_{sa} + d_{Bb}v_{sb} + d_{Bc}v_{sc}, \\ v_{oC} &= d_{Ca}v_{sa} + d_{Cb}v_{sb} + d_{Cc}v_{sc}. \end{aligned} \tag{7.7}$$

Meanwhile, the input currents of the matrix converter can be expressed using the presence function in Eq. (7.8).

$$\begin{aligned} i_{sa} &= s_{Aa}i_{oA} + s_{Ba}i_{oB} + s_{Ca}i_{oC}, \\ i_{sb} &= s_{Ab}i_{oA} + s_{Bb}i_{oB} + s_{Cb}i_{oC}, \\ i_{sc} &= s_{Ac}i_{oA} + s_{Bc}i_{oB} + s_{Cc}i_{oC}. \end{aligned} \tag{7.8}$$

If the switching period T_s is sufficiently small, the input currents of the matrix converter are expressed in Eq. (7.9)

$$\begin{aligned} i_{sa} &= d_{Aa} i_{oA} + d_{Ba} i_{oB} + d_{Ca} i_{oC}, \\ i_{sb} &= d_{Ab} i_{oA} + d_{Bb} i_{oB} + d_{Cb} i_{oC}, \\ i_{sc} &= d_{Ac} i_{oA} + d_{Bc} i_{oB} + d_{Cc} i_{oC}. \end{aligned} \tag{7.9}$$

As a result, Eqs. (7.6), (7.7) indicate the voltage transfer interaction equations of the matrix converter and Eq. (7.9) gives the current transfer interaction equation. These can be defined using the matrix form as Eq. (7.10).

$$\begin{aligned} v_o &= T \cdot v_i, \\ i_i &= T^T \cdot i_o, \end{aligned} \tag{7.10}$$

where each variable in Eq. (7.10) is expressed in Eq. (7.11).

$$\begin{aligned} &T = \begin{bmatrix} d_{Aa} & d_{Ab} & d_{Ac} \\ d_{Ba} & d_{Bb} & d_{Bc} \\ d_{Ca} & d_{Cb} & d_{Cc} \end{bmatrix}, \\ &v_o = \begin{bmatrix} v_{oA} \\ v_{oB} \\ v_{oC} \end{bmatrix}, \quad v_i = \begin{bmatrix} v_{sa} \\ v_{sb} \\ v_{sc} \end{bmatrix}, \quad i_i = \begin{bmatrix} i_{sa} \\ i_{sb} \\ i_{sc} \end{bmatrix}, \quad i_o = \begin{bmatrix} i_{oA} \\ i_{oB} \\ i_{oC} \end{bmatrix}, \\ &0 \le d_{jk} \le 1, \quad j \in \{A, B, C\}, \quad k \in \{a, b, c\}, \\ &d_{ja} + d_{jb} + d_{jc} = 1, \quad j \in \{A, B, C\}. \end{aligned} \tag{7.11}$$

Each matrix element (d_{jk}) of the duty cycle matrix T of the matrix converter is represented for the duty ratio of the applied switches. For example, d_{Aa}, d_{Ab}, and d_{Ac} are represented for the turn-on time ratio of replied switches S_{Aa}, S_{Ab}, and S_{Ac} during one switching period. Additionally, each switch is turned on for the calculated time through the three portions of the time in one period depending on the ratio of d_{Aa}, d_{Ab}, and d_{Ac}.

7.2.3 Modulation Technique of Direct Matrix Converter

In this section, the duty cycle matrix for controlling each switch of the three-phase direct matrix converter and the modulation method of the three-phase direct matrix converter using the duty cycle matrix will be described. The input phase voltage and the output phase current of the direct matrix converter are given as the independent variables in Eq. (7.12).

$$\begin{aligned} v_i &= \begin{bmatrix} v_{sa} \\ v_{sb} \\ v_{sc} \end{bmatrix} = V_{im} \begin{bmatrix} \cos(\omega_i t) \\ \cos(\omega_i t - 2\pi/3) \\ \cos(\omega_i t + 2\pi/3) \end{bmatrix}, \\ i_o &= \begin{bmatrix} i_{oA} \\ i_{oB} \\ i_{oC} \end{bmatrix} = I_{om} \begin{bmatrix} \cos(\omega_o t - \phi_o) \\ \cos(\omega_o t - \phi_o - 2\pi/3) \\ \cos(\omega_o t - \phi_o + 2\pi/3) \end{bmatrix}. \end{aligned} \tag{7.12}$$

In this case, suppose the operation generates output phase voltage and input phase current in Eq. (7.13) by control.

$$v_o = \begin{bmatrix} v_{oA} \\ v_{oB} \\ v_{oC} \end{bmatrix} = V_{om} \begin{bmatrix} \cos(\omega_o t) \\ \cos(\omega_o t - 2\pi/3) \\ \cos(\omega_o t + 2\pi/3) \end{bmatrix},$$
$$i_i = \begin{bmatrix} i_{sa} \\ i_{sb} \\ i_{sc} \end{bmatrix} = I_{im} \begin{bmatrix} \cos(\omega_i t - \phi_i) \\ \cos(\omega_i t - \phi_i - 2\pi/3) \\ \cos(\omega_i t - \phi_i + 2\pi/3) \end{bmatrix}, \tag{7.13}$$

where $\cos(\phi_o)$ and $\cos(\phi_i)$ are the power factors of the load and the input stage, respectively, and ω_i and ω_o are the input and output angular frequencies, respectively. The reference potential of the output phase voltage v_{oA}, v_{oB}, and v_{oC} is a neutral point of the three-phase source voltage of the input stage as shown in Fig. 7.3.

The input power of the direct matrix converter must be equal to the output power. Therefore, Eq. (7.14) is defined from $v_i^T i_i = v_o^T i_o$.

$$V_{im} I_{im} \cos(\phi_i) = V_{om} I_{om} \cos(\phi_o). \tag{7.14}$$

When the voltage gain of the direct matrix converter is defined as $q = V_{om}/V_{im}$, Eq. (7.15) is defined as

$$V_{om} = qV_{im},$$
$$I_{im} = qI_{om}\frac{\cos(\phi_o)}{\cos(\phi_i)}. \tag{7.15}$$

When Eqs. (7.12), (7.13) are substituted into Eq. (7.10), the duty matrix T that satisfies the limited condition of the duty ratio as in Eq. (7.11) is calculated by using Eq. (7.16).

$$T = \begin{bmatrix} d_{Aa} & d_{Ab} & d_{Ac} \\ d_{Ba} & d_{Bb} & d_{Bc} \\ d_{Ca} & d_{Cb} & d_{Cc} \end{bmatrix} = \frac{p_1}{3} \begin{bmatrix} d_1 & d_2 & d_3 \\ d_3 & d_1 & d_2 \\ d_2 & d_3 & d_1 \end{bmatrix} + \frac{p_2}{3} \begin{bmatrix} d'_1 & d'_2 & d'_3 \\ d'_2 & d'_3 & d'_1 \\ d'_3 & d'_1 & d'_2 \end{bmatrix}, \tag{7.16}$$

where d_1, d_2, d_3, d'_1, d'_2, and d'_3 are expressed in Eq. (7.17).

$$\begin{cases} d_1 = 1 + 2q\cos(\omega_1 t), \\ d_2 = 1 + 2q\cos\left(\omega_1 t + \dfrac{2\pi}{3}\right), \\ d_3 = 1 + 2q\cos\left(\omega_1 t - \dfrac{2\pi}{3}\right), \end{cases} \quad \begin{cases} d'_1 = 1 + 2q\cos(\omega_2 t), \\ d'_2 = 1 + 2q\cos\left(\omega_2 t - \dfrac{2\pi}{3}\right), \\ d'_3 = 1 + 2q\cos\left(\omega_2 t + \dfrac{2\pi}{3}\right), \end{cases} \tag{7.17}$$

where ω_1 and ω_2 are $\omega_o - \omega_i$ and $\omega_o + \omega_i$, respectively, and p_1 and p_2 are positive and negative direction power factor control variables, respectively, which are expressed in Eq. (7.18).

$$p_1 = \frac{1}{2}(1+p), \quad p_2 = \frac{1}{2}(1-p),$$
$$p = \frac{\tan(\phi_i)}{\tan(\phi_o)}. \tag{7.18}$$

From Eq. (7.18), $p_1+p_2=1$ and $p_1-p_2=p$. Additionally, p is the input-to-output phase transfer ratio of the direct matrix converter. Among the variables that determine p, ϕ_o is determined by the characteristics of the load and ϕ_i is determined by the desired command value.

If the input stage of the matrix converter is operated at unity power factor ($\phi_i=0$), Eq. (7.16) can be rewritten simply as given by Eq. (7.19).

$$d_{jk}=\frac{1}{3}\left[1+\frac{2v_{oj}v_{sk}}{V_{im}^2}\right] \quad (j=A,B,C, \quad k=a,b,c). \tag{7.19}$$

Fig. 7.10 shows the magnitude range of the three-phase input source voltage and the output phase voltage of the direct matrix converter. The three-phase output phase voltage cannot exceed the range of the input phase voltage because the output phase voltage of the direct matrix converter is synthesized from the input voltage. Therefore, the maximum magnitude of the output phase voltage is restricted to 50% of that of the input phase voltage. In other words, the maximum value of control parameter q is 0.5 in the duty matrix of Eq. (7.16).

Fig. 7.11 shows method to obtain a larger output phase voltage than the output phase voltage of Fig. 7.10 by adding the common-mode voltage to the output phase voltage of Eq. (7.13). As mentioned earlier, the common-mode voltage applied to the output phase voltage does not make effect to the output stage line-to-line voltage of the direct matrix converter because the reference

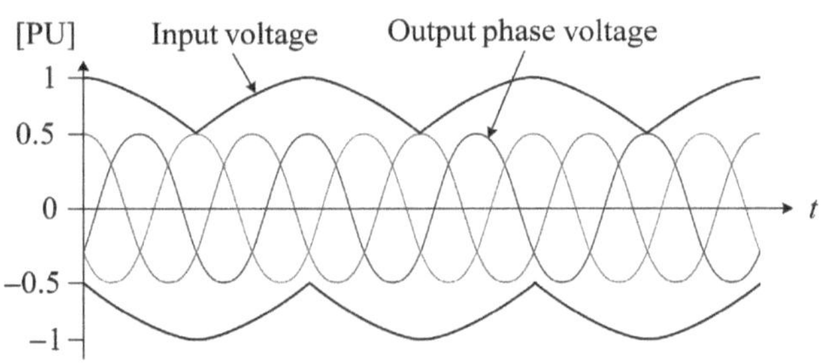

FIG. 7.10 Input voltage and output phase voltage ($q_{max}=0.5$).

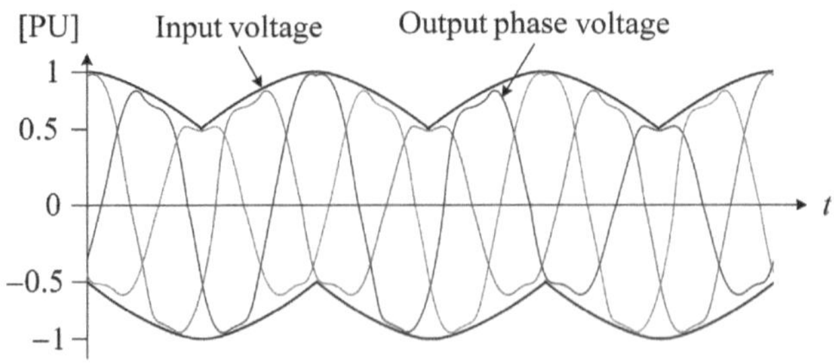

FIG. 7.11 Input voltage and output phase voltage ($q_{max}=0.866$) using a common-mode voltage in the modulation.

potentials of the output phase voltage v_{oA}, v_{oB}, and v_{oC} are neutral points of the input stage three-phase source voltage.

Therefore, the output phase voltages are expressed in Eq. (7.20) as

$$v_o = \begin{bmatrix} v_{oA} \\ v_{oB} \\ v_{oC} \end{bmatrix} = V_{om} \begin{bmatrix} \cos(\omega_o t) + v_{cm}(t) \\ \cos(\omega_o t - 2\pi/3) + v_{cm}(t) \\ \cos(\omega_o t + 2\pi/3) + v_{cm}(t) \end{bmatrix}, \tag{7.20}$$

where v_{cm} is the common-mode voltage and it is expressed in Eq. (7.21) as

$$v_{cm}(t) = -\frac{1}{6}\cos(3\omega_o t) + \frac{\sqrt{3}}{6}\cos(3\omega_i t). \tag{7.21}$$

As a result, the maximum value of q is increased to $\sqrt{3}/2$ (=0.866). Additionally, $q_{max}=0.866$ is a unique characteristic of the direct matrix converter, which is determined regardless of the control modulation method of the direct matrix converter.

If the output phase voltage of Eq. (7.20) instead of Eq. (7.13) is applied, the final solution is generally expressed to complex equation obtained from Venturini's optimum method. In addition, this method is necessary to many calculations for the actual application. However, if the input stage of the direct matrix converter is operated at unit power factor ($\phi_i=0$), the final solution can be realized easily as shown in Eq. (7.22).

$$d_{jk} = \frac{1}{3}\left[1 + \frac{2v_{oj}v_{sk}}{V_{im}^2} + \frac{4q}{3\sqrt{3}}\sin(\omega_i t + \beta_k)\sin(3\omega_i t)\right], \tag{7.22}$$
$$j = A,B,C,\ \ k = a,b,c,\ \ \beta_a = 0,\ \ \beta_b = -2\pi/3,\ \ \beta_c = 2\pi/3.$$

Depending on Venturini's optimum method analysis, the relationship between input-output phase transfer ratio p of the direct matrix converter and voltage gain q is selected from Eq. (7.23).

$$2q\left[|p|\cdot\left(1 - \frac{\mathrm{sgn}(\lambda)}{\sqrt{3}}\right) + \frac{\mathrm{sgn}(\lambda)}{\sqrt{3}}\right] \le 1, \tag{7.23}$$

where λ and sgn(λ) are expressed as the following in Eq. (7.24).

$$\lambda = \frac{2q}{\sqrt{3}}(1 - |p|),\ \ \mathrm{sgn}(\lambda) = \begin{cases} 1, & \lambda \ge 0 \\ -1, & \lambda < 0 \end{cases}. \tag{7.24}$$

Fig. 7.12 shows the change of the maximum voltage gain q_{max} depending on p value. If p is controlled for power factor control of the input stage of the direct matrix converter, caution is necessary because the maximum voltage gain q_{max} changes as shown in Fig. 7.12.

If q_{max} is required to be >0.5, the range of p must be restricted in the range of $-1<p<1$. Additionally, in the range of $-1<p<1$, the range of the power factor angle control of the input stage is restricted as $-|\phi_o|<\phi_i<|\phi_o|$ from Eq. (7.18).

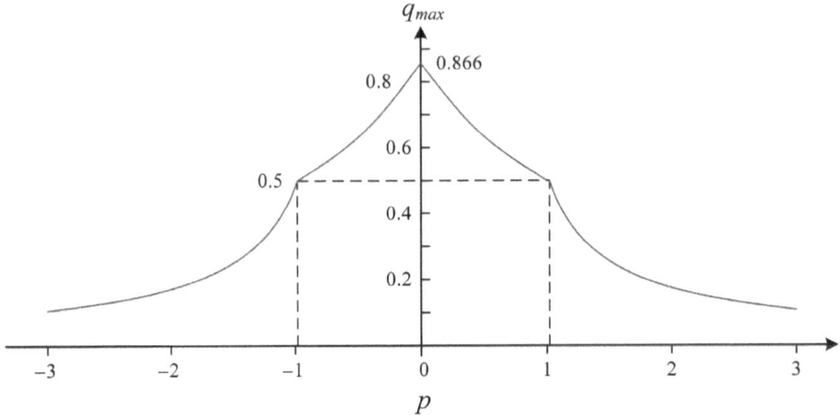

FIG. 7.12 Maximum voltage gain q_{max} depending on p value.

Fig. 7.13 shows an example of the method that generates the gating signals, which are the presence function of switch (S_{jk}), using each matrix element (d_{jk}) of the duty matrix T of the matrix converter. The gating signals of switches S_{Aa}, S_{Ab}, and S_{Ac} connected to the A-phase output stage are determined by comparing the carrier signal v_{tri} of the triangular shape with d_{Aa} and ($d_{Aa}+d_{Ab}$) instantaneously. In addition, they are expressed as follows in Eq. (7.25):

$$[s_{Aa}\ s_{Ab}\ s_{Ac}] = \begin{cases} [100], & 0 \le v_{tri} < d_{Aa}, \\ [010], & d_{Aa} \le v_{tri} < (d_{Aa}+d_{Ab}), \\ [001], & (d_{Aa}+d_{Ab}) \le v_{tri} < 1, \end{cases} \tag{7.25}$$

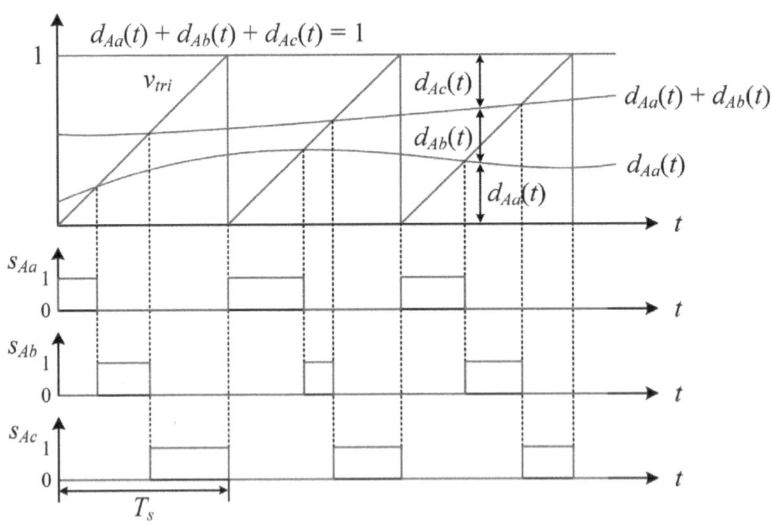

FIG. 7.13 Generation of gating signals from duty signal (switch of A-phase).

where $s_{ij}=0$ represents the off state of the switch and $s_{ij}=1$ represents the on state. The methods that generate the gating signals of switches (S_{Ba}, S_{Bb}, and S_{Bc}) connected to the B-phase output stage and switches (S_{Ca}, S_{Cb}, and S_{Cc}) connected to the C-phase output stage are similar to the method for switches connected to the A-phase output stage.

7.2.4 Modulation Technique of Indirect Matrix Converter

Fig. 7.14 shows the circuit configuration of the indirect matrix converter. In this section, the modulation techniques of the rectifier stage and the inverter stage are described.

7.2.4.1 Modulation of Rectifier Stage

The modulation technique of the rectifier stage will be discussed in this paragraph. Among the switches of the rectifier stage shown in Fig. 7.14, the prime (′)-signed switches have equal state (on state or off state) with nonprime-signed switches. In the rectifier stage, one switch among the upper switches and one switch among the lower switches are turned on instantaneously. Depending on the on state of the switches, the vectors are classified into six active vectors and three null vectors as shown in Fig. 7.15. In case the switches of the same phase leg are turned on, the null vector is generated, and the fictitious DC link is shorted.

The phase angle of the reference current is expressed as θ_a in region 1 as shown in Fig. 7.15. In addition, this vector can be expressed by the most adjacent vectors *SC1* and *SC6*, and it can be expressed as follows in Eq. (7.26).

$$\begin{aligned} &i_a^*=I_{im}\cos\theta_a,\quad i_b^*=I_{im}\cos\theta_b,\quad i_c^*=I_{im}\cos\theta_c,\\ &\theta_a=\omega_i t,\quad \theta_b=\theta_a-2\pi/3,\quad \theta_c=\theta_a+2\pi/3, \end{aligned} \tag{7.26}$$

where I_{im} is the amplitude of the current and θ_a, θ_b, and θ_c are the phase angles of the current.

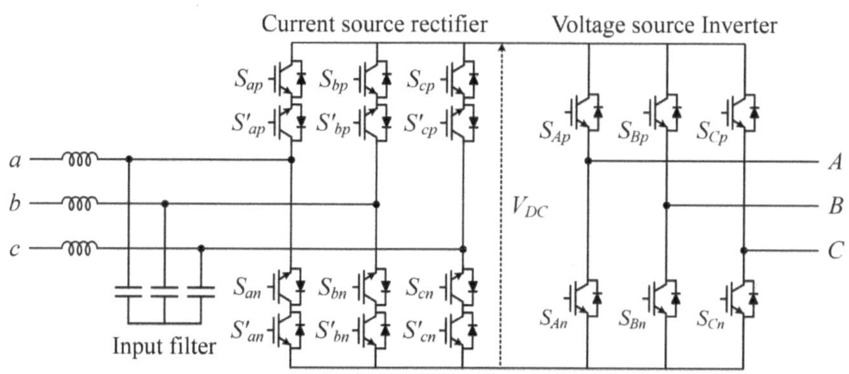

FIG. 7.14 Circuit configuration of an indirect matrix converter.

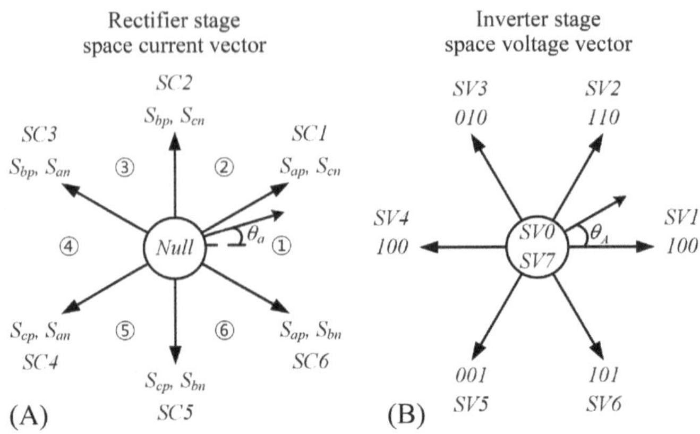

FIG. 7.15 Space vector diagram of (A) rectifier stage and (B) inverter stage.

In this case, switch S_{ap} maintains the on state in the switching period. Therefore, the upper side of the DC link is connected to the a-phase. The switches of the other phases are alternately turned on using the duty ratio calculated, as shown in Eq. (7.27) in order to connect to the lower side of the DC link.

$$\cos\theta_a + \cos\theta_b + \cos\theta_c = 0, \quad -\frac{\cos\theta_b}{\cos\theta_a} - \frac{\cos\theta_c}{\cos\theta_a} = 1,$$
$$d_x = -\cos\theta_b/\cos\theta_a, \quad d_y = -\cos\theta_c/\cos\theta_a. \tag{7.27}$$

Other regions shown in Fig. 7.15 can be interpreted as the equal method. Additionally, the multiplication of the duty ratio and the input line-to-line voltage are equal to the average DC-link voltage, and it is expressed as given in Eq. (7.28).

$$\begin{aligned} V_{DC(av)} &= d_x v_{ab} + d_y v_{ac} \\ &= \frac{3V_{im}}{2\cos\theta_{ia}} \cdot \cos\phi_i, \quad -\frac{\pi}{6} \le \theta_{ia} \le \frac{\pi}{6}, \end{aligned} \tag{7.28}$$

where V_{im} is the amplitude of the voltage and ϕ_i is the input power factor.

The purpose of the modulation of the rectifier stage is to maintain positive DC-link voltage, sinusoidal input current, and unity input power factor. In the rectifier stage, the input voltage having maximum magnitude as the absolute value is connected to the positive pole or negative pole at each π/3 region as shown in Fig. 7.16 in order to generate the maximum DC-link voltage.

For example, in the 3π/6–5π/6 region, the instantaneous value of input voltage v_b is positive and the upper switch of Phase-b is turned on. Otherwise, input voltages v_a and v_c are negative and the lower switches of Phase-a and Phase-c are turned on. In this region, all other switches are turned off. Therefore, the DC-link voltage is generated by the switching of the largest input voltage and the second largest input voltage in the rectifier stage.

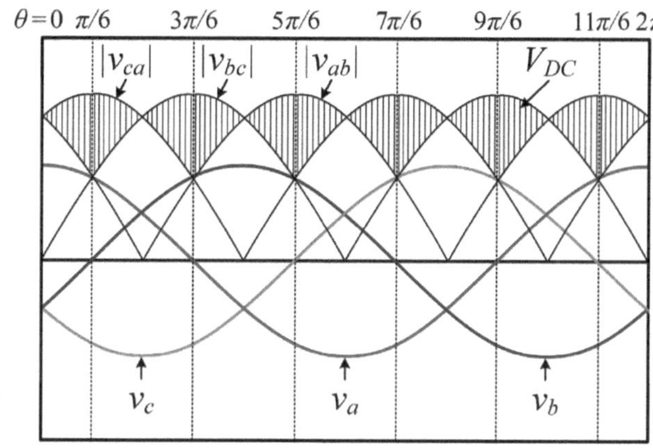

FIG. 7.16 Input voltage and DC-link voltage.

7.2.4.2 *Modulation of the Inverter Stage*

The modulation technique of the inverter stage shown in Fig. 7.14 will be discussed in this section. In the inverter stage, the vectors are classified into six active vectors and two zero vectors as shown in Fig. 7.15, depending on the on state of the switches. If the upper switches (S_{Ap}, S_{Bp}, S_{Cp}) of all phase legs or the lower switches (S_{An}, S_{Bn}, S_{Cn}) of all phase legs are turned on, the zero vectors *SV0* or *SV7* are generated, respectively. In order to modulate the rectifier stage, the modulation signals are calculated by using Eq. (7.29) from the duty ratios and the average DC-link voltage calculated from Eqs. (7.27), (7.28).

$$\begin{aligned} v_{A(upper)} &= -2d_y \cdot \frac{v_A + v_{offset}}{V_{DC(av)}} + d_x, \\ v_{A(lower)} &= 2d_x \cdot \frac{v_A + v_{offset}}{V_{DC(av)}} - d_y, \end{aligned} \tag{7.29}$$

where v_A is the amplitude of A-phase voltage and v_{offset} is expressed in Eq. (7.30).

$$v_{offset} = -0.5 \cdot \{\max(v_A, v_B, v_C) + \min(v_A, v_B, v_C)\}. \tag{7.30}$$

As a result, the rectifier stage and the inverter stage are modulated by the carrier-based PWM method using the duty ratios and modulation signals obtained from Eqs. (7.27), (7.29), respectively.

7.3 CONTROL STRATEGY OF MATRIX CONVERTER

In this section, the configuration of the matrix converter and the commutation method of the bidirectional switches will be described. Finally, a control strategy of the matrix converter will be described using PSIM (power simulation) as a simulation tool.

7.3.1 Configuration of Direct Matrix Converter System

The main circuit for driving the matrix converter, shown in Fig. 7.17, is composed of the input filter, the matrix converter including the bidirectional power switches, and a clamp circuit. The LC filter of the input stage is configured to reduce the harmonics of the input current as a secondary low-pass filter [19,20]. The matrix converter consists of nine bidirectional switches that are connected to the input and output. The clamp circuit is included in the circuit to prevent damage to the system from the overvoltage of the input stage and the output stage using the two diode bridges and the clamp capacitor.

7.3.1.1 Input Filter

In general, the harmonics of the switching frequency can be efficiently reduced by adding the low-pass LC filter in the matrix converter. To achieve this, the input filter must be designed to reduce the ripple of the input current, while the reactive power generated by the filter is minimized. The LC filter is generally used for the input filter. In case a higher damping rate is required, a multistage filter is used for the input filter. However, the large volume and the high cost of the input filter are the key problems to be addressed. The cut-off frequency of the input filter, which is expressed as ω_0, must be designed to have a damping ratio determined by the switching frequency, and it must have a lower frequency than the switching frequency. In addition, the inductance of the input filter is determined by the voltage drop at the rated current. The input filter is configured to the inductor connected to the input stage as series and the capacitor connected to the Δ- or Y-connection as shown in Fig. 7.17. Therefore, the cut-off frequency of the input filter is calculated using Eq. (7.31); the inductance of the input filter has to be determined to a value that satisfies Eq. (7.32) at the rated current.

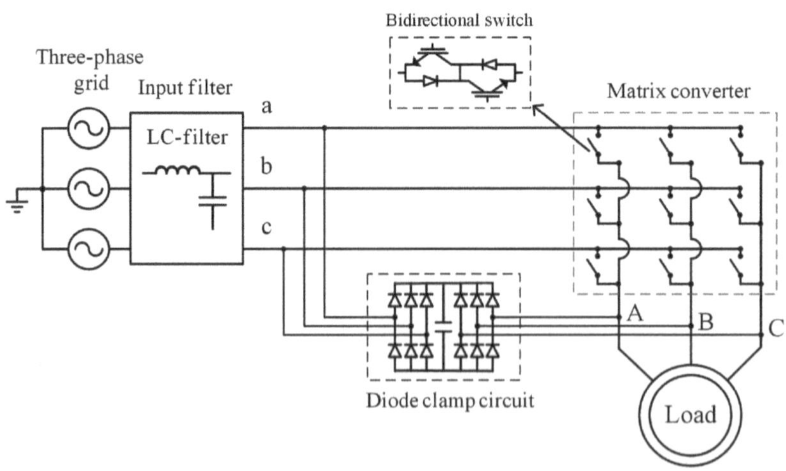

FIG. 7.17 Three-phase to three-phase matrix converter system with a motor as a load.

$$\omega_0 = \frac{1}{\sqrt{L_f C_f}}, \tag{7.31}$$

$$\frac{\Delta V}{V_n} = 1 - \sqrt{1 - (\omega \cdot L_f)^2 \cdot \left(\frac{I_n}{V_n}\right)^2}, \tag{7.32}$$

where ω_0 is the cut-off frequency, ω is the frequency of power supply, and L_f and C_f are the inductance and capacitance of the input filter.

7.3.1.2 Clamp Circuit

The clamp circuit is used to operate the matrix converter safely by limiting the overvoltage generated by the input and output stages of the matrix converter. The capacitance of the clamp circuit is determined by Eq. (7.33).

$$C_{clamp} = \frac{3i_{max}^2 (L_{\sigma s} + L_{\sigma r})}{2\left[V_{max}^2 - \left(\sqrt{2}V_{line}\right)\right]}, \tag{7.33}$$

where i_{max} is the overcurrent protection level as an effective value, $L_{\sigma s}+L_{\sigma r}$ is the total leakage inductance of the motor, C_{clamp} is the capacitance of the clamp circuit, and V_{max} and V_{line} are the overvoltage protection level and line-to-line voltage.

7.3.2 Commutation of Bidirectional Switches

The matrix converter has always to be controlled to prevent the short circuit of the input phase and open circuit of the output phase in order to prevent damage during operation. The actual switching device cannot perform ideal switching because it has switching delay times. Therefore, when the two switches operate alternately, the risk of short circuit of the input phase arises. Several methods have been proposed to solve these problems and to implement the matrix converter safely. The methods used most frequently are two-stage commutation and a four-step commutation. In these two methods, the switching sequence is determined depending on the direction of load current, and the switches are operated to perform semisoft switching operation. In this section, four-step commutation will be described because it is used in general.

The four-step commutation process is shown in Fig. 7.18. First, switch S_{Aa+} is turned off immediately to prevent the reverse conduction of the switch. In addition, when the delay time has elapsed, switch S_{Ab-} is turned on so that the path of the load current is changed to the path through switch S_{Ab-} from the path through switch S_{Aa-}. Therefore, bidirectional switches such as S_{Aa} and S_{Ab} operate so that the current flows to the load. In the bidirectional switch S_{Aa} that will be turned off, the current does not flow naturally under reversed bias condition of the diode depending on the polarity of the line-to-line voltage. If the reversed bias condition of the diode does not occur, the conducting switch S_{Aa-} is forcibly turned off. Finally, switch S_{Ab+} that does not conduct is turned on

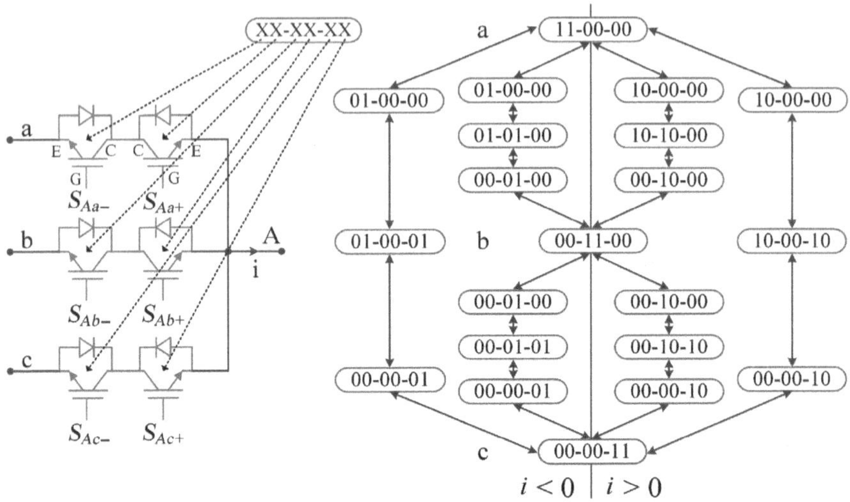

FIG. 7.18 Process of four-step commutation.

after the delay time that can be determined depending on the characteristics of the switch. Additionally, these commutation sequences must be changed depending on the direction of the load current. Therefore, in an actual system, a high-performance sensor and additional circuit in the hardware must be used for quick and accurate zero-crossing or a compensation algorithm in software is required.

The switching sequence depending on the four-step commutation process is shown in Fig. 7.19. In this process, the switching state is shifted depending on the polarity of the load phase current in a leg. The four-step commutation safely accomplishes the current commutation between the switches, while the short circuit of the input stage and open circuit of the output stage are prevented [21,22]. However, the four-step commutation generates distortion, which is similar to the dead time of the three-phase voltage source inverter, in the PWM pulse width calculated and it causes distortion of the output voltage. The delay time must be as short as possible within the permitted range of the switching device to reduce the distortion.

7.3.3 Simulation Results of Matrix Converter

In this section, an operation and control strategy of the matrix converter will be described using PSIM as simulation tool.

7.3.3.1 Circuit Configuration Using a PSIM Simulation Tool

Fig. 7.20 shows the circuit configuration for the space vector modulation method of the matrix converter using the PSIM tool. The PSIM tool and Visual C Studio are used to simulate the vector control of the matrix converter.

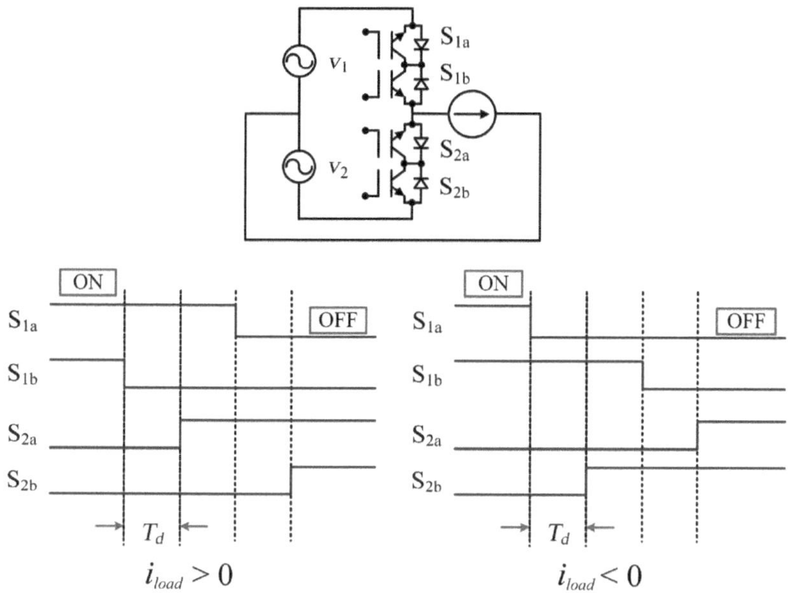

FIG. 7.19 Switching sequence depending on the process of four-step commutation.

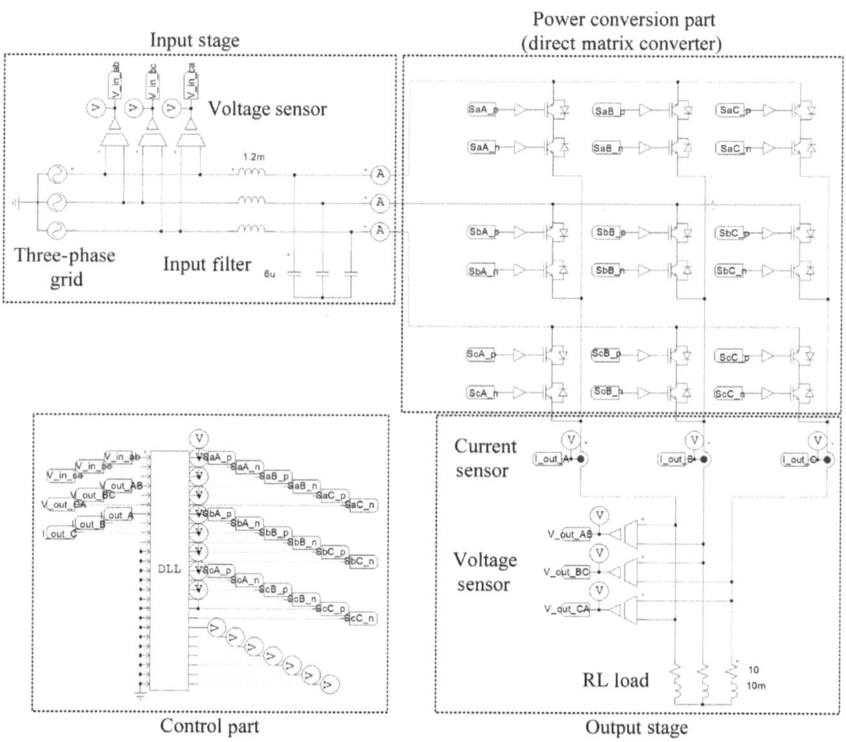

FIG. 7.20 Circuit configuration of matrix converter system using PSIM.

The circuit configuration is divided into the input stage, power conversion part, output stage, and control part. The input stage consists of a three-phase AC source, voltage sensors, and an LC filter. The three-phase AC source is connected to the input stage of the matrix converter via an LC filter, and it generates the three-phase 60 Hz/380 V_{rms}. Additionally, the elements of the LC filter are determined as $L = 1.2\,mH$, $C = 6\,\mu F$. The power conversion part is the direct matrix converter composed of the bidirectional switches. The output stage consists of the three-phase *R-L* load and voltage and current sensors. In the simulation, a three-phase *R-L* load is used and they are $R = 10\,\Omega$, $L = 10\,mH$. The control part is used to execute the dynamic link library (DLL) using C code. Additionally, the control period of the simulation is determined as 200 μs.

Fig. 7.21 shows the simulation results where the input voltage sector of the matrix converter is detected. The line-to-line voltages generated by the

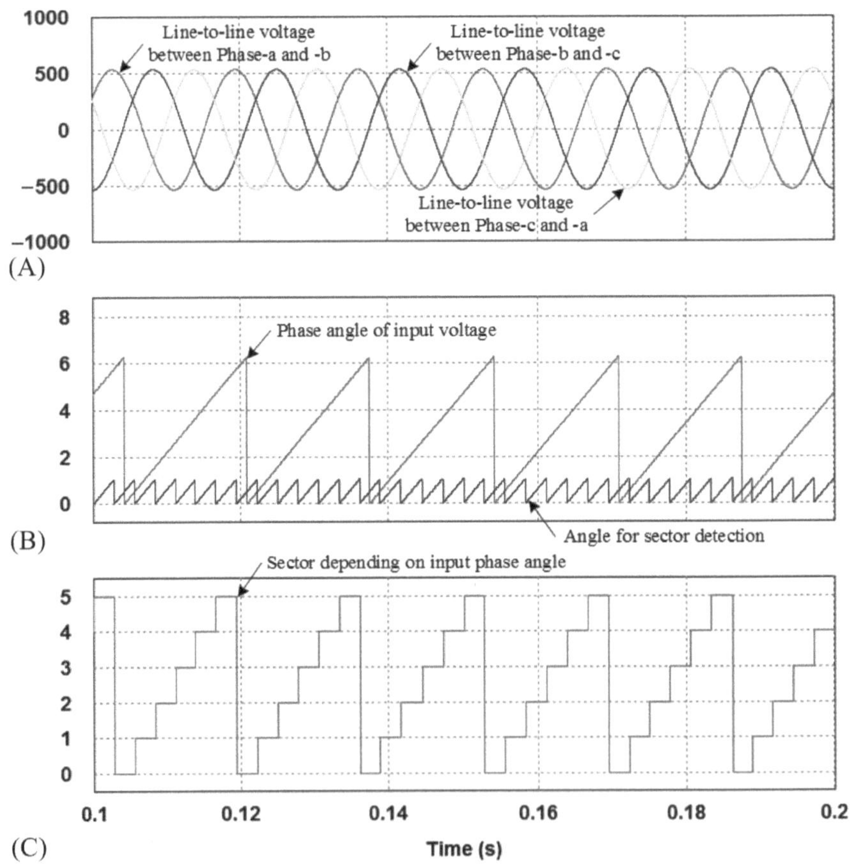

FIG. 7.21 Detection of input voltage sector. (A) Three-phase grid line-to-line voltages, (B) phase angle of input voltage and angle for sector detection, and (C) sector depending on input phase angle.

three-phase grid are shown in Fig. 7.21A. Through the reference frame transformation of the voltages, the phase angle of the input voltage is detected. In addition, the angle for sector detection is separated by a 60° interval. Finally, the sector depending on the input phase angle is detected as shown in Fig. 7.21D.

Figs. 7.22 and 7.23 show the simulation results where the output voltage sector of the matrix converter is detected. First, the output stage currents flowing in the three-phase *R-L* load of the matrix converter are controlled to the command values. In Fig. 7.21A, the *d-q* axis rotating reference frame currents are controlled to 0 and 15 A, respectively. Through the current control, *d-q* axis reference voltages with respect to the rotating reference frame are calculated.

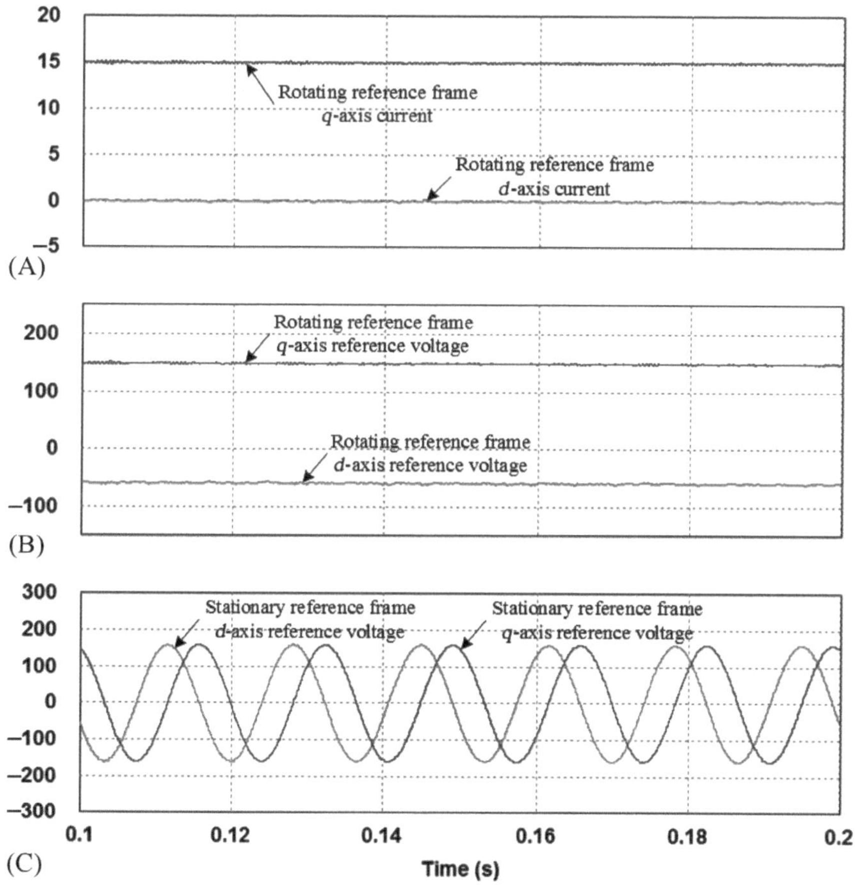

FIG. 7.22 Calculation of *d-q* axis reference voltages with respect to stationary reference frame. (A) Rotating reference frame *d-q* axis current, (B) rotating reference frame *d-q* axis reference voltage, and (C) stationary reference frame *d-q* axis reference voltage.

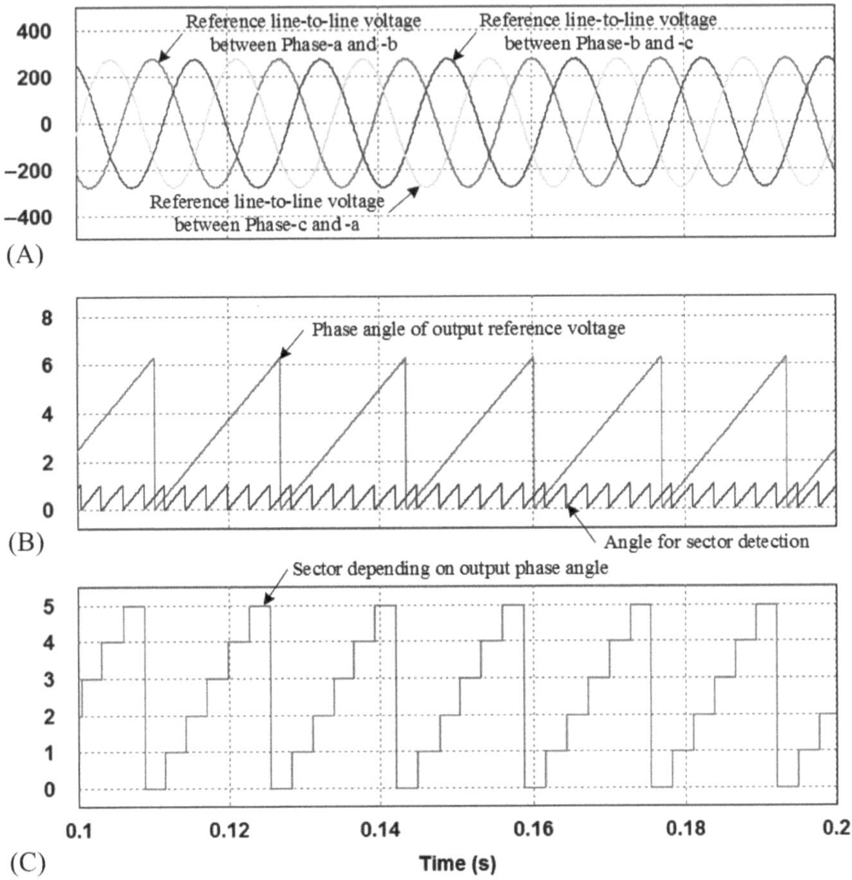

FIG. 7.23 Detection of output voltage sector. (A) Three-phase reference line-to-line voltages, (B) phase angle of output reference voltage and angle for sector detection, and (C) sector depending on output phase angle.

They are transformed to *d-q* axis stationary reference frame voltages as shown in Fig. 7.22C.

Additionally, the *d-q* axis reference voltages with respect to the stationary reference frame are transformed to three-phase line-to-line voltages as shown in Fig. 7.23A. Similar to the detection of the input voltage sector, the output voltage sector is detected by the three-phase line-to-line voltages.

As a result, the input and output voltage sectors are used to determine the switching states of the matrix converter.

Fig. 7.24 shows the simulation results where the *d-q* axis rotating reference frame currents are controlled to 0 and 15 A, respectively. The three-phase line-to-line voltages generated by the AC-source have 60 Hz/380 V_{rms},

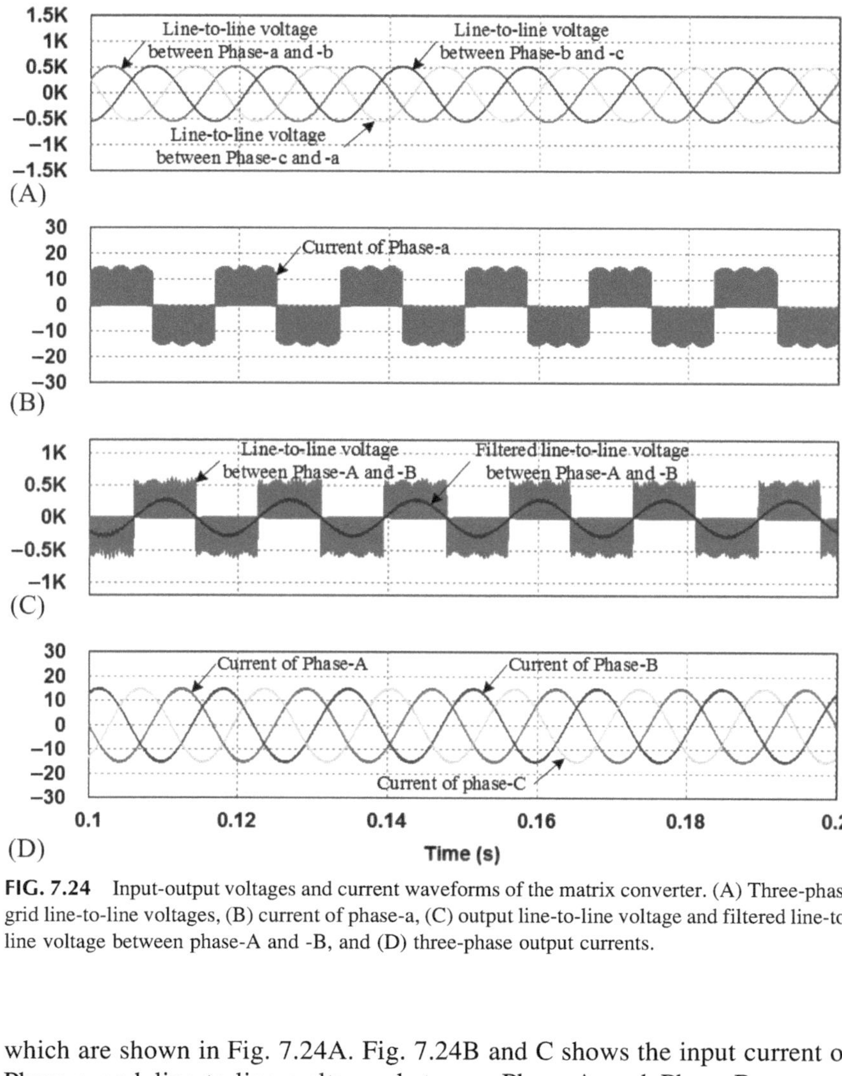

FIG. 7.24 Input-output voltages and current waveforms of the matrix converter. (A) Three-phase grid line-to-line voltages, (B) current of phase-a, (C) output line-to-line voltage and filtered line-to-line voltage between phase-A and -B, and (D) three-phase output currents.

which are shown in Fig. 7.24A. Fig. 7.24B and C shows the input current of Phase-a and line-to-line voltages between Phase-A and Phase-B, respectively. They are generated by the switching of the matrix converter. Finally, the three-phase output currents are controlled to 15 A, and they are shown in Fig. 7.24D.

Additionally, the switches of the direct matrix converter, shown in Fig. 7.20, are operated with four-step commutation depending on the sign of the output phase current. As an example, the switching sequence depending on the four-step commutation process regarding on/off states of the switches such as SaA_p, SaA_n, SbA_p, and SbA_n is shown in Fig. 7.25.

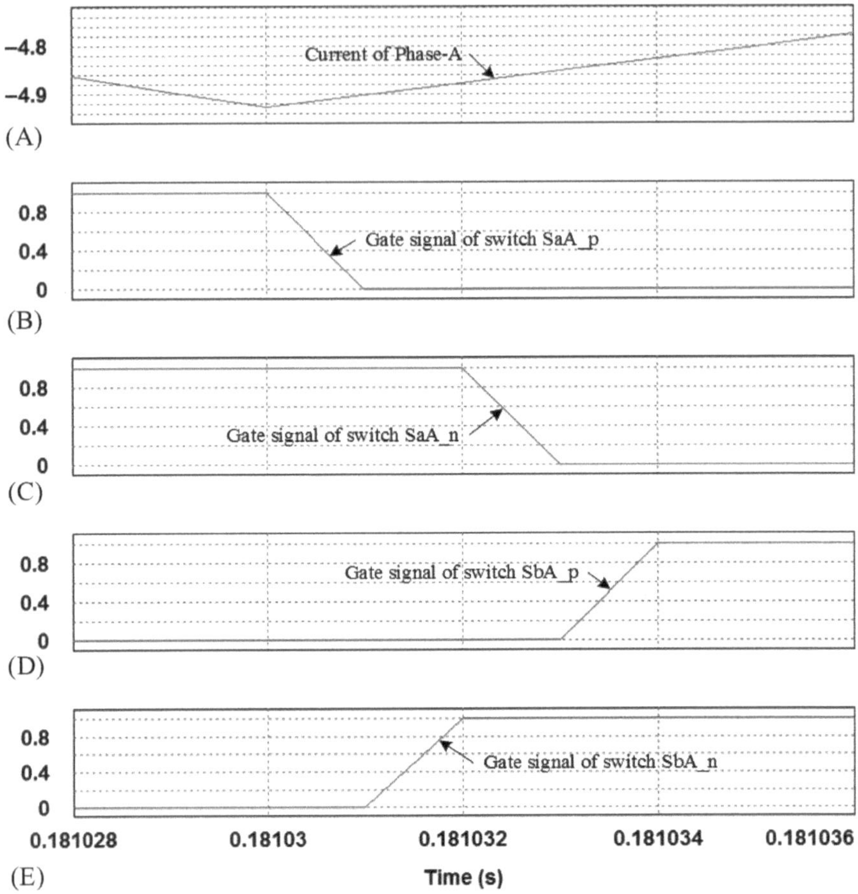

FIG. 7.25 Switching sequence of switches SaA and SbA depending on process of four-step commutation. (A) Current of phase-A, (B) gate signal of switch SaA_p, (C) gate signal of switch SaA_n, (D) gate signal of switch SbA_p, and (E) gate signal of switch SbA_n.

Fig. 7.26 shows the simulation results where the output currents in the *d-q* axis rotating reference frame are controlled to various command values. The command value of the *d*-axis current is changed to 5 A from 0 A at 0.3 s. The command value of the *q*-axis current is changed to 20 A from 15 A at 0.2 s, and it is also changed to 15 A from 20 A at 0.4 s. In the simulation results, the *d-q* axis rotating reference frame currents are controlled to the command values, respectively.

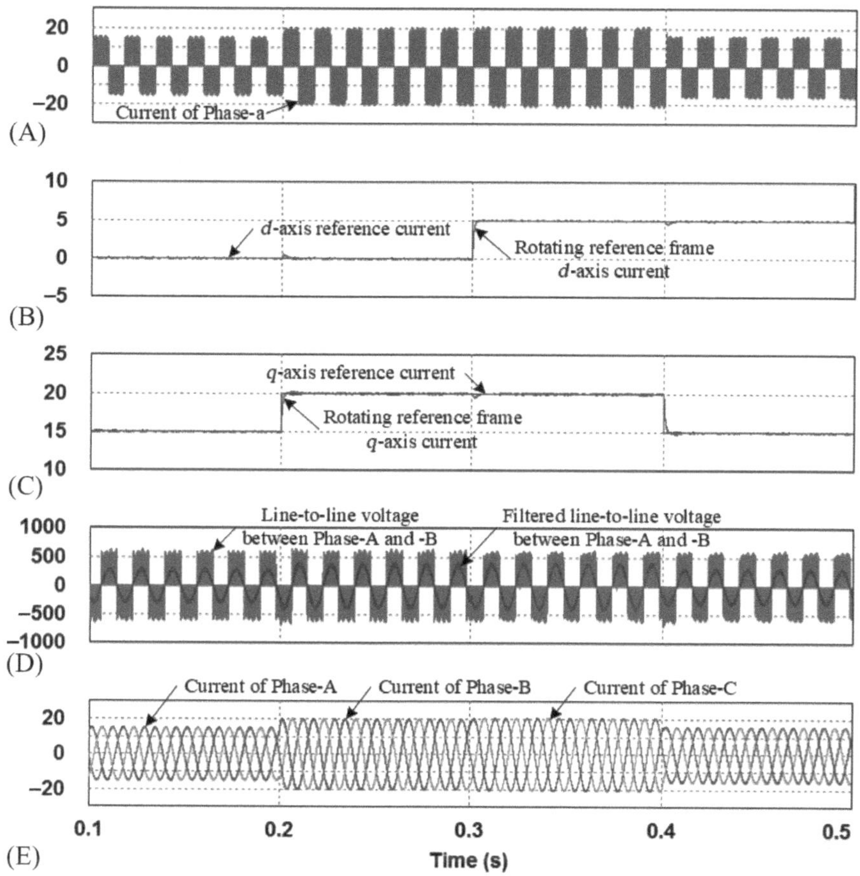

FIG. 7.26 Output currents of the matrix converter are controlled to various command values. (A) Current of phase-a, (B) *d*-axis reference current and rotating reference frame *d*-axis current, (C) *q*-axis reference current and rotating reference frame *q*-axis current, (D) output line-to-line voltage and filtered line-to-line voltage between phase-A and -B, and (E) three-phase output currents.

7.4 CONCLUSION

In this chapter, a matrix converter for direct AC-AC power conversion is described. The matrix converters have been researched since the 1970s. The research field of the matrix converter has mostly dealt with various modulation methods, control algorithms, and hardware design of the matrix converter. More recently, researches being conducted to overcome the drawbacks of the matrix converter have been studied and the applications fields are also varied. The matrix converter is one-step direct power conversion system of AC-AC. It has advantages of driver reliability, long lifetime, small size, and low weight because the DC link does not use electrolytic capacitors, which are large in size and they have also reliability issues. Additionally, the matrix converter offers advantages such as efficiency improvement and reduced maintenance. Therefore, it is used in industrial applications that require high reliability, miniaturization, and sinusoidal input and output currents as well as power flow capabilities.

REFERENCES

[1] L. Gyugyi, B.R. Pelly, Static Power Frequency Changers: Theory, Performance and Applications, Wiley, New York, 1976.

[2] L. Gyugyi, B.R. Pelly, An Electrical Converter System Employing a Control Converter for Controlling the Operation of a Power Converter, US patent 3,585,486, to Westinghouse Electric Corp., Patent and Trademark Office, 1971.

[3] M. Venturini, A. Alesina, The generalized transformer: a new bidirectional sinusoidal waveform frequency converter with continuously adjustable input power factor, Proc. PESC, 1980, pp. 242–252.

[4] M. Venturini, A new sine wave in sine wave out, conversion technique which eliminates reactive elements, Proc. POWERCON, 1980. E3_1–E3_15.

[5] A. Alesina, M.G.B. Venturini, Analysis and design of optimum-amplitude nine-switch direct AC-AC converters, IEEE Trans. Power Electron. 4 (1) (1989) 101–112.

[6] J. Rodriguez, A new control technique for AC-AC converters, Proc. IFAC Control in Power Electronics and Electrical Drives, 1983, pp. 203–208.

[7] G. Kastner, J. Rodriguez, A forced commutated cycloconverter with control of the source and load currents, Proc. EPE Conf, 1985, pp. 1141–1146.

[8] C.L. Neft, C.D. Schauder, Theory and design of a 30-hp matrix converter, IEEE Trans. Ind. Appl. 28 (1992) 546–551.

[9] N. Burany, Safe control of four-quadrant switches, Conf. Rec. IEEE-IAS Annual Meeting, 1989, pp. 1190–1194.

[10] T. Friedli, J.W. Kolar, J. Rodriguez, P.W. Wheeler, Comparative evaluation of three-phase AC-AC matrix converter and voltage DC-link back-to-back, IEEE Trans. Ind. Electron. 59 (12) (2012) 4487–4510.

[11] P. Mutschler, M. Marcks, A direct control method for matrix converters, IEEE Trans. Ind. Electron. 49 (2) (2002) 362–369.

[12] B. Metidji, N. Taib, L. Baghli, T. Rekioua, S. Bacha, Novel single current sensor topology for venturini controlled direct matrix converters, IEEE Trans. Power Electron. 28 (7) (2013) 3509–3516.

[13] I. Sato, J.-I. Itoh, H. Ohguchi, A. Odaka, H. Mine, An improvement method of matrix converter drives under input voltage disturbances, IEEE Trans. Power Electron. 22 (1) (2007) 132–138.

[14] K.-B. Lee, F. Blaabjerg, Reduced-Order extended luenberger observer based sensorless vector control driven by matrix converter with nonlinearity compensation, IEEE Trans. Ind. Electron. 53 (1) (2006) 66–75.

[15] K.-B. Lee, F. Blaabjerg, Improved sensorless vector control for induction motor drives fed by a matrix converter using nonlinear modeling and disturbance observer, IEEE Trans. Energy Convers. 21 (1) (2006) 52–59.
[16] K.-B. Lee, F. Blaabjerg, T.-W. Yoon, Speed-sensorless DTC-SVM for matrix converter drives with simple nonlinearity compensation, IEEE Trans. Ind. Appl. 43 (6) (2007) 1639–1649.
[17] J.W. Kolar, F. Schafmeister, S.D. Round, H. Ertl, Novel three-phase AC-AC sparse matrix converters, IEEE Trans. Power Electron. 22 (5) (2007) 1649–1661.
[18] K. Park, K.-B. Lee, F. Blaabjerg, Improving output performance of a Z-source sparse matrix converter under unbalanced input-voltage Conditions, IEEE Trans. Power Electron. 27 (4) (2012) 2043–2054.
[19] A. Trentin, P. Zanchetta, J. Clare, P. Wheeler, Automated optimal design of input filters for direct AC/AC matrix converters, IEEE Trans. Ind. Electron. 59 (7) (2012) 2811–2823.
[20] A. Dasgupta, P. Sensarma, Filter design of direct matrix converter for synchronous applications, IEEE Trans. Ind. Electron. 61 (12) (2014) 6483–6493.
[21] A. Ecklebe, A. Lindemann, S. Schulz, Bidirectional switch commutation for a matrix converter supplying a series resonant load, IEEE Trans. Power Electron. 24 (5) (2009) 1173–1181.
[22] J. Mahlein, J. Igney, J. Weigold, M. Braun, O. Simon, Matrix converter commutation strategies with and without explicit input voltage sign measurement, IEEE Trans. Ind. Electron. 49 (2) (2002) 407–414.

Part II

Renewable Energy Systems

Chapter 8

Design and Control of Voltage Source Converters With *LCL*-Filters

Rafael Peña-Alzola* and Frede Blaabjerg†
**University of Strathclyde, Glasgow, United Kingdom,* †*Aalborg University, Aalborg, Denmark*

8.1 INTRODUCTION

Active rectifier using a voltage-sourced three-phase full-bridge converter may provide bidirectional energy flow and sinusoidal current with not only unity power factor but also other power factors on demand. As perfect sinusoidal waveforms are never possible with power electronic converters, the standards (being IEEE 519-1992 [1] the most commonly referred) limit the total THD (total harmonic distortion) typically to 5% and the amplitude of the individual harmonics of the current injected to the grid. Considering this harmonic limitation, an L-filter (a simple inductor) as an interface between the full-bridge and the grid may require a too large inductance value. The required L-filter is heavy, bulky, and expensive. It may also increase losses in the voltage source converter due to the large voltage drops and even to a poor dynamic response considering the DC-link ceiling. In contrast, the LCL-filter allow selecting a lower value for the inductances and thereby a more compact design and lower losses. The main disadvantage is that the present resonance in the LCL-filter may turn the current-control loop unstable. Passive damping solves this issue by using dissipative elements (resistors) which produce additional losses and thereby lower efficiency. Active damping modifies the control algorithm to make the current-control loop stable without using dissipative elements.

Fig. 8.1A shows a grid-connected LCL-filter-based three-phase converter using active damping along with the nomenclature used in the chapter. The converter inductor has an inductance L and a resistance R, the filter capacitor has a capacitance C_f, and finally, the grid inductor has an inductance L' and

Control of Power Electronic Converters and Systems. https://doi.org/10.1016/B978-0-12-805245-7.00008-1

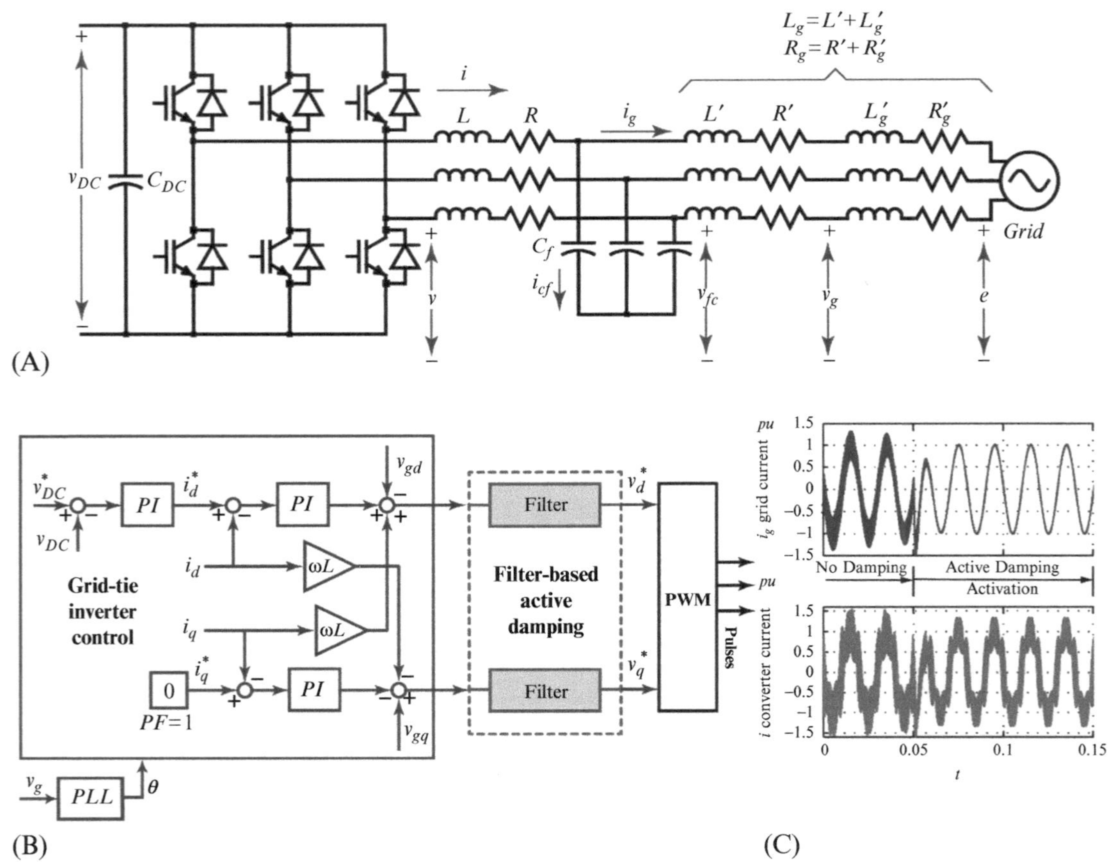

FIG. 8.1 (A) Grid-connected *LCL*-filter-based three-phase converter using active damping, (B) block diagram for the control: grid-tie inverter controller in the *dq*-frame and active damping mechanism (filter based as an example), and (C) grid and converter current waveforms with and without the active damping mechanism activated.

a resistance R'. The grid inductor may include the leakage inductance of a transformer. The line has an inductance L_g' and a resistance R_g', and these values are usually unknown. In order to obtain a more compact notation the inductance $L_g = L' + L_g'$ and the resistance $R_g = R' + R_g'$ are defined. The converter current is i, the grid current is i_g, and finally, the filter capacitor current is i_c. The converter voltage (PWM output) is v, the filter capacitor voltage is v_c, the measured grid voltage is v_g, and finally, the grid voltage is e. Fig. 8.1B shows the block diagram for the control of the *LCL*-filter-based grid-tie inverter. This control consists of the nested loops for the current and DC-link voltage in the *dq*-frame and the damping mechanism. The nested loops are the same as used in *L*-filter-based three-phase converters. Fig. 8.1B shows the current-controller for the converter current i, but the grid current i_g can also be controlled. Fig. 8.1B shows a damping mechanism based on a filter, but there are many options as it will be explained in the following sections. Fig. 8.1C shows the converter and grid currents for the *LCL*-filter-based converter using active damping. When the active damping mechanism is not activated, the inductor resistances do not provide sufficient damping. The current-control may be unstable, and the current has an unacceptable high ripple. Once the active damping mechanism is activated, the current-control loop is stable and the ripple size fulfills the norms. It is clear that the converter current ripple is higher than the grid current ripple as the filter capacitor acts as a sink for high-frequency current by providing a low-impedance path.

Fig. 8.2A shows a grid-connected *LCL*-filter-based three-phase converter using passive damping (simple resistor case). Fig. 8.2B shows the block diagram for the control of the *LCL*-filter-based grid-tie inverter. This control consists only of the nested loops for the current and DC-link voltage in the *dq*-frame. The nested loops are exactly the same as used in *L*-filter-based three-phase converters so that no control software modification is required when using passive damping for the *LCL*-filter. Fig. 8.2B also shows the current controller for the converter current i, but the grid current i_g can also be controlled. Finally, Fig. 8.2C shows the characteristic converter and grid currents of the *LCL*-filter-based converter for the passive damping case.

LCL-filters are a standard solution for grid-tie inverters, and many papers on the topic are published. This chapter covers the basic control principles for the *LCL*-filter design and its damping. The procedures shown in the chapter are well proven and simple to implement. They are based on articles published in the specialized literature combined with the experience of the authors in order to achieve the most satisfactory results. The chapter is organized as follows: after this brief introduction in Section 8.1, Section 8.2 explains the most popular procedure for *LCL*-filter design. Section 8.3 reviews the overall control of the three-phase full-bridge converter. Section 8.4 explains the different options for the passive damping of the *LCL*-filter. Section 8.5 explains simple procedures for active damping of the *LCL*-filter, and finally, Section 8.6 concludes the chapter.

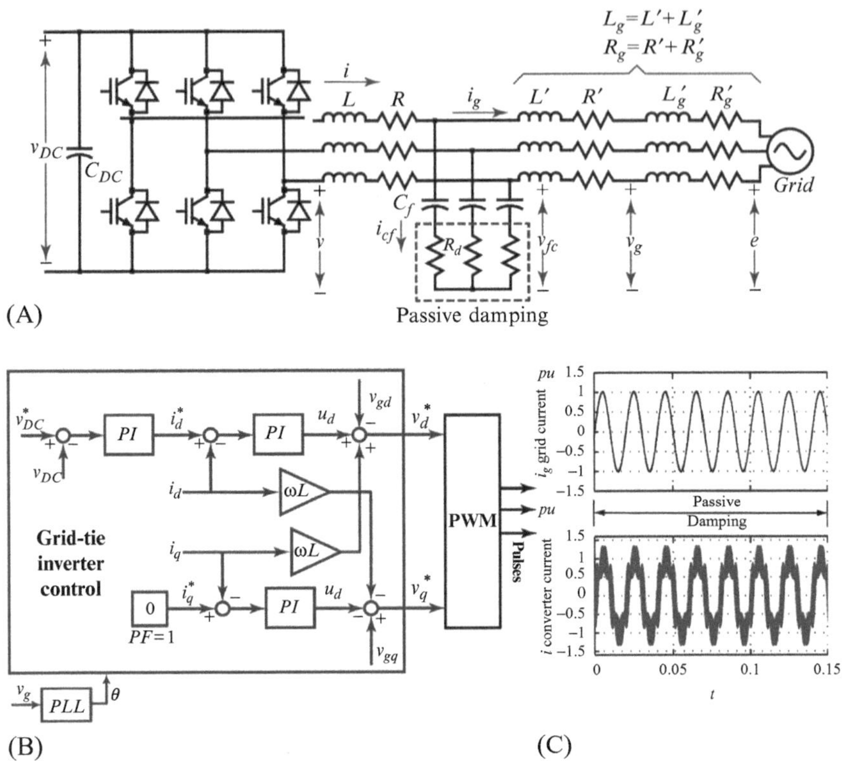

FIG. 8.2 (A) Grid-connected *LCL*-filter-based three-phase converter using passive damping (simple resistor case), (B) block diagram for the control: grid-tie inverter controller in the *dq*-frame, and (C) grid and converter waveforms.

8.2 DESIGN OF *LCL*-FILTERS

Refs. [2,3] have explained the basic guidelines for the *LCL*-filter design. In Ref. [4], the wye and delta connections of the capacitor branch were also considered. Ref. [5] uses an analytical solution of the converter harmonics, and the design is optimized for minimum energy stored in the *LCL*-filter elements. The capacitor branch that splits into two parallel capacitors for lower passive damping losses is considered in Refs. [6,7]. Ref. [6] makes use of simple approximations to obtain the *LCL*-filter configuration fulfilling the grid requirements while incurring minimal losses. In Ref. [7], after modeling in detail all the component losses, a search algorithm calculates the optimal *LCL*-filter parameters resulting in the smallest volume or minimal losses. Ref. [8] integrates both inductors in one single core and Ref. [9] uses a graphic approach for an *LCL*-filter design. Finally, Ref. [10] includes the ratio between the switching frequency and the resonance frequency as a parameter, which results in designs which are robust against the grid inductance variations. For this section, the procedure in Ref. [2] will be considered with some additional formulas that can help in the iteration process. The method [2] is simple to follow and it always results in a proper design trade-off. This is the reason for its popularity in spite of being one of the first proposed procedures.

The aim of the *LCL*-filter is reducing the equivalent inductance $L_{eq}=L+L_g$ of the otherwise large inductance value L_f of the simple *L*-filter inductance (typically to a half approximately [2]). The rms value of the harmonic current i_{hLf} circulating through an inductor (*L*-filter) connected to a three-phase converter when using the space vector modulation was derived in Ref. [11] and it is as follows:

$$i_{hL_f} = \frac{1}{2\sqrt{3}}\frac{1}{\sqrt{48}}\frac{V_{DC}}{f_{sw}L_f}\sqrt{\left[\frac{3}{2}m^2 - \frac{4\sqrt{3}}{\pi}m^3 + \frac{9}{8}\left(\frac{3}{2}-\frac{9}{8}\frac{\sqrt{3}}{\pi}\right)m^4\right]} \tag{8.1}$$

with m the modulation index, f_{sw} the switching frequency, L_f the inductance of the *L*-filter, and V_{DC} the DC-link voltage level. A simple model for the harmonic current in the converter inductor L is assuming that all the harmonic current circulates through the filter capacitor branch C_f [10]. As the modulation index is designed to be around $m \approx 1$, an estimation of the harmonic current through the converter inductor L is as follows:

$$i_{hl} \approx 0.022\frac{V_{DC}}{f_{sw}L} \tag{8.2}$$

The ripple attenuation, passing from the converter side to the grid side, can be calculated approximately as follow by considering only the most predominant harmonic at the switching frequency [2]:

$$\frac{i_g(j\omega_{sw})}{i(j\omega_{sw})} = \frac{z_{LC}^2}{\left|\omega_{res}^2 - \omega_{sw}^2\right|} \tag{8.3}$$

where $z_{LC}{}^2=(L_gC_f)^{-1}$ and the resonance frequency f_{res} is $2\pi f_{res}=\omega_{res}{}^2=(1+L_g/L)z_{LC}{}^2$ [12]. When designing an *LCL*-filter, the parameters are subject to the following practical constraints [2]:

(a) The capacitor value is limited by the decrease of the power factor at the rated power to generally less than 5% of the base capacitance $C_b=(\omega_n Z_b)^{-1}$, with $Z_b=E_n{}^2/P_n$ being the impedance base, ω_n is the rated grid frequency, E_n is the rms line-to-line rated voltage, and finally, P_n is the rated power.

(b) The total value of the inductance should be less than 0.1 pu ($L_b=Z_b/\omega_n$) to limit the AC voltage drop during the operation of the converter. Otherwise, a higher DC-link voltage level will be required, which will result in higher switching losses [7] and also a slower step response.

(c) The resonance frequency should be located between the current-control bandwidth and one-half of the switching frequency to avoid resonance interfering in the lower and upper parts of the harmonic spectrum.
(d) Passive damping or active damping must be sufficiently effective to avoid oscillations that would render the *LCL*-filter useless.

Taking these constraints into account, the *LCL* filter can be designed by using the following step-by-step procedure [2]:

(1) Select the required current ripple on the converter side to determine the converter inductance L. Another choice is selecting the inner inductor to have approximate 80% of the required grid current THD. The converter inductance L_g is determined as a function of L using the following ratio:

$$L_g = rL \tag{8.4}$$

The ratio r affects the per unit variation of resonance frequency related to the per unit variation of the *LCL*-filter inductance as follows [10]:

$$\frac{\frac{df_{res}}{f_{res}}}{\frac{dL_g}{L_g}} = -\frac{1}{2}\frac{1}{1+r} \tag{8.5}$$

As expected a ratio for $r \gg 1$ makes the system intrinsically more robust to grid inductance variations.

(2) Select the reactive power absorbed at rated conditions in order to determine the filter capacitance C_f:

$$C_f = xC_b \tag{8.6}$$

As explained previously, x is limited by restriction (a) to be around 5%.

(3) Select the desired current ripple reduction in order to calculate the ratio r. The desired attenuation should be given some margin to take into account the losses that decrease the filter effectiveness. Eq. (8.3) can be rewritten by taking into account the previous derivations as follows [2]:

$$\frac{i_g(j\omega_{sw})}{i(j\omega_{sw})} = \frac{1}{|1 + r(1 - ax)|} \tag{8.7}$$

where $a = LC_b\omega_{sw}$ is a constant. If the sum of the two inductances does not fulfill constraint (b), another attenuation level should be chosen, or another value for the absorbed reactive power should be selected in step 2).

(4) Verify the resonance frequency to fulfill restriction (c), if this is not correct then the absorbed reactive power returned in step (2) or the attenuation returned in step (3) should be changed.

(5) If the filter attenuation is not adequate, the design procedure returns to step (3) to increase the desired current ripple reduction. If this is not sufficient the design procedure should go back to step number (2) and select a higher value of the reactive power.

(6) As the design formulas come from simplified models and there is a decrease of the filtering action due to nonmodeled losses, it is necessary to verify the filter attenuation by using simulation.

As an example, the step-by-step procedure was applied to a system in Ref. [2] with a rated voltage of $V_n = 380\,\text{V}$ (rms line to line), a rated power of $P_n = 4.1\,\text{kW}$, and a switching frequency of 5–8 kHz. The base impedance is approximately $Z_b = 35\,\Omega$, and the base capacitance is $C_b = 90\,\mu\text{F}$.

(1) A 10% current ripple is obtained by selecting a 2.7% impedance for the converter inductor. Adding the *LC* aims to reduce the current ripple to 2%.

(2) The maximum capacitance C_f is $4.7\,\mu\text{F}$ under restriction (a). If a too low capacitor value is selected, a too high value of total inductance could be necessary, and so the *LCL*-filter advantage over the simple *L*-filter is lost. A good starting point for the iteration with C_f is a half of this value $2.2\,\mu\text{F}$ ($0.025C_b$). If the restrictions are not met, this value should be progressively increased.

(3) A value of $r = 1$ is calculated using Eq. (8.7) by selecting a current ripple attenuation of 20% with respect to the ripple on the converter side. Selecting $r = 1.6$ takes into account the reduction in filter effectiveness caused by dissipative losses. The final parameters for the *LCL*-filter by following this simple but highly effective procedure are [2] $L = 3\,\text{mH}$, $L_g = 5\,\text{mH}$, and $C_f = 2.2\,\mu\text{F}$.

(4) The resulting resonance frequency is $f_{res} = 2.5\,\text{kHz}$ which is located between the current-control bandwidth (a tenth of the switching frequency at most) and the different switching frequencies.

(5) Procedures for passive and active damping will be explained in the following sections.

(6) Simulations in Ref. [2] demonstrate the correctness of the designed *LCL* filter over the full range of switching frequencies.

8.3 CURRENT-CONTROL OF VOLTAGE SOURCE CONVERTERS WITH *LCL*-FILTERS

The converter control is studied using the space vector notation and an average model in the *dq*-frame rotating at grid frequency [2,12,13]. The synchronization with the grid voltage uses a PLL (phase-locked loop). Figs. 8.1B and 8.2B show the block diagrams of the inner and faster loops controlling the *dq*-currents and the outer and slower loop controlling the DC-link voltage [3]. Figs. 8.1B and 8.2B also show the cross-coupling of the dq-currents and the feedforward of the grid voltage e_{dq}. The PWM modulator can be regarded as a zero-order hold (ZOH) [14]. The parameters for the digital PI controller according to the technical optimum criterion (damping factor of dominant pole pair $\zeta_{lf}=0.707$ for 4% overshoot) are [15] as follows:

$$K_p = \frac{L_{eq}}{3T_s} \tag{8.8}$$

$$T_i = \frac{L_{eq}}{R_{eq}} \tag{8.9}$$

where T_s is the sampling period. The PWM carrier is synchronized with the ADC (analog-to-digital converters) and simple ($T_s = T_{sw}$ where T_{sw} is the switching period) or double ($2T_s = T_{sw}$) update mode can be used for the sampling of all the measurement signals [15,16]. Using the previous parameters for tuning the PI controllers the following bandwidth is achieved:

$$\omega_{bw} = \frac{K_p}{L_{eq}} \tag{8.10}$$

The digital PI controller for the DC-link voltage is usually adjusted following the symmetrical optimum criterion [15] and by considering the inner and faster d-current control.

The gain margin GM_{lfm} and phase margin PM_{lfm} achieved are, respectively, as follows [17]:

$$GM_{lfm} \approx \pi\,(9.94\,\text{dB})\ \text{with}\ \omega_{pc} = \pi/(3T_s) \tag{8.11}$$

$$PM_{lfm} \approx 90 - \frac{270}{\pi}\frac{K_p T_s}{L_{eq}}\ \text{with}\ \omega_g = K_p/T_s \tag{8.12}$$

where ω_{pc} and ω_{gc} are the phase and gain crossover frequencies. As stated previously, the proportional gain is usually adjusted for approximately 4% overshoot with $K_p = L_{eq}/(3T_s)$ and so $PM_{lfm} = 61.4°$.

In an industrial converter, the current sensors are integrated on the converter side as they are also used for protection [3]. In this chapter, the converter current will be sensed [18]; the case for the grid current measurement is completely analogous. The grid current-control allows accurate control of the power factor at the PCC. In addition, grid current-control requires no active damping mechanism for stability when the sampling frequency is lower than six times the resonance frequency [19].

In this chapter, the control structure for the LCL-filter-based converter is the same as that of the L-filter-based converter [20]. The active damping case requires additional blocks to modify the control algorithm in order to achieve stability. The passive damping case does not require any additional block with no software modification required. Neglecting the parasitic resistances in all the

LCL-filter elements, the transfer function related to the converter voltage, v, and the converter current, i, is [12]:

$$G_{ud}(s) = \frac{i(s)}{v(s)} = \frac{1}{Ls} \frac{s^2 + z_{LC}^2}{s^2 + \omega_{res}^2} \tag{8.13}$$

The parasitic converter and grid coil resistances, R and R_g, respectively, provide extra damping and make Eq. (8.13) to have a finite DC gain. The main problem for the *LCL*-filter stability is that the resonance frequency is never accurately known. This is because the unknown line-inductance L_l, which may be large, should be included in the known grid inductance L_g so that the real grid inductance is $L_{greal} = L_g + L_l$ [21]. This situation gets worse when N converters are working in parallel as each converter will see an equivalent grid inductance equal to $L_{greal} = L_g + NL_l$ [21].

The tuning formulas are given for the *L*-filter plant with a first-order transfer function [15]. The transfer function of the *LCL*-filter Eq. (8.13) is linear, but it is second order. Thus, an equivalent model of the *LCL*-filter for the low-frequency is needed to use the simple formulas established for the *L*-filter. The Padé approximant [22] can be used to obtain the *LCL*-filter equivalent model at low-frequency. The Padé approximant $p_{N,M}(x)$ of a function $f(x)$ consists of a quotient of two polynomials with numerator degree N and denominator degree M. The Taylor series expansion of Padé approximant $p_{N,M}(x)$ equals to the Taylor series expansion of $f(x)$ up to degree $M+N$ [22]. The Padé approximant $p_{N,M}(s)$ of Eq. (8.13) with $N=0$ and $M=1$, taking into account the parasitic coil resistances, is as follows:

$$p_{0,1}(s) = \frac{1}{\left(L + L_g - C_f R_g^2\right)s + \left(R + R_g\right)} \approx \frac{1}{\left(L + L_g\right)s + \left(R + R_g\right)} \tag{8.14}$$

The factor $C_f R_g{}^2$ in Eq. (8.14) is negligible when comparing to the values of $L + L_g$ [23]. This demonstrates mathematically that the equivalent model for low frequency is approximately equivalent to eliminate the capacitor branch in the *LCL*-filter [2]. The parameters $R_{eq} = R + R_g$ and $L_{eq} = L + L_g$ of the equivalent model for low frequency are substituted in Eqs. (8.8), (8.9) for the digital PI controllers of the current loops in the *LCL*-filter-based converter.

8.4 DESIGN OPTIONS FOR PASSIVE-DAMPED *LCL*-FILTERS

Passive damping is still the most widely adopted method in the industry because of its simplicity. Compared to the *L*-filter case, neither new sensors nor changes in the control software are necessary. However, there are additional encumbrances and the additional losses could claim for forced cooling [2]. This section reviews the most practical design options [24,25] for passive damping along with analytical estimations of the incurring losses. More sophisticated techniques for passive damping design rely on the procedures explained in this section, and more details can be found in Ref. [26]. Stability analysis, including grid inductance variations, uses simple root locus design in the z-plane.

8.4.1 Design of the Passive Damping

Fig. 8.2A shows a three-phase LCL-filter-based converter with passive damping for the simple resistor case. The resistor is located in the capacitor branch where the harmonic current circulates almost exclusively with a very little circulation of fundamental current. The transfer function $G_{pd}(s)$, related to the converter voltage v and the converter current i, is [12]:

$$G_{pd}(s) = \frac{i(s)}{v(s)} = \frac{1}{Ls}\frac{s^2 + 2\zeta' z_{LC} s + z_{LC}^2}{s^2 + 2\zeta\omega_{res} s + \omega_{res}^2} \tag{8.15}$$

The damping factors in the previous transfer function of the LCL-filter with passive damping are:

$$\zeta = \frac{C_f \omega_{res}}{2} R_d \tag{8.16}$$

and $\zeta' = R_d C_f z_{LC}/2$ where R_d is the damping resistor value. From Eq. (8.16) the damping resistor R_d has an order of magnitude of the LCL-filter capacitor impedance at $\omega = \omega_{res}$. The transfer function related to the grid current i_g and v has a zero at $z = -1/R_d C_f$. This decreases the attenuation for frequencies higher than the resonance frequency $\omega \gg \omega_{res}$ from 60 dB/decade for the case with no damping down to 40 dB/decade, see Fig. 8.3. Hence, the damping resistor may reduce the LCL-filter effectiveness when it is not properly sized.

The addition of the damping resistor modifies the LCL-filter transfer function $G_{pd}(s)$ and the current-control loop based on simple digital PI controllers results in stability. The open-loop transfer function of the overall system is as follows [27]:

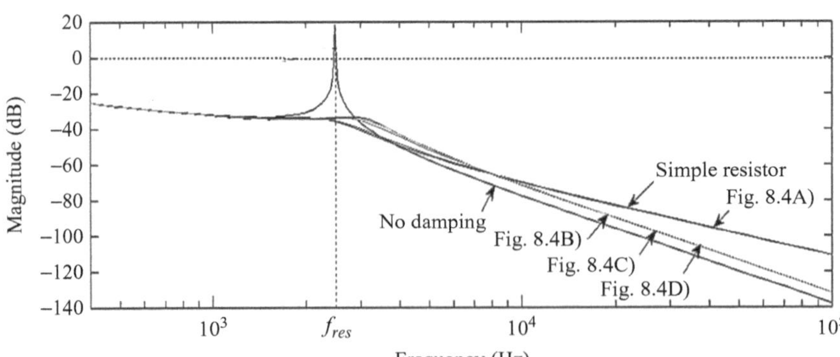

FIG. 8.3 Frequency response of the grid current i_g versus the converter voltage v (case with no damping, simple resistor case, and configurations shown in Fig. 8.4 [27]).

$$\left|G_{ol}(s)\right|_{s=j\omega_{res}} = \left|G_{PI}(s)G_{delays}(s)G_{pd}(s)\right|_{s=j\omega_{res}} = 1\ (0\ \text{dB}) \tag{8.17}$$

where $G_{delays}(s)$ models the computational delay (one sampling period $T_s = 1/f_s$) and the PWM delay (half the sampling period) and $G_{PI}(s)$ is the PI controller. The phase shift for $-180°$ in $G_{ol}(s)$ is near $\omega = \omega_{res}$ [12,29], and the minimum damping resistor R_{dmin} for a stable overall system must comply with Eq. (8.17) in order to achieve a positive gain margin. An approximate value for the minimum damping resistor was calculated in Ref. [27], and it is as follows,

$$R_{dmin} = \frac{1}{3} f_s \frac{L_g^2}{(L + L_g)} \tag{8.18}$$

From the assumed simplifications in Ref. [27], Eq. (8.18) will be accurate for the elevated switching frequencies $f_s \gg f_{res}$ and $L_g/L \gg 1$. As the passive damping losses are directly proportional to the damping resistor value, Eq. (8.18) is also useful to determine the minimum losses that the passive damping will cause.

As passive damping is only necessary at the resonance frequency, the damping losses P_d can be reduced by diverting the current from the dissipative elements at other frequencies. For this aim, new passive elements must be located in parallel/series to the *LCL*-filter capacitor branch. Fig. 8.4 shows different design options for reducing the passive damping losses.

In Fig. 8.4A, the inductance L_d in parallel with the resistor must provide a low-impedance path $L_d\omega_f \ll R_d$ [25] at the fundamental frequency $\omega = \omega_f$ so that the fundamental losses P_{df} are cancelled. Conversely, the resistor must have the dominant current flow path at the resonant frequency $\omega = \omega_{res}$, $R_d \ll L_d\omega_{res}$ in order to have a proper damping. A trade-off can be achieved by doing the impedance ratios as follows [27]:

$$\frac{R_d}{L_d\omega_f} = \frac{L_d\omega_{res}}{R_d} \tag{8.19}$$

In Fig. 8.4B, the capacitance C_d in parallel with the resistor must provide a low-impedance path $1/(C_d\omega_{sw}) \ll R_d$ [25] at the switching frequency $\omega = \omega_f$ so

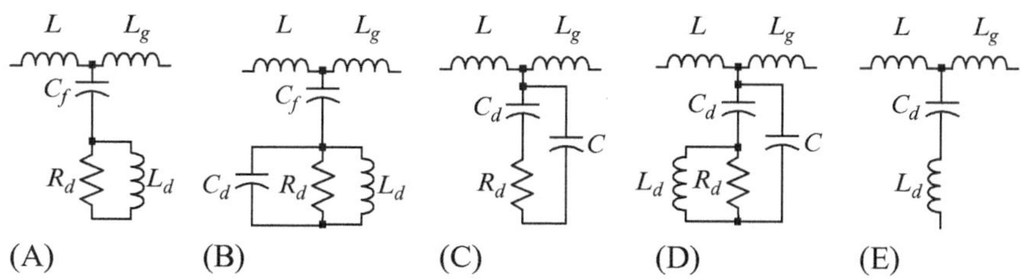

FIG. 8.4 Different configurations for passive damping [27,28] used for the voltage source converter shown in Fig. 8.2, (A) parallel inductor, (B) parallel inductor and capacitor, (C) double capacitor branch, (D) double capacitor branch with parallel inductor, and (E) *LLCL*-filter.

that the switching losses P_{dH} are highly reduced. Conversely, the resistor must have the dominant current flow path at the resonant frequency $R_d \ll 1/(C_d\omega_{res})$ for proper damping. Similar to previously, a proper trade-off can be achieved by doing the impedance ratios as follows [27]:

$$\frac{1/(C_d\omega_{res})}{R_d} = \frac{R_d}{1/(C_d\omega_{sw})} \tag{8.20}$$

Fig. 8.4E shows an interesting topology, the *LLCL*-filter, proposed in Ref. [28] that includes an additional inductor L_d in series with the capacitor filter C_d. The impedance at the switching frequency in the capacitor branch is exactly null, so $1/C_{fd}j\omega_{sw} = L_{d}j\omega_{sw}$. Therefore, the harmonic component at the switching frequency (the most preponderant) will circulate entirely through the capacitor branch and not through the grid inductance, which will improve the grid current THD.

Fig. 8.5 shows the z-plane root locus for an increasing value of the damping resistor R_d. The data of the *LCL*-filter are as those resulting from the design example in Section 8.2: $L_g = 5\,\text{mH}$, $L = 3\,\text{mH}$, $C_f = 2.2\,\mu\text{F}$, $P_{nom} = 4.1\,\text{kW}$, and $V_{nom} = 380\,\text{V}$. The minimum value of the damping resistor resulting in stable behavior is $R_{dmin} = 8.3\,\Omega$, which can be estimated by using Eq. (8.18). A proper damping results from achieving a damping factor around $\zeta_{cl} \approx 0.1$ [12], which requires a damping resistor value of $R_d = 16\,\Omega$.

Fig. 8.6 shows the root locus in the z-plane when varying the grid inductance L_{greal} for the simple resistor case. Increasing the value of the grid inductance

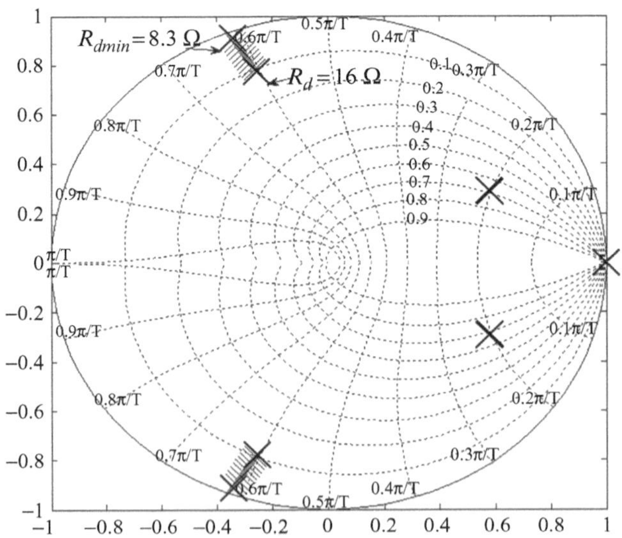

FIG. 8.5 Closed-loop poles in the z-plane of the converter current-control ($f_s = f_{sw} = 8\,\text{kHz}$) by varying the damping resistor value between $R_{dmin} = 8.3\,\Omega$ and $R_d = 16\,\Omega$ [27].

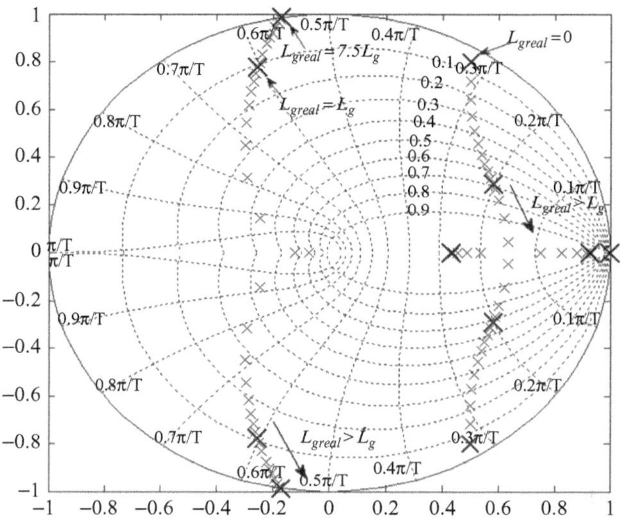

FIG. 8.6 Closed loop poles in the z-plane of the converter current-control ($f_s = f_{sw} = 8$ kHz) for $R_d = 16\,\Omega$ by varying the grid inductance between $L_{greal} = 0$ and $L_{greal} = 7.5L_g$ [27].

L_{greal} increases the damping of the dominant poles at low-frequency, which results in lower overshoots in the step response for the control. However, the damping of the resonance poles Eq. (8.16) is progressively decreased until the system becomes unstable. This analysis can be generalized for all the configurations shown in Fig. 8.4.

8.4.2 Estimation of the Passive Damping Losses

The damping losses consist of the fundamental losses produced by the fundamental current I_c and the harmonic losses produced by the harmonic current $\hat{\imath}_c$ circulating through the LCL-filter capacitor. In order to calculate the fundamental current, the damping resistors R_d can be neglected and the fundamental losses P_{df} for unity power factor are as follows [27]:

$$P_{df} \approx 3I_c^2 R_d = 3\left(\frac{V_{cf}}{Z_{cf}}\right)^2 R_d = 3\frac{\left[\left(V_{nom}/\sqrt{3}\right)^2 + \left(\omega_f L_g I_{nom}\right)^2\right]}{1/\left(\omega_f C_f\right)^2} R_d \tag{8.21}$$

with V_{nom} and I_{nom} are the rated voltage and current, respectively, and V_{cf} is the fundamental capacitor voltage. The harmonic currents find a path of negligible impedance in the capacitor branch. Thus, the converter connected to the inductance L can be used as an equivalent circuit for estimating the capacitor harmonic current. The rms value of the harmonic current circulating through an inductor connected to a three-phase converter when using the space vector modulation was derived in Ref. [11]. Using this calculation, the rms value of the harmonic capacitor current $\hat{i}_c$ can be estimated as follows:

$$\hat{i}_c = \frac{1}{2\sqrt{3}} \frac{1}{\sqrt{48}} \frac{V_{DC}}{f_{sw}L} \sqrt{\left[\frac{3}{2}m^2 - \frac{4\sqrt{3}}{\pi}m^3 + \frac{9}{8}\left(\frac{3}{2} - \frac{9}{8}\frac{\sqrt{3}}{\pi}\right)m^4\right]} \tag{8.22}$$

where m is the modulation index, which can be estimated as:

$$m = \frac{2\sqrt{2}}{V_{DC}} \sqrt{\left(\frac{V_{nom}}{\sqrt{3}}\right)^2 + \left[\omega_f\left(L + L_g\right)I_{nom}\right]^2} \tag{8.23}$$

In Ref. [27] it is shown that Eq. (8.22) is a lower bound of the *LCL*-filter capacitor rms harmonic current and an upper approximation is also considered. Thus, a lower approximation P_{dH}^{lower} for the harmonic damping losses P_{dH} is:

$$P_{dH} = 3\left(\hat{i}_c\right)^2 R_d \tag{8.24}$$

Eq. (8.24) leads to a reasonable prediction of the power losses useful for a first approximation in the passive damping design [27].

In Fig. 8.4B the current through the whole capacitor branch can be estimated with Eq. (8.22). As the most important harmonics are around the switching frequency $\omega = \omega_{sw}$, the current through the damping resistor can be estimated by the ratio of impedances at the switching frequency $\omega = \omega_{sw}$. Hence,

$$P_d \approx P_{dH} \left|\frac{Z_d}{R_d}\right|^2 \tag{8.25}$$

where $Z_d = Z_{C_d} || Z_{R_d} || Z_{L_d} \approx Z_{C_d} || Z_{R_d}$ are impedances for $\omega = \omega_{sw}$. In Fig. 8.4C most of the current harmonics is bypassing the damping resistor branch by circulating through the parallel capacitor C [6]. Ref. [6] proposes to use the same capacitance in both branches $C_d = C = C_f/2$. The fundamental capacitor current is halved, and to calculate the harmonic damping losses P_{dH} for Fig. 8.4C, it must be proceeded as given in Eq. (8.25) by considering the main harmonics around $\omega = \omega_{sw}$. Therefore, the damping losses P_d result as in the following,

$$P_d \approx \frac{P_{df}}{4} + P_{dH} \left|\frac{Z_{CT}}{Z_d}\right|^2 \tag{8.26}$$

where $Z_d = R_d + Z_{Cd}$ and $Z_{CT} = Z_C || Z_d$ are also the impedances for $\omega = \omega_{sw}$. In Fig. 8.4D [24] the inductor L_d is in parallel to the damping resistor shown in Fig. 8.4C to cancel the fundamental losses P_{df}.

The losses calculated by the previous estimations and by simulation (Matlab/Simulink) are shown in Table 8.1 for different switching frequencies in the single-resistor case. The losses calculated by the previous estimations and simulation are shown in Table 8.2 for a single switching frequency (8 kHz) and the different passive damping configuration. The results for

TABLE 8.1 Damping Losses Obtained by Simulation and Estimation for the Simple Resistor Case ($R_d = 10\,\Omega$) Shown in Fig. 8.2A [27].

f_{sw}	5 kHz	6 kHz	7 kHz	8 kHz
Simulation (W)	41.5	25.8	18.0	13.4
Estimation (W)	41.1	25.4	17.7	13.2

TABLE 8.2 Damping Losses, Simulation, and Estimation for the Different Configurations ($f_{sw} = 8$ kHz).

	Simple Resistor	Fig. 8.4A	Fig. 8.4B	Fig. 8.4C	Fig. 8.4D
Simulation (W)	21.1	19.8	4.1	4.9	3.7
Estimation (W)	20.9	19.8	5.3	6.2	4.8

TABLE 8.3 Damping Losses Obtained by Experiments for the Simple Resistor Case ($R_d = 10\,\Omega$) Shown in Fig. 8.2A [27]

f_{sw}	5 kHz	6 kHz	7 kHz	8 kHz
Lower estimation Eq. (8.21) + Eq. (8.24)	29.5	20.7	15.4	12.0
Experiments (W)	32	20	13	10

estimation and simulation are very close to each other because the simulation models of the power devices and the passive elements are ideal and the switching frequency is relatively high.

Table 8.3 shows the passive losses P_d calculated by using experiments and using Eqs. (8.21) and (8.24) for the simple resistor case. The results are sufficiently close to each other so that the formulas can be safely used for a first approximation when designing the passive damping of the *LCL*-filter. The values for the passive losses P_d shown in Table 8.3 are lower than those shown in Table 8.1 because the nonlinearities of the passive elements (saturation, core losses, etc.), which further limit the harmonics, were not considered in the simulations.

8.5 PRACTICAL METHODS FOR ACTIVE DAMPING

8.5.1 Introduction

Much literature is devoted to the active damping for *LCL*-filter-based converter with many proposed procedures. The method of using a virtual resistor, originally proposed in Ref. [30], modifies the control algorithm to emulate the behavior of a damping resistor without physically adding it [31,32]. Depending on the location of the virtual resistor, the procedure may require additional sensors. State observers can be used to avoid using extra sensors [33] at the expense of higher design complexity and more computations. The delays present in the feedback loop due to the computations and the PWM modulator require a careful design of the digital controller. The effects of the computational delay in the feedback loop of *LCL*-filter based converters were studied in Refs. [34,35]. The control in space-state domain allows selecting the locations of closed-loop poles [36]. However, it is difficult to achieve locations that result in acceptable phase and gain margins [37]. The use of an LQR regulator [38] may alleviate this issue by displacing the choice to the proper weighting coefficients. In addition, multivariable controllers involve an elevated computational load, and they are difficult to retune in time-varying conditions [39]. To overcome the effect of the line-inductance variations, which modifies the resonance frequency, adaptive control techniques have been proposed in Refs. [40,41]. Sliding mode control for *LCL*-filter-based converters was used in Refs. [42,43], which includes the Kalman filter for magnitude estimation. The feedback of the *LCL*-filter capacitor-current to the voltage reference produces resonance damping [44]. One of the first proposed method for active damping consists of adding the capacitor voltage filtered by a lead-lag network [29] to the modulator voltage reference [23]. In Ref. [45] a high-pass filter was used instead of the lead-lag filter. Passing the reference voltage to the modulators through a notch filter is a single-loop strategy in order to achieve active damping [23,46]. The Posicast controller [47] is also an interesting technique using a single loop. The main disadvantage of these single loop strategies is that they are sensitive to changes in the resonance frequency due to line-impedance variations and may require parameter identification techniques [48]. Neural networks [49] and genetic algorithms [50] have also been proposed for active damping. Fulfilling the practical philosophy of this book, in the following the most widely used procedures for active damping will be explained in their simplest implementation. These procedures use transfer functions, and the design flow makes use of common methods such as root locus and bode plots.

8.5.2 Feedback-Type Active Damping

Fig. 8.7 shows, enclosed by the dashed lines, the feedback-type block for active damping of the *LCL*-filter-based converter.

The rationale of this active damping procedures comes from the fact that the capacitor-current feedback results in resonance damping [44]. Fig. 8.8

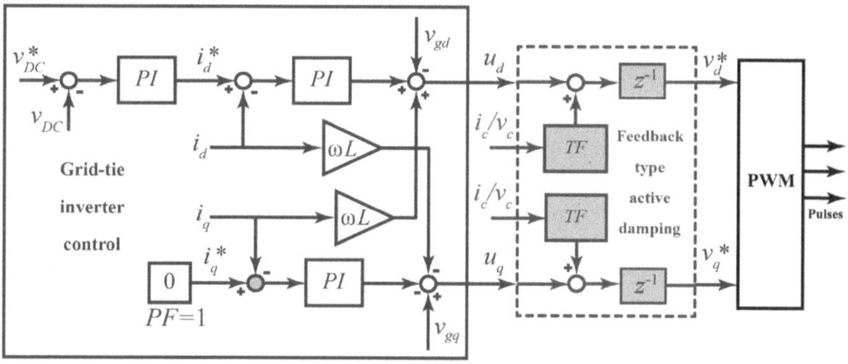

FIG. 8.7 Block diagram of the digital control for the *LCL*-filter-based rectifier (the active damping is shown enclosed by *dashed lines*).

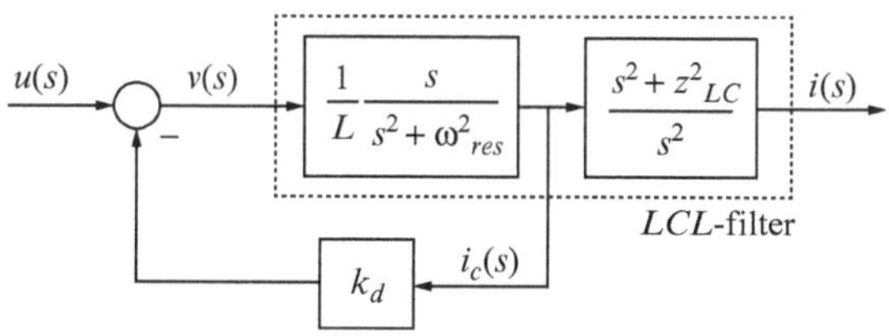

FIG. 8.8 *LCL*-filter capacitor-current feedback to obtain damping of the system.

illustrates this concept with the negative feedback of the *LCL*-filter capacitor-current to the converter voltage reference [44,51]. Assuming an ideal feedback, the transfer function related to the grid current i_g and current-controller output voltage u (see Fig. 8.7) results as in the following:

$$G_{gad}(s) = \frac{i_g(s)}{u(s)} = \frac{1}{Ls}\frac{z_{LC}^2}{s^2 + 2\left[\frac{k_d}{2L\omega_{res}}\right]\omega_{res} + \omega_{res}^2} \tag{8.27}$$

The proportional gain k_d is related to the damping factor ζ_{ad} in the following way:

$$\zeta_{ad} = \frac{k_d}{2L\omega_{res}} \tag{8.28}$$

Unlike the passive damping case, the transfer function (Eq. 8.27) does not result in a zero in the numerator so that the effectiveness of the *LCL*-filter is not degraded [52] by the active damping procedure. The transfer function related to the converter current i and voltage u is as follows [52]:

$$G_{ad}(s) = \frac{i(s)}{u(s)} = \frac{1}{Ls}\frac{\left(s^2 + z_{LC}^2\right)}{s^2 + 2\left[\frac{k_d}{2L\omega_{res}}\right]\omega_{res}s + \omega_{res}^2} \tag{8.29}$$

Considering the open-loop transfer function of the converter current-control in the continuous-time domain:

$$G_{adol}(s) = G_{PI}(s)G_{delays}(s)G_{ad}(s) \tag{8.30}$$

where $G_{ad}(s)$ is the plant (Eq. 8.29), $G_{delays}(s)$ is a pure delay modeling the computational and PWM delays, and finally, $G_{PI}(s)$ is a PI controller. The frequency that corresponds to $-180°$ phase shift for the open-loop transfer function is also near the resonant frequency [12,29]. The resonant peak should be below unity (0 dB) [29] for the current-control stability with positive gain margin. Thus, the minimum k_{dmin} resulting in stability can be obtained by equating the open-loop transfer function to unity (0 dB) at the resonance frequency ($s = j\omega_{res}$):

$$|G_{adol}(s)|_{s=j\omega_{res}} = |G_{PI}(s)G_{delays}(s)G_{ad}(s)|_{s=j\omega_{res}} = 1 \ (0\text{dB}) \tag{8.31}$$

Substituting and solving Eq. (8.31) as well as neglecting the smaller terms the following expression is obtained for the minimum gain k_d that results in an overall stable system [23]:

$$k_{dmin} = \frac{1}{3}\frac{L_g}{T_s} \tag{8.32}$$

The digital implementation of the capacitor-current feedback adds phase lags due to the computational and PWM delays, see Fig. 8.9, which can be critical.

In the s-domain, the transfer function modeling the effects of the computational delay and the PWM is [10]:

$$G_d(s) = e^{-T_s s}\frac{1 - e^{-T_s s}}{T_s s} \tag{8.33}$$

The first product of Eq. (8.33) is a pure delay and models the computational delay. The second product of Eq. (8.33) models a ZOH equivalent to the PWM. The effect of the computational delay and PWM at the resonance frequency can

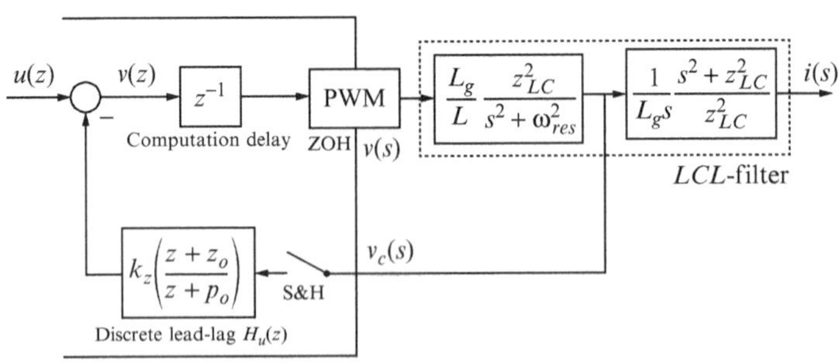

FIG. 8.9 Digital implementation of the capacitor-current feedback.

be studied by substituting $s = j\omega_{res}$ in Eq. (8.33). After some algebraic manipulations, Eq. (8.33) results in the following [53]:

$$G_d(j\omega_{res}) = \text{sinc}\left(\frac{T_s}{2}\omega_{res}\right)e^{-\frac{3T_s}{2}j\omega_{res}} \tag{8.34}$$

Hence, the effect of the PWM and computations at the resonance frequency is a pure gain, the first factor in Eq. (8.34), and a pure delay, which is the second factor Eq. (8.34). Selecting $\omega_{res} = 2\pi/3T_s$ in Eq. (8.34) cancels the pure delay produced by the PWM and computations, and the overall gain results in a real negative value:

$$G_d(j\omega_{res}) = -3\frac{\sqrt{3}}{2\pi} \tag{8.35}$$

Hence, the digital implementation delays at the resonance frequency, where damping is necessary, can be compensated by simply inverting Eq. (8.35). Selecting $\omega_{res} = 2\pi/3T_s$, Eq. (8.34) corresponds to the ratio between the resonance and the sampling frequency $f_{sw}/f_{res} = 3$. For the general case, separating the real and imaginary components of Eq. (8.34) must be considered [10]. Substituting in Eq. (8.34) and calculating the maximum results in the optimal value for $f_s/f_{res} \approx 3.12$ [10]. Hence, it can be stated that the capacitor-current feedback is adequate for:

$$\frac{f_s}{f_{res}} \approx 3 \tag{8.36}$$

Simulation and experiments show that fulfilling this condition results in robustness against large grid inductance variations [10].

As most harmonic current circulates through the capacitor branch, the additional sensor for the capacitor current must be accurate and thus it is expensive. In order to avoid the use of new current sensors [52–54], the *LCL*-filter capacitor-current i_c can be estimated from the derivative of the *LCL*-filter capacitor voltage v_c [51], see Fig. 8.10, as follows:

$$H_d(s) = \frac{i_c}{v_c} = k_d C_f s \tag{8.37}$$

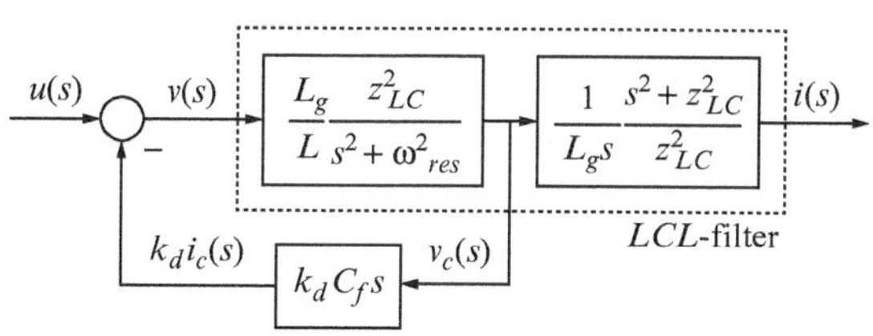

FIG. 8.10 Feedback of the capacitor voltage derivative to obtain damping in the *LCL*-filter.

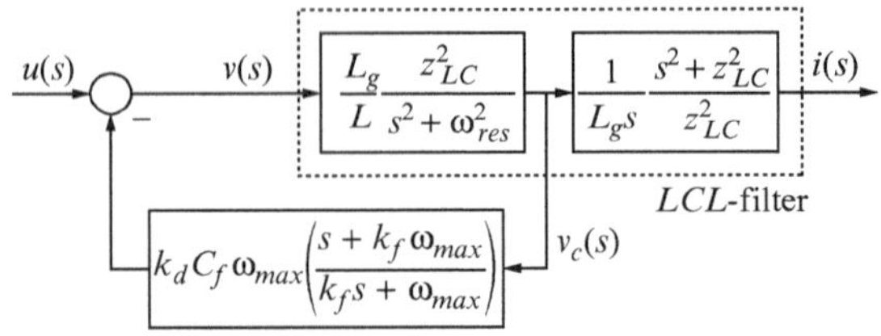

FIG. 8.11 Feedback of *LCL*-filter capacitor voltage filtered by a lead-lag network to obtain damping.

However, a perfect differentiator for the capacitor voltage would amplify the measurement signal noise and a lead-lag network (a lead network, strictly speaking) should be used instead, see Fig. 8.11.

In order to emulate the transfer function Eq. (8.37), the lead-lag network can be parametrized in a very convenient way as follows:

$$H_{ll}(s) = k_d C_f \omega_{res}\left(\frac{s + k_f\omega_{res}}{k_f s + \omega_{res}}\right) \text{ with } k_f = \sqrt{\frac{1-\sin\phi_{max}}{1+\sin\phi_{max}}} \tag{8.38}$$

where φ_{max} is the maximum phase shift ($|\varphi_{max}| < 90°$) at the resonance frequency, see Fig. 8.12. The lead-lag network behaves as a differentiator at frequencies around the resonance frequency, where the damping is necessary, by selecting φ_{max} close to 90°. By doing this, Eq. (8.38) has an amplitude equal to

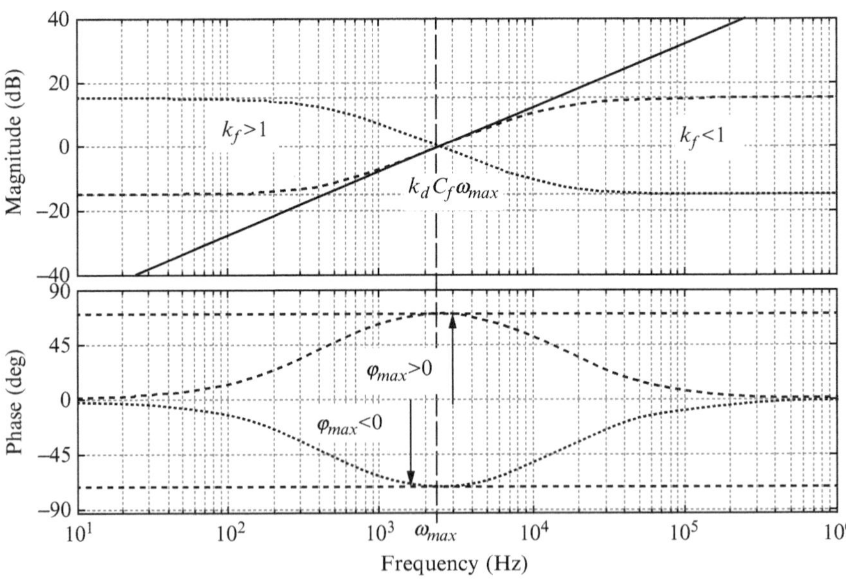

FIG. 8.12 Bode diagram of the perfect differentiator in Eq. (8.37) and the lead-lag network in Eq. (8.38) for $k_f > 1$ and $k_f < 1$ [23].

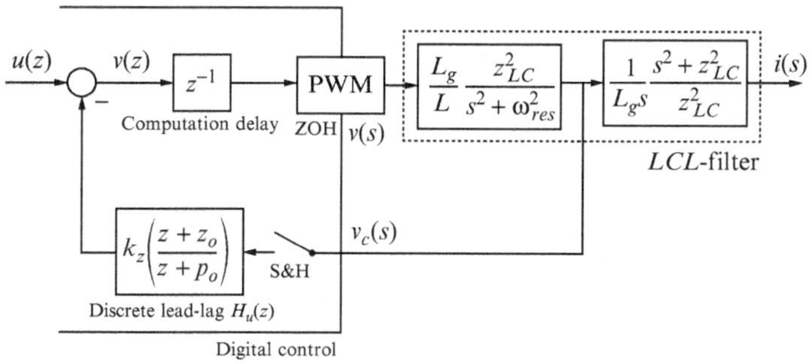

FIG. 8.13 Implementation of the lead-lag network for active damping in the digital controller [23].

$k_d C\omega_{res}$ and a phase close to 90° with zero slope at the resonance frequency just like it happens in Eq. (8.37). In Fig. 8.12, it can also be seen that the lead-lag network does not behave as a differentiator for $-90° < \varphi_{max} < 0$ ($k_f > 1$).

In practice, the lead-lag network for active damping is implemented in the digital domain, see Fig. 8.13. The synchronous sampling of the capacitor voltage requires symmetry in the PWM pulses to obtain a ripple free waveform, and so simple update mode ($T_s = T_{sw}$) is only used.

The digital implementation adds phase lags due to the computational and PWM delays. The discrete lead-lag network must behave as a perfect differentiator around the resonance frequency taking into account this extra phase lag. Hence, the phase shift at the resonance frequency must lead to 90° once the phase shifts due to the PWM and the computational delays are compensated:

$$90^{\circ} = \phi_{max} - 1.5T_s\omega_{res}\frac{360}{2\pi} \tag{8.39}$$

If the gain k_d in Eq. (8.38) is selected negative, a phase lag of 180° must be added in Eq. (8.39). A negative gain k_d in Eq. (8.38) is consistent with the original formulation of the lead-lag method [29], which used the positive feedback of the capacitor voltage filtered by the lead-lag network. A practical upper limit is $\varphi_{max} = 80°$ [12], corresponding to $f_s = 3.2 f_{res}$ Eq. (8.39). In addition, there is a lower limit $\varphi_{max} = 70°$ to make the lead-lag to have a perfect differentiator slope [23]. Hence, the frequency ratio for the lead-lag method should be selected in the following range:

$$\frac{f_s}{f_{res}} \approx 3.2 \div 3.4 \tag{8.40}$$

When this condition is fulfilled, simulations and experiments show that the resulting overall design results are very robust against the grid line variations [10]. To preserve the phase and amplitude characteristics at the resonance frequency, Eq. (8.38) must be discretized by using the bilinear transformation (Tustin method) with prewarping at $\omega = \omega_{res}$ [12,55].

For both cases, the capacitor-current feedback and the lead-lag network, root locus analysis in the z-plane must be used to select the gain k_d that results in an adequate damping for the closed-loop poles. A proper value for k_d consists in selecting the gain that results in a damping factor around $\zeta_{lf}=0.1$ approximately. If the lower-frequency poles are moved for this value, the system may result in excessive overshoots. In such case, a smaller gain k_d can be selected or the proportional gain K_p in Eq. (8.8) can be reduced provided that the bandwidth (Eq. 8.10) is still acceptable.

Fig. 8.14A shows the root locus in the z-plane for the closed loop poles of the current-control by varying k_d in the case of the capacitor-current feedback. The selected gain k_d results in a damping factor equal to $\zeta=0.1$ as previously explained. Fig. 8.15A shows the root locus in the z-plane for varying L_g from

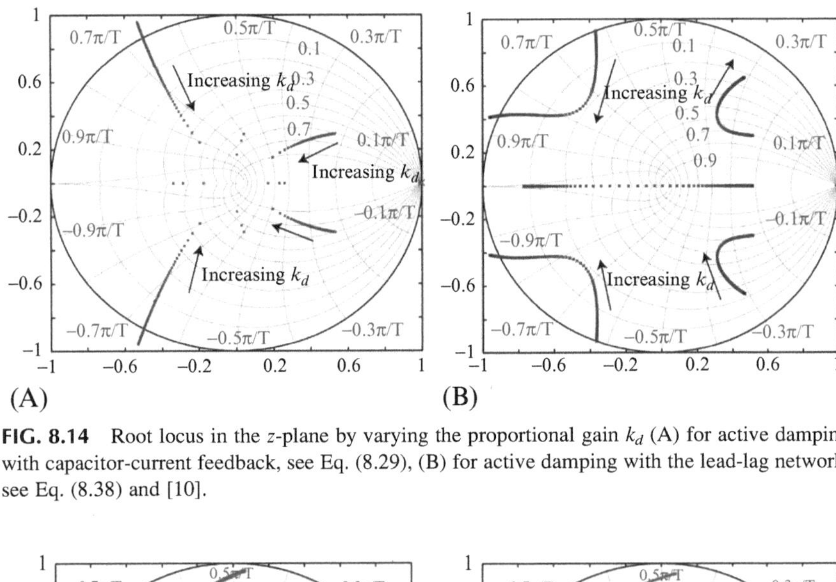

(A) (B)

FIG. 8.14 Root locus in the z-plane by varying the proportional gain k_d (A) for active damping with capacitor-current feedback, see Eq. (8.29), (B) for active damping with the lead-lag network, see Eq. (8.38) and [10].

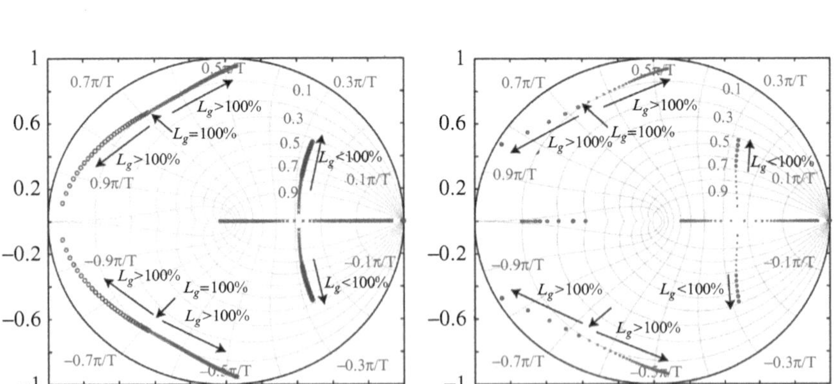

(A) (B)

FIG. 8.15 Root locus in the z-plane by varying L_g from 40% to 1000% for (A) active damping with capacitor-current feedback, (B) active damping with the lead-lag network [10].

40% to 1000%. The minimum value of 40% for L_g was selected considering inaccuracies in the measurements, temperature variations, and aging effects of the components. The maximum value of 1000% for L_g was selected considering the line impedance in a weak grid and with the presence of many parallel converters. The system remains stable for the full variation range. Fig. 8.14B shows the root locus in the *z*-plane for the closed-loop system in the case of the lead-lag network. Fig. 8.15B shows the root locus in the *z*-plane for varying L_g from 40% to 1000%, with the system also remaining stable for the full range.

It can also be seen in Fig. 8.14A and B that the active damping affects the low-frequency poles. For the case of the capacitor-current feedback the low-frequency poles result in larger damping. For the case of the lead-lag network, elevated values of k_d can cause the low-frequency poles to have little damping with large overshoot. In such cases, it may be necessary to reduce the proportional gain K_p at the expense of reducing the bandwidth. However, by limiting the damping factor of the resonance poles to a damping factor of $\zeta = 0.1$ the variations in the locations of the low-frequency poles are very small. If needed, the proportional gain K_p reductions are also very small.

Table 8.4 shows the parameters for the experimental setup shown in Fig. 8.16. The DC link is supplied by a programmable power source. The DC source by Delta Elektronika is unidirectional so that the power flows from the DC link to the grid. A Danfoss FC302 converter is directly connected to the grid through an isolating transformer, the leakage inductance of which corresponds to the grid inductance L_g. The overall system is controlled using a dSpace DS1103 controller board.

TABLE 8.4 **Parameters of the Experimental Setup**

Parameter	Symbol	Value
Rated power of VSI	S_n	2.2 kVA
Rated AC voltage	V_n	380 V
Rated frequency	f_n	50 Hz
DC link voltage	v_{DC}	650 V
Sampling frequency	f_s	8 kHz
PWM frequency	f_{sw}	8 kHz
Converter inductor	L	1.8 mH
Filter capacitor	C_f	4.7 μF
Grid inductor	L_g	1.8 mH
Resonance frequency of the filter	f_{res}	2.447 kHz

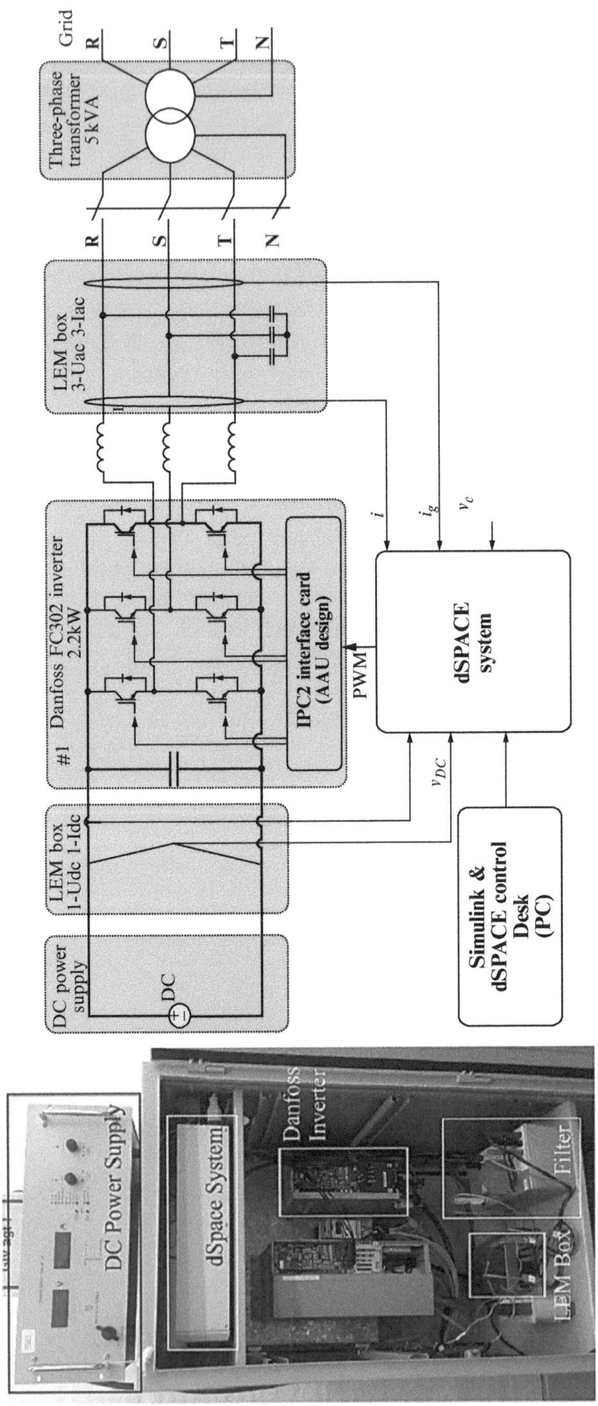

FIG. 8.16 Setup used for the testing the active damping experimentally.

The resistive component in the distribution-grid lines provides some damping so that the closed-loop current-control is initially stable without connecting the active damping mechanism. However, the oscillatory component at the resonance frequency is very prominent indicating the lack of sufficient damping. Upon the connection of the capacitor-current feedback, the resonance oscillation is canceled due to the active damping mechanism, see Fig. 8.17. To test the robustness of the proposed procedure, additional inductors of 6.9 mH are inserted between the converter and the isolated transformer. In this condition, the damping provided by the resistive distribution-grid is unable to stabilize the current-control when the active damping mechanism is not connected and the system protection trips. Therefore, the overall system must be started with the active damping mechanism connected. The experiment shown in Fig. 8.18 goes from inactive gate drivers to full rated generation in the converter. It can be seen that the capacitor-current feedback makes the current-control loop to be stable in this difficult condition.

Fig. 8.19 shows the experimental results for the lead-lag network upon the connection of the capacitor-current feedback where the initial resonance oscillation is also canceled due to the active damping. For the case of the lead-lag network, the experiment with additional 6.9 mH inductors is presented in Fig. 8.20. Also for this case, without the lead-lag network the damping provided by the resistive distribution-grid is unable to stabilize the current-control and the system protection trips. With the active damping mechanism connected the grid current displays considerable oscillation at the resonance frequency, but the current-control remains stable and the system protection does not trip. More analysis on this experiment can be found in Ref. [10].

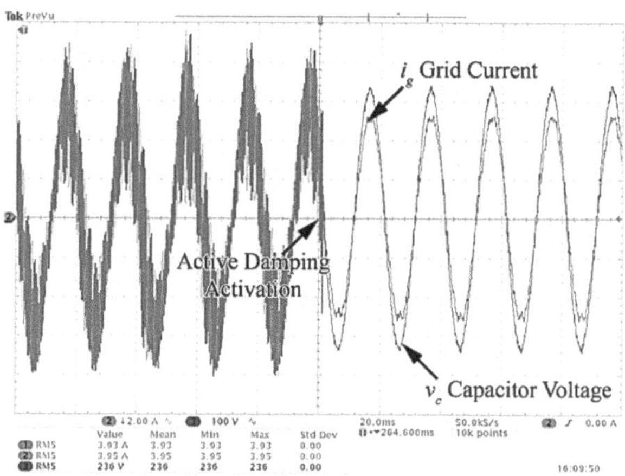

FIG. 8.17 Capacitor voltage (20 ms/div and 100 V/div) and grid current (20 ms/div and 2 A/div) upon the connection of the capacitor-current feedback [10].

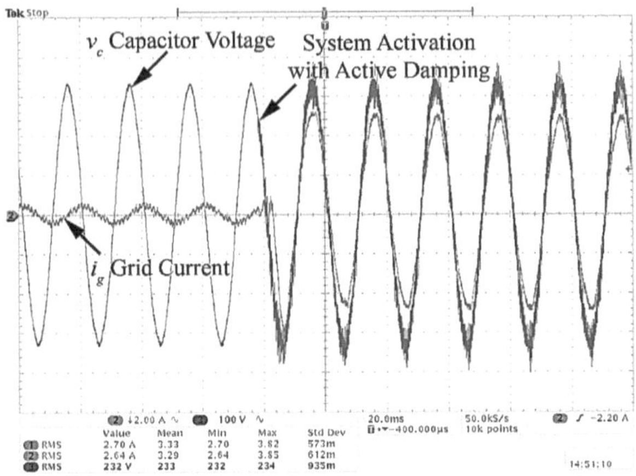

FIG. 8.18 Capacitor voltage (20ms/div and 100V/div) and grid current (20ms/div and 2A/div) from disconnection to connection of the *LCL*-filter-based converter for active damping with capacitor-current feedback [10].

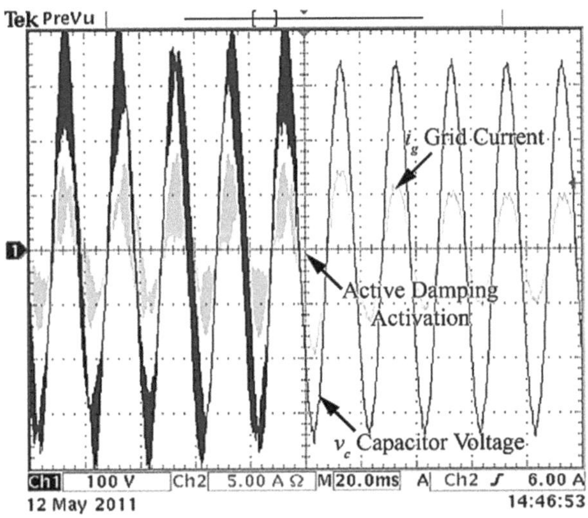

FIG. 8.19 Measured capacitor voltage (100V/div) and grid current (5 A/div) upon the connection of the lead-lag network [23].

8.5.3 Filter-Based Active Damping

Another kind of active damping explained in this chapter is based on simple filters on the reference voltage for the modulator, specifically notch filters. This procedure is simple to implement and does not require any additional sensors [3]. However, the interactions with the current-control loop make the notch filter tuning more difficult. The notch filter tuning procedure proposed in Ref. [50] uses complex genetic algorithms, and the formulas proposed in Ref. [56] require

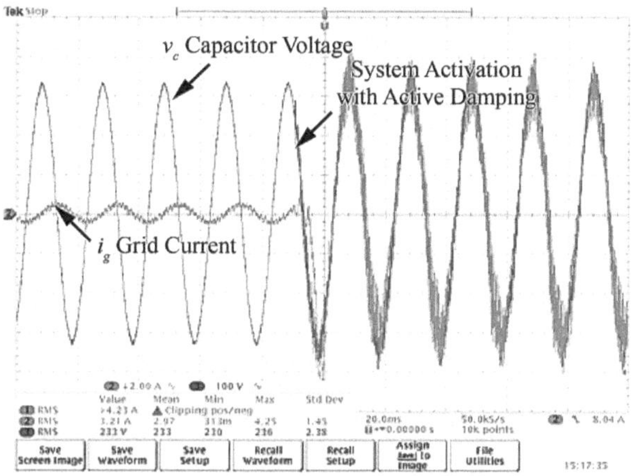

FIG. 8.20 Capacitor voltage (20ms/div and 100V/div) and grid current (20ms/div and 2A/div) from disconnection to connection of the *LCL*-filter-based converter for active damping with the lead-lag network [10].

numerous trial and error iterations. Once the notch frequency is properly tuned at the resonance frequency, the voltage reference does not contain any component susceptible of exciting the *LCL*-filter resonance. Grid impedance variations change the resonance frequency rendering the notch filter mistuned, which may compromise the *LCL*-filter stability. Ref. [17] uses a simple and straightforward tuning procedure for the notch filter and proposes to estimate the resonance frequency before commissioning in order to overcome the line-inductance variations.

Fig. 8.21 shows the usual cascaded control structure of the grid converter in the *dq*-frame including the notch filter on the reference voltage for the modulator. The generic transfer function for an analog notch filter is given as follows [46]:

$$N_n(s) = \left(\frac{s^2 + 2D_z\omega_{nf}s + \omega_{nf}^2}{s^2 + 2D_p\omega_{nf}s + \omega_{nf}^2} \right)^n \quad (8.41)$$

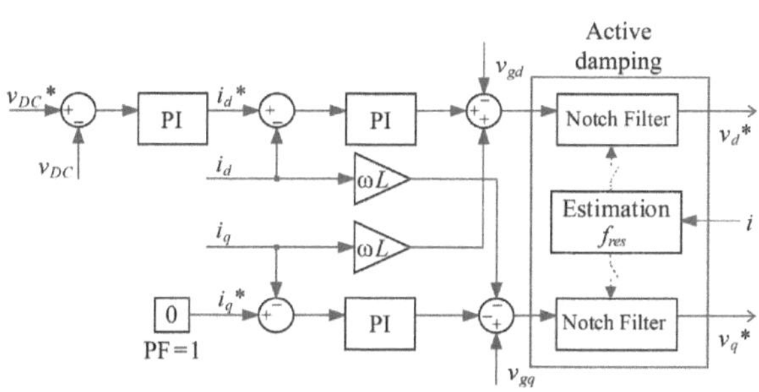

FIG. 8.21 Block diagram of the converter control with active damping based on the notch filter.

where ω_{nf} is the notch frequency, D_z and D_p are the damping factors for the complex conjugates poles and zeros, respectively, and n is the number of sections. The open-loop transfer function of the converter current-control in Fig. 8.21 for the continuous-time domain and without considering the notch filter is:

$$G_{ol}(s) = G_{PI}(s)G_d(s)G(s) \tag{8.42}$$

where $G(s)$ is the transfer function related to the converter voltage v and current i, $G_d(s)$ models the computational and PWM delays, and finally, $G_{PI}(s)$ is the PI controller. The frequency for $-180°$ phase shift in the open-loop transfer function is near the resonance frequency [29]. To achieve stability with a positive gain margin, the resonant peak should be below unity (0 dB). Therefore, the inclusion of a notch filter $N_n(s)$ in cascade to the PI controller output should result in:

$$|G_{ol}(s)N_n(s)|_{s=j\omega_{res}} < 1\,(0\,\text{dB}) \tag{8.43}$$

It is clear that the notch frequency must be the resonant frequency $\omega_{nf}=\omega_{\text{res}}$ and the condition for stability is $D_z/D_p < |G_{ol}(s)|_{s=j\omega_{res}}^{-1}$. As a digital implementation will be used for the notch filter, absolute cancellation at the resonance frequency $|N_n(j\omega_{\text{res}})|=0$ $(-\infty$ dB) is perfectly possible by setting $D_z=0$. Hence, there will be no component present in the voltage reference able to excite the *LCL*-filter resonance and Eq. (8.43) is always fulfilled. To preserve the null amplitude when discretizing Eq. (8.41) the bilinear transformation [12] (Tustin method) with prewarping at $\omega=\omega_{res}$ must be used.

For the low-frequency range $\omega \ll \omega_{\text{res}}$ the behavior of the notch filter with $D_z=0$ can be approximated to n first-order systems with the time constant $\tau=2D_p/\omega_{nf}$, and the module of $G_{ol}(s)$ Eq. (8.42) is much less affected than the phase angle due to the presence of $N_n(s)$ [17]. Therefore, the gain crossover frequency ω_{gc} of the open-loop transfer function will be very close to that of the low-frequency model in Eq. (8.12). Conversely, the phase margin will be inferior to that of the low-frequency model (Eq. 8.12) due to the phase delay of $N_n(s)$. The parameter D_p can be adjusted to result in a phase margin reduction ΔPM as follows:

$$D_p = \frac{1}{2}\tan\left(\frac{\Delta PM}{n}\frac{2\pi}{360}\right)\left(\frac{\omega'_{gc}}{\omega_{nf}} - \frac{\omega_{nf}}{\omega'_{gc}}\right) \tag{8.44}$$

where ω_{gc}' is the cross-over frequency (Eq. 8.12) taking into account the bilinear transformation using prewarping:

$$\omega'_{gc} = \omega_{nf}\frac{\tan\left(\omega'_{gc}T_s/2\right)}{\tan\left(\omega_{nf}T_s/2\right)} \tag{8.45}$$

If the notch filter width is very narrow, the system is too sensible to the variations in the resonance frequency. Fig. 8.22 shows the notch filter amplitude

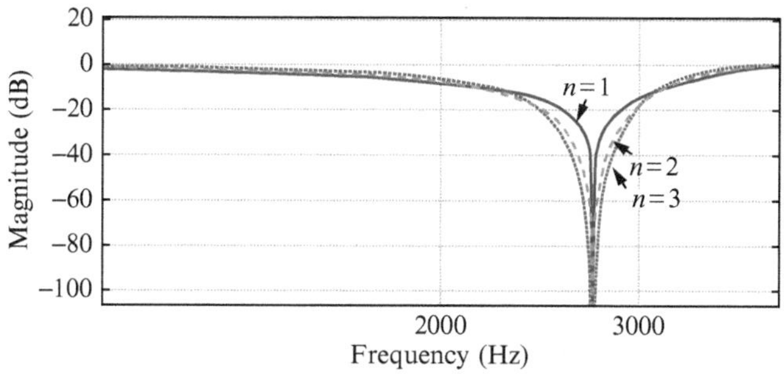

FIG. 8.22 Notch filter magnitudes for increasing number of sections n; see Eq. (8.41) [17].

for an increasing number of sections n and with the same ΔPM in Eq. (8.44). The notch width is larger for an increasing number of sections n, and so the system is more robust to the resonance frequency variation. However, increasing the number of sections n requires additional computations. Selecting $n=2$ [17] results in a proper compromise between robustness and computational burden.

The low-frequency dynamics of the closed-loop system behaves almost like a second-order system with the phase margin PM_{lf} approximately related to the damping factor according to $\xi = PM_{lf}/100$. A reduction in phase margin leads to a reduction in the damping factor and so an increased overshoot that may not be acceptable. Reducing the proportional gain K_p increases the phase gain achieving an acceptable overshoot at the expense of reducing the bandwidth. This reduction in bandwidth can be overcome by using a double update mode in the control system. A lower bound for the necessary gain reduction can be obtained from Eq. (8.12) by precompensating the resulting decrease in the phase gain ΔPM:

$$\%K_p = 100\left(1 - \frac{\pi}{90}\Delta PM\right) \tag{8.46}$$

Hence, the phase gain reduction ΔPM should not be selected higher than 15 degrees in order not to reduce the bandwidth more than 50%. Thus, selecting a phase margin reduction of no more than $\Delta PM = 15^{\circ}$ results in a proper trade-off between robustness to the resonance variations and a proper low-frequency behavior.

When a large grid impedance variation occurs [57], the resonance frequency changes and the notch filter is tuned at an erroneous frequency. The notch filter is not able to cancel the voltage reference components susceptible of exciting the *LCL*-filter and the stability will be compromised. In Ref. [58], it is proposed to estimate the resonance frequency first and later to use it for tuning the notch filter to take into account the grid inductance variations.

There are many procedures to estimate the grid inductance, but it must be taken into account that for this case the power electronic converter is only available after the notch filter is properly tuned. Therefore, the resonance is excited in a controlled manner, and its frequency is identified in the spectrum by using Fourier analysis [58]. Without the notch filter, the inductor resistances R and R_g provide normally passive damping which is sufficient to achieve stability for very low proportional gain K_p in the PI controller. The provided damping is so little that resonance excitation will result in large oscillations. Considering again $G_{ol}(s)$ in Eq. (8.42) and the inductor resistances, also for this case the phase for the resonance frequency is near $-180°$. Therefore, the condition for stability resulting into a positive gain margin is:

$$|G_{ol}(s)|_{s=j\omega_{res}} = |G_{PI}(s)G_d(s)G(s)|_{s=j\omega_{res}} < 1\,(0\text{dB}) \tag{8.47}$$

After some algebraic manipulations and simplifications, the maximum proportional gain that results in a stable control is:

$$K_{pre} = R + R_g\left(\frac{L}{L_g}\right)^2 \tag{8.48}$$

This estimation (Eq. 8.48) can be used as a guideline to gradually increase K_p until exciting the resonance but without the system to become unstable. Once the presence of the resonance is evident, its frequency can be calculated by using the FFT. The FFT is usually included in the standard libraries for DSP coding. However, it may consume large computational resources. The Goertzel algorithm is used in Ref. [17] to calculate the FFT coefficients necessary for identifying the resonance frequency reducing and thereby computational resources and memory.

Fig. 8.23 shows the root loci in the z-plane of the overall system with notch filter for $n=2$ when varying the grid inductance L_g. For small values of L_g the

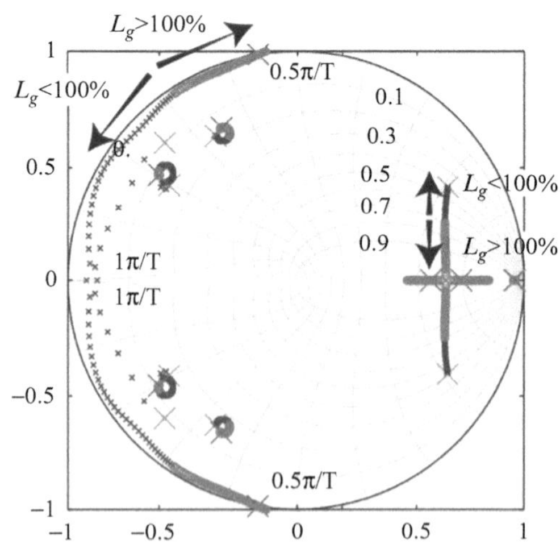

FIG. 8.23 Root locus in the z-plane of the overall system using notch filter with $n=2$, see Eq. (8.41), when varying the grid inductance L_g [17].

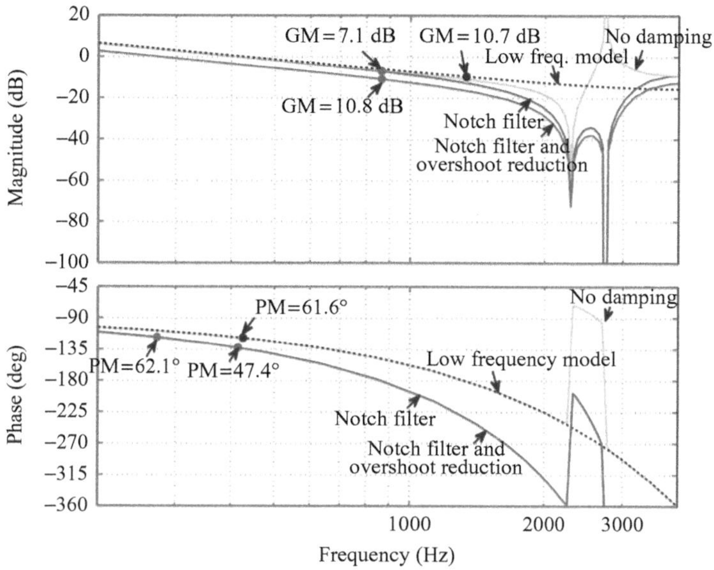

FIG. 8.24 Bode plots of the converter current-control [17].

overall system is stable up to $L_g > 210\%$ where instability occurs. For larger variations in the grid inductance L_g, the resonance frequency should be measured before commissioning the converter.

Fig. 8.24 shows the Bode plots for the converter current-control with no damping, the notch filter ($\Delta PM = 15^{\circ}$) and the notch filter ($\Delta PM = 15^{\circ}$) with proportional gain K_p for overshoot reduction. Fig. 8.24 shows that the resonant peak is completely canceled by the notch filter. The K_p reduction for 4% overshoot was 65% consistent with the lower bound (Eq. 8.46).

Fig. 8.25 shows a screen capture for the converter current and its Fourier spectrum when the resonance is being excited. The detected resonance

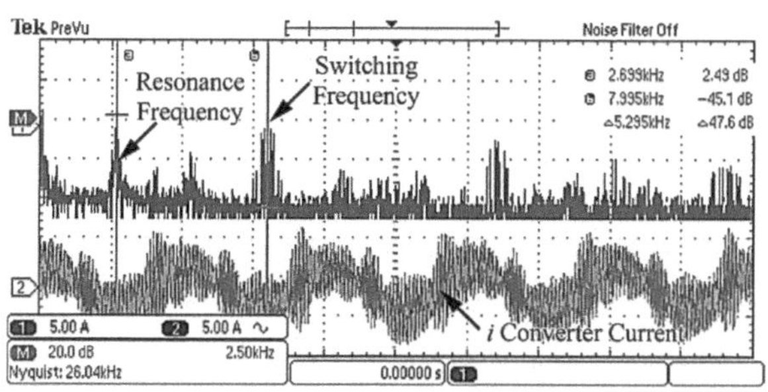

FIG. 8.25 Oscilloscope screen capture showing the inverter current (lower channel, time base 10 ms/div and 5 A/div) and its FFT (upper channel) [17] when the resonance frequency detection is in operation.

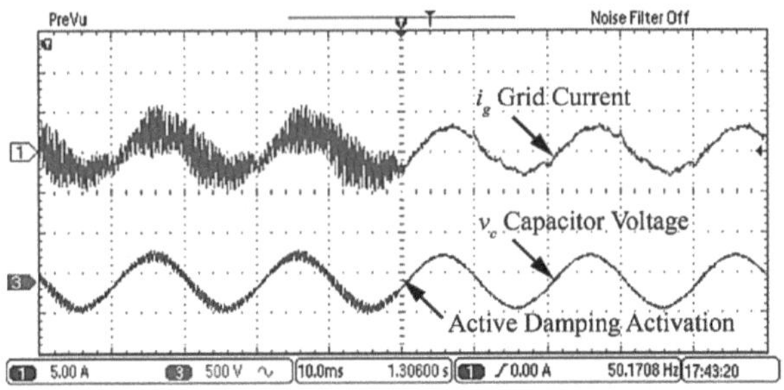

FIG. 8.26 Oscilloscope screen capture for the grid current (upper channel, 10 ms/div and 5 A/div) and capacitor voltage (lower channel, 10 ms/div and 500 V/div) upon the connection of the notch filter [17].

frequency is 2695 Hz according to the Goertzel algorithm implemented in the dSpace system, and 2699 Hz according to the oscilloscope using FFT (see Fig. 8.25). Once the resonance frequency is detected, all the parameters of the notch filter are calculated. The oscillations are effectively damped upon connection the active damping mechanism based on the notch filter, see Fig. 8.26.

Finally, more recent advances on notch filter tuning for robust design of *LCL*-filters using direct digital techniques can be found in Ref. [59].

8.6 CONCLUSIONS

This chapter deals with the voltage source converters connected to the grid through an *LCL*-filter. The main purpose is to reduce the high inductance that would otherwise be present in a conventional *L*-filter. The chapter is dealing with practical procedures based on the author experience instead of an exhaustive listing of all the methods available in the literature. The method for the *LCL*-filter design is well proven and straightforward. After considering the constraints in the sizing of the passive elements, the *LCL*-filter is designed by following several steps and requiring only a few iterations to be determined.

Because of its simplicity, passive damping is still widely used in the industry despite the additional losses and thereby lower efficiency. In order to reduce the passive damping losses, extra passive elements can be used to divert the current from the damping resistor. The formulas explained in the chapter allow obtaining a first approximation for the losses incurred in the passive damping methods.

Literature also contains many procedures for active damping, but those well-known and based on simple transfer functions are only considered in this chapter. The capacitor-current feedback needs an accurate additional sensor, but it is very simple to implement requiring only a sum and a gain. The lead-lag network method does not need additional sensors as it uses the capacitor voltage sensors required for synchronization for the capacitor-current estimation. When selecting the proper ratio between switching and resonance frequencies, both procedures result in robustness against the grid inductance variations. The last simple active damping procedure explained in the chapter uses a notch filter at the voltage reference output to the modulator. The notch filter blocks all the components that may excite the *LCL*-filter resonance. The tuning procedure for the notch filter is straightforward, but the overall system may not result in sufficient robustness for large grid inductance variations. For such situation, the resonance frequency is estimated before commissioning the converter.

Finally, the basic techniques explained in this chapter can serve as a stepping stone for more sophisticated algorithms present in the literature and for future algorithms resulting from the ongoing research on *LCL*-filter-based grid converters.

REFERENCES

[1] IEEE Recommended Practice and Requirements for Harmonic Control in Electric Power Systems—Redline, in: IEEE Std 519-2014 (Revision of IEEE Std 519-1992), June 11, 2014, pp. 1–213.

[2] M. Liserre, F. Blaabjerg, S. Hansen, Design and control of an *LCL*-filter based three-phase active rectifier, IEEE Trans. Ind. Appl. 41 (5) (2005) 1281–1291.

[3] M. Liserre, F. Blaabjerg, A. DellAquila, Step-by-step design procedure for a grid-connected three-phase PWM voltage source converter, Int. J. Electron. 91 (8) (2004) 445–460.

[4] A. Reznik, M. Simoes, A. Al-Durra, S. Muyeen, LCL filter design and performance analysis for grid-interconnected systems, IEEE Trans. Ind. Appl. 50 (2) (2014) 1225–1232.

[5] K. Jalili, S. Bernet, Design of LCL filters of active-front-end two-level voltage-source converters, IEEE Trans. Ind. Electron. 56 (5) (2009) 1674–1689.

[6] P. Channegowda, V. John, Filter optimization for grid interactive voltage source inverters, IEEE Trans. Ind. Electron. 57 (12) (2010) 4106–4114.

[7] J. Muhlethaler, M. Schweizer, R. Blattmann, J. Kolar, A. Ecklebe, Optimal design of LCL harmonic filters for three-phase PFC rectifiers, IEEE Trans. Power Electron. 28 (7) (2013) 3114–3125.

[8] D. Pan, X. Ruan, C. Bao, W. Li, X. Wang, Magnetic integration of the LCL filter in grid-connected inverters, IEEE Trans. Power Electron. 29 (4) (2014) 1573–1578.

[9] Q. Liu, L. Peng, Y. Kang, S. Tang, D. Wu, Y. Qi, A novel design and optimization method of an LCL filter for a shunt active power filter, IEEE Trans. Ind. Electron. 61 (8) (2014) 4000–4010.

[10] R. Peña-Alzola, M. Liserre, F. Blaabjerg, M. Ordonez, Y. Yang, LCL-filter design for robust active damping in grid-connected converters, IEEE Trans. Ind. Inf. 10 (4) (2014) 2192–2203.

[11] D.G. Holmes, T.A. Lipo, Pulse Width Modulation for Power Converters: Principles and Practice, Wiley-IEEE Press, Hoboken, NJ, USA, 2003, . ISBN 978-0-471-20814-3pp. 291–294 (Chapter 6).

[12] M. Liserre, A. Dell'Aquila, F. Blaabjerg, Stability improvements of an *LCL*-filter based three-phase active rectifier, Proc. of 2002 IEEE 33rd Annu. Power Electron. Specialists Conf., PESC'02, vol. 3, 2002, , pp. 1195–1201.

[13] R. Sebastian, R. Peña-Alzola, Simulation of an isolated Wind Diesel System with battery energy storage, Electr. Power Syst. Res. 81 (2) (2011) 677–686.

[14] S. Buso, P. Mattavelli, Digital Control in Power Electronics, Morgan and Claypool Publishers, San Rafael, CA, USA, 2006 1-598-29112-2.

[15] V. Blasko, V. Kaura, A new mathematical model and control of a three-phase AC-DC voltage source converter, IEEE Trans. Power Electron. 12 (1) (1997) 116–123.

[16] M.P. Kazmierkowski, M. Jasinski, G. Wrona, DSP-based control of grid-connected power converters operating under grid distortions, IEEE Trans. Ind. Inf. 7 (2) (2011) 204–211.

[17] R. Peña-Alzola, M. Liserre, F. Blaabjerg, M. Ordonez, T. Kerekes, A self-commissioning notch filter for active damping in a three-phase LCL-filter-based grid-tie converter, IEEE Trans. Power Electron. 29 (12) (2014) 6754–6761.

[18] J. Yin, S. Duan, B. Liu, Stability analysis of grid-connected inverter with LCL-filter adopting a digital single-loop controller with inherent damping characteristic, IEEE Trans. Ind. Inf. 9 (2) (2013) 1104–1112.

[19] S.G. Parker, B.P. McGrath, D.G. Holmes, Regions of active damping control for LCL filters, IEEE Trans. Ind. Appl. 50 (1) (2014) 424–432.

[20] J. Dannehl, C. Wessels, F.W. Fuchs, Limitations of voltage-oriented PI current control of grid-connected PWM rectifiers with LCL filters, IEEE Trans. Ind. Electron. 56 (2) (2009) 380–388.

[21] J.L. Agorreta, M. Borrega, J. López, L. Marroyo, Modeling and control of n-paralleled grid-connected inverters with LCL filter coupled due to grid impedance in PV plants, IEEE Trans. Power Electron. 26 (3) (2011) 770–785.

[22] W.Y. Yang, W. Cao, T.S. Chung, J. Morris, Applied Numerical Methods Using Matlab, (B8.5.1st edition)Wiley-Interscience, Hoboken, NJ, USA, 2005. ISBN 978-0-471-69833-3.

[23] R. Peña-Alzola, M. Liserre, F. Blaabjerg, R. Sebastian, J. Dannehl, F.W. Fuchs, Systematic design of the lead-lag network method for active damping in LCL-filter based three phase converters, IEEE Trans. Ind. Inf. 10 (1) (2014) 43–52.

[24] K.H. Ahmed, S.J. Finney, B.W. Williams, Passive filter design for three-phase inverter interfacing in distributed generation, Proc. of Compatibility in Power Electron., CPE '07, 2007, , pp. 1–9.

[25] T.C.Y. Wang, Z. Ye, G. Sinha, X. Yuan, Output filter design for a grid-interconnected three-phase inverter, Proc. of IEEE 34th Annu. Power Electron. Specialist Conf., PESC '03, 15–19 June, vol. 2, 2003, , pp. 779–784.

[26] R.N. Beres, X. Wang, F. Blaabjerg, M. Liserre, C.L. Bak, Optimal design of high-order passive-damped filters for grid-connected applications, IEEE Trans. Power Electron. 31 (3) (2016) 2083–2098.

[27] R. Peña-Alzola, M. Liserre, F. Blaabjerg, R. Sebastian, J. Dannehl, F. Fuchs, Analysis of the passive damping losses in LCL-filter-based grid converters, IEEE Trans. Power Electron. 28 (6) (2013) 2642–2646.

[28] W. Wu, Y. He, F. Blaabjerg, An LLCL power filter for single-phase grid-tied inverter, IEEE Trans. Power Electron. 27 (2) (2012) 782–789.

[29] V. Blasko, V. Kaura, A novel control to actively damp resonance in input LC filter of a three-phase voltage source converter, IEEE Trans. Ind. Appl. 33 (2) (1997) 542–550.

[30] P.A. Dahono, A control method to damp oscillation in the input LC filter of AC-DC PWM converters, Proc. PESC'02, June, 2002, , pp. 1630–1635.

[31] X. Wang, F. Blaabjerg, P.C. Loh, Virtual RC damping of LCL filtered voltage source converters with extended selective harmonic compensation, IEEE Trans. Power Electron. 30 (9) (2015) 4726–4737.

[32] X. Wang, F. Blaabjerg, P.C. Loh, Grid-current-feedback active damping for LCL resonance in grid connected voltage-source converters, IEEE Trans. Power Electron. 31 (1) (2016) 213–223.

[33] V. Miskovic, V. Blasko, T.M. Jahns, A.H.C. Smith, C. Romenesko, Observer-based active damping of LCL resonance in grid-connected voltage source converters, IEEE Trans. Ind. Appl. 50 (6) (2014) 3977–3985.

[34] Y. Lyu, H. Lin, Y. Cui, Stability analysis of digitally controlled LCL-type grid-connected inverter considering the delay effect, IET Power Electron. 8 (9) (2015) 1651–1660.

[35] C. Chen, J. Xiong, Z. Wan, J. Lei, K. Zhang, A time delay compensation method based on area equivalence for active damping of an LCL-type converter, IEEE Trans. Power Electron. 32 (1) (2017) 762–772.

[36] R. Guzman, L.G. de Vicua, J. Morales, A. Momeneh, J. Miret, J. Torres-Martinez, Active damping based on Ackermann's formula for a three-phase voltage source inverter with LCL filter, Proc. of IECON 2015 41st Annual Conference of the IEEE Ind. Electron. Soc., November, 2015, , pp. 1217–1222.

[37] D. Pérez-Estévez, J. Doval-Gandoy, A.G. Yepes, Ó. López, Positive- and negative-sequence current controller with direct discrete-time pole placement for grid-tied converters with LCL filter, IEEE Trans. Power Electron. 32 (9) (2017) 7207–7221.

[38] F. Huerta, D. Pizarro, S. Cobreces, F.J. Rodriguez, C. Giron, A. Rodriguez, LQG servo controller for the current control of LCL grid connected voltage-source converters, IEEE Trans. Ind. Electron. 59 (11) (2012) 4272–4284.

[39] J. Roldan-Perez, A. Garcia-Cerrada, J.L. Zamora-Macho, M. Ochoa-Gimenez, Helping all generations of photo-voltaic inverters ride-through voltage sags, IET Power Electron. 7 (10) (2014) 2555–2563.

[40] M. Andresen, M. Liserre, F.W. Fuchs, N. Hoffmann, Design of a grid adaptive controller for PWM converters with LCL filters, Proc. of IECON 2015—41st Annual Conf. of the IEEE Ind. Electron. Soc. November, 2015, , pp. 3664–3671.

[41] J. Kukkola, M. Hinkkanen, State observer for grid-voltage sensorless control of a converter equipped with an LCL filter: direct discrete-time design, IEEE Trans. Ind. Appl. 52 (4) (2016) 3133–3145.

[42] H. Komurcugil, S. Ozdemir, I. Sefa, N. Altin, O. Kukrer, Sliding mode control for single-phase grid-connected LCL-filtered VSI with double-band hysteresis scheme, IEEE Trans. Ind. Electron. 63 (2) (2016) 864–873.

[43] R. Guzman, L.G. de Vicuña, J. Morales, M. Castilla, J. Miret, Model-based active damping control for three-phase voltage source inverters with LCL filter, IEEE Trans. Power Electron. 32 (7) (2017) 5637–5650.

[44] J. Dannehl, F.W. Fuchs, S. Hansen, P.B. Thogersen, Investigation of active damping approaches for pi-based current control of grid connected pulse width modulation converters with LCL filters, IEEE Trans. Ind. Appl. 46 (4) (2010) 1509–1517.

[45] X. Bao, F. Zhuo, Y. Tian, P. Tan, Simplified feedback linearization control of three-phase photovoltaic inverter with an LCL filter, IEEE Trans. Power Electron. 28 (6) (2013) 2739–2752.

[46] J. Dannehl, M. Liserre, F.W. Fuchs, Filter-based active damping of voltage source converters with LCL filter, IEEE Trans. Ind. Electron. 58 (8) (2011) 3623–3633.

[47] J.Y. Hung, Feedback control with posicast, IEEE Trans. Ind. Electron. 50 (1) (2003) 94–99.

[48] A. Vidal, A.G. Yepes, F.D. Freijedo, O. Lopez, J. Malvar, F. Baneira, J. Doval-Gandoy, A method for identification of the equivalent inductance and resistance in the plant model of current-controlled gridtied converters, IEEE Trans. Power Electron. 30 (12) (2015) 7245–7261.

[49] X. Fu, S. Li, Control of single-phase grid-connected converters with LCL filters using recurrent neural network and conventional control methods, IEEE Trans. Power Electron. 31 (7) (2016) 5354–5364.

[50] M. Liserre, A. Aquila, F. Blaabjerg, Genetic algorithm-based design of the active damping for an LCL-filter three-phase active rectifier, IEEE Trans. Power Electron. 19 (1) (2004) 76–86.

[51] W. Gullvik, L. Norum, R. Nilsen, Active damping of resonance oscillations in LCL-filters based on virtual flux and virtual resistor, Proc. of European Conf. on Power Electron. and Appl., 2–5 September, 2007, , pp. 1–10.

[52] C.P. Dick, S. Richter, M. Rosekeit, J. Rolink, R.W. De Doncker, Active damping of LCL resonance with minimum sensor effort by means of a digital infinite impulse response filter, Proc. of 2007 European Conf. on Power Electron. and Appl., 2–5 September, 2007, , pp. 1–8.

[53] C.L. Phillips, R.D. Harbor, Feedback Control Systems, fourth ed., Prentice Hall, Upper Saddle River, NJ, USA, 2000 ISBN: 0139490906.

[54] A. Xu, Z. Xu, S. Xie, M. Zou, Study on dual-loop grid current control scheme for grid-connected inverter with an *LCL*-filter, Proc. of 4th IEEE Conf. on Ind. Electron. and Appl., 2009. ICIEA 2009, 25–27 May, 2009, , pp. 3200–3203.

[55] G.F. Franklin, J.D. Powell, M.L. Workman, Digital Control of Dynamic Systems, third ed., Addison Wesley Longman, Boston, MA, USA, 1997. ISBN 0-201-82054-4.

[56] J. Dannehl, M. Liserre, F. Fuchs, Filter-based active damping of voltage source converters with LCL filter, IEEE Trans. Ind. Electron. 58 (8) (2011) 3623–3633.

[57] M. Liserre, R. Teodorescu, F. Blaabjerg, Stability of grid-connected PV inverters with large grid impedance variation, Proc. Of IEEE 35th Annu. Power Electron. Spec. Conf. PESC'04, vol. 6, 2004, , pp. 4773–4779.

[58] M. Liserre, F. Blaabjerg, R. Teodorescu, Grid impedance estimation via excitation of LCL-filter resonance, IEEE Trans. Ind. Appl. 43 (5) (2007) 1401–1407.

[59] W. Yao, Y. Yang, X. Zhang, F. Blaabjerg, P.C. Loh, Design and analysis of robust active damping for LCL filters using digital notch-filters, IEEE Trans. Power Electron. 32 (3) (2017) 2360–2375.

Chapter 9

Modeling and Control of PV Systems

Yongheng Yang*, Katherine A. Kim† and Tao Ding‡
**Aalborg University, Aalborg, Denmark, †Ulsan National Institute of Science and Technology (UNIST), Ulsan, Korean, ‡Xi'an Jiaotong University, Xi'an, China*

9.1 INTRODUCTION

Along with the fast growth in the global economy, the consumption of conventional fossil-based energy resources (e.g., coal, gas, and oil) has also been increasing. Energy consumption was high in the past due to modernization and industrialization, and it is still increasing in today's society. Because of the increased energy demand coupled with limited fossil-based energy resources, many nations are turning to sustainable energy. Many national and international strategic energy plans are being developed so that societal energy consumption can be sustainably maintained while not causing adverse impacts on the environment. Seen from this standpoint, renewable energy is very promising, and it has already gained much popularity worldwide. In 2016, the global total renewable energy capacity was more than 2000 GW [1], as shown in Fig. 9.1. It can also be seen that solar energy is the third place in terms of total capacity. At present, the two main technologies for converting the solar energy into electricity are solar photovoltaic (PV) and concentrated solar power. Between the two types, solar PV power is more often seen in today's energy conversion systems. This chapter will focus on solar PV technology and how to control its operation to fully utilize its produced power.

Abundant solar energy can be directly converted into DC electricity using semiconducting materials (e.g., crystalline Silicon), called a PV cell, which operates based on the photovoltaic effect [2]. However, the voltage level of a single PV cell is quite low (typically about 0.5 V) compared with the voltage required for many loads (e.g., batteries in an electric car, a lamp, and the electric ac grid). Hence, a number of PV cells are connected in series to increase the voltage level to construct a PV module, as illustrated in Fig. 9.2. In practice, the PV modules and panels are also connected in series to increase voltage, in parallel to increase current, or in a combination of both, which enables the powering of heavy loads

Control of Power Electronic Converters and Systems. https://doi.org/10.1016/B978-0-12-805245-7.00009-3

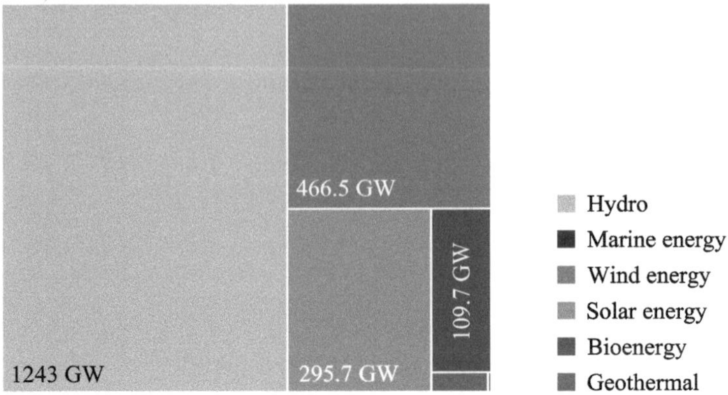

FIG. 9.1 Global renewable power capacity composition in 2016 (more than 2000 GW in total) [1], where hydro power also includes pumped storage and mixed plants; solar energy includes solar photovoltaic (290.8 GW) and concentrated solar power (4.7 GW).

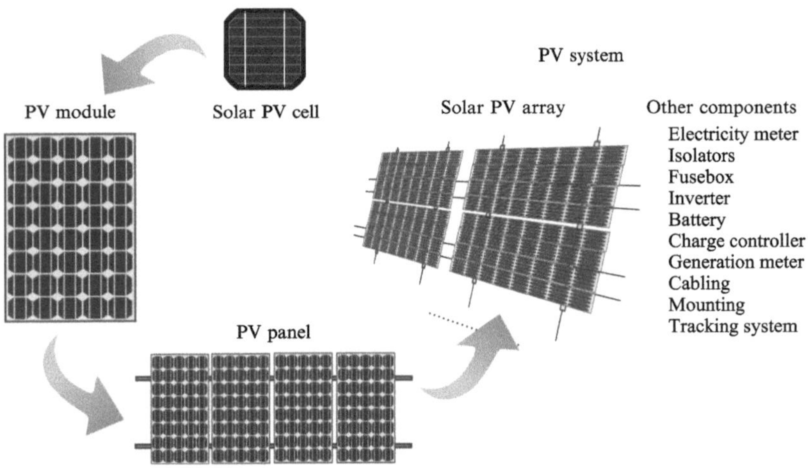

FIG. 9.2 Components of a typical PV system: from solar PV cells to a PV system. The figure is adapted from the figure published on Wikimedia. Online available: https://commons.wikimedia.org/wiki/File:From_a_solar_cell_to_a_PV_system.svg.

(high-voltage and high-power applications), as shown in Fig. 9.2. A PV system consists of, not only solar PV arrays, but many components like power converters (e.g., inverter), AC and DC isolators, and controllers.

Although the PV modules can directly power some DC loads, in most cases, power electronic converters are adopted in PV systems to improve power utilization [3]. There are two main reasons for utilizing power electronic converters: maximum power point tracking and dc-ac power conversion. PV output power is dependent on the varying environmental conditions (i.e., solar irradiance and ambient temperature), such that a maximum power point tracking (MPPT) scheme should be employed to optimize energy capture. Additionally, for PV systems with AC loads, power electronic converters act as the link between the DC input power and the AC loads. That is, the power electronic converters are crucial to the proper operation of PV systems and control strategies for those converters should be developed to ensure the above functionalities of PV systems.

Because power electronic converters consume some power to operate properly, the efficiency of the entire PV power conversion will be affected. Topological innovations and advancements in power electronic semiconductors can contribute to an improved efficiency of PV system power converters. However, this may also complicate the overall control for the PV systems. Nevertheless, dedicated control algorithms should be investigated for power converters in PV systems to ensure efficient and reliable operation [4]. To this end, modeling is a required step to appropriately develop and evaluate control schemes.

Hence, this chapter presents the modeling of PV systems (mainly the modeling of PV modules). Commonly-used MPPT algorithms are also introduced to show how they maximize PV power production. As examples, MPPT control is implemented in both single- and double-stage PV systems based on the models of the power converters (i.e., the DC/AC inverter) described in Chapter 6 of this book. Control design is demonstrated in a practical PV system example.

9.2 PV MODULE MODELING

9.2.1 PV Cell Basics—Operating Principle

PV cells, or solar cells, are semiconductor devices. In the 1950s, PV cells were initially introduced to power space satellites, but since the 1970s, they have been used in terrestrial applications as well [5,6]. Along with the continuous price drop of semiconductor materials, today, PV cells can be seen in a wide variety of applications (e.g., grid-connected systems, buildings, off-grid areas). Fig. 9.3 exemplifies a roof- mounted PV system in Aalborg, Denmark, where a number of PV panels are shown and they have been connected in series or in

FIG. 9.3 Example of a roof-mounted PV system at Nordjysk Kollegium in Aalborg, Denmark.

parallel. It also shows that the surrounding environment (e.g., partial shading by trees) can negatively affect the PV power production.

PV cells can be fabricated from many different types of materials using various techniques. In terms of PV cell materials, crystalline silicon (Si) is the most commercialized. This is partially due to the high availability of low-cost silicon PV materials that has prevented new and emerging cell types from gaining significant presence in the PV market. Nevertheless, there are various emerging technologies like dye-sensitized, perovskite, organic, and quantum PV cells that have been tested in laboratories; however, initially, the efficiency is low and the cost is high [7]. Hence, intensive research is undergoing with a focus on both increasing efficiency and decreasing manufacturing cost [8]. This section will introduce the basic characteristics and operating principles of crystalline Si PV cells along with some basic considerations for designing PV systems.

A solar PV cell is actually fabricated as a large-area p-n semiconductor (e.g., Si) junction. The power generation from a PV cell is achieved by exploiting the *photovoltaic effect* that converts solar energy into electrical energy as illustrated in Fig. 9.4. The p-n junction creates a depletion region with an electric field through migration (i.e., electrons into the p-layer and holes into the n-layer), as demonstrated in Fig. 9.4. The electric field then provides a force opposing the continuous exchanging process. When a photon with sufficient energy (i.e., sunlight) hits the material, an electron will be excited by the photon and then enters into the conduction band, leaving a hole in the valence band. Due to the built-in electric field of the depletion region, the electron and hole will be swept from the junction in opposite directions. That is, holes will further concentrate toward the anode; electrons will move toward the cathode, as illustrated in Fig. 9.4, which are separated by the depletion region (the electric field). When an external path is created (e.g., the anode and cathode are connected through a bulb), the electrons will flow to "fill" the holes, and then a DC current is generated.

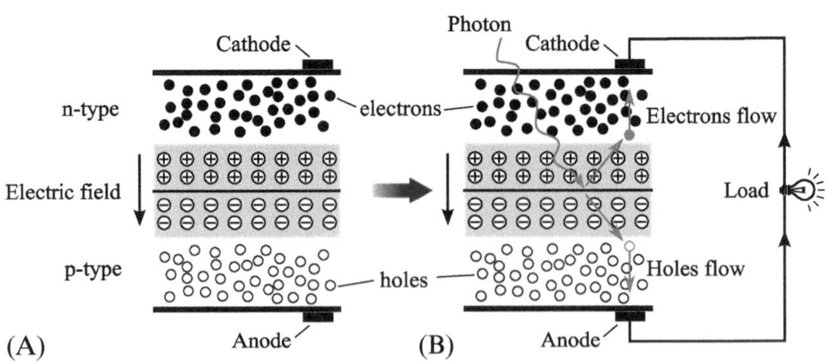

FIG. 9.4 Photovoltaic effect: (A) structure of a photovoltaic cell creating a depletion region and (B) a photon generating an electron-hole pair in a photovoltaic cell.

9.2.2 PV Cell Modeling

Based on the above discussion, the PV cell is a large area photodiode, and thus its structure of a PV cell can be modeled as various electrical components, similar to a photodiode [9–11]. Ideally, the photon-to-electron-flow process can be modeled as a current source, I_{ph}, where the generated current depends on the intensity of the sunlight hitting the cell. The p-n semiconductor junction is modeled as an ideal diode, D, as shown in Fig. 9.5. However, in practical cases, there are additional components. More specifically, the p-n junction has associated parallel capacitance, $C_{\rm p}$, and parallel resistance, $R_{\rm p}$. Additionally, the wire leads attached to the PV cell have some series resistance, $R_{\rm s}$, and series inductance, $L_{\rm s}$, as depicted in Fig. 9.5, and normally $R_{\rm p} >> R_{\rm s}$. For simplicity, these parasitic components are ignored in the modeling of PV cells, but they should be reconsidered if more accurate PV models are required.

As previously discussed, the power generation from PV cells is based on the *photovoltaic effect*, where light with sufficient energy is the key to exciting electrons in the solar cell material (e.g., Si layers). This physical process is affected by many factors. That is, many environmental factors affect the characteristics of a PV cell and, thus, its power generation. There are mainly two parameters: the solar irradiance, G, and the solar cell temperature, T. Thus, the PV operating characteristics can be modeled mathematically by linking the two factors with the generated photocurrent $i_{\rm ph}$, as illustrated in the following.

For simplicity, the parasitic parameters (i.e., capacitance $C_{\rm p}$ and inductance $L_{\rm s}$) in Fig. 9.5 are ignored. Accordingly, following the Kirchhoff's Current Law (KCL), the output current of the solar PV cell $i_{\rm pv}$ can be obtained as

$$i_{\rm pv} = i_{\rm ph} - i_{\rm d} - i_{\rm p} \tag{9.1}$$

where $i_{\rm d}$ is the diode current and $i_{\rm p}$ is the current flowing through the parallel resistance $R_{\rm p}$. The photocurrent is dependent of the solar irradiance level and the solar cell temperature as

$$i_{\rm ph} = [I_{\rm scn} + k_i(T - T_n)]\frac{G}{G_n} \tag{9.2}$$

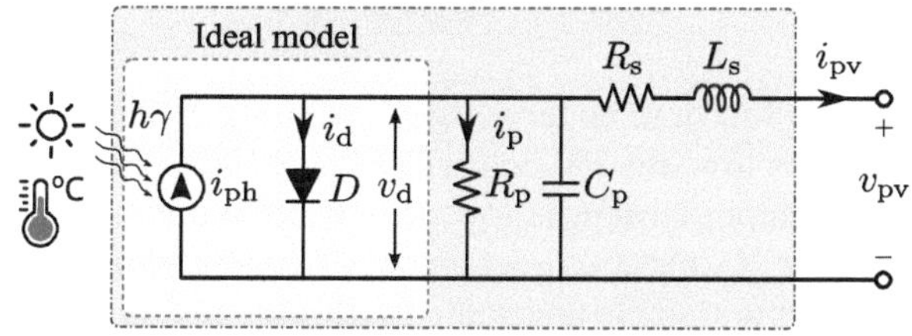

FIG. 9.5 Electricals model of a photovoltaic cell, where $h\gamma$ represents photons.

with I_{scn} being the nominal short-circuit current (at the standard test condition, i.e., air mass (AM): AM1.5, cell temperature: 25°C, and solar irradiance level: 1 kW/m^2), k_i being the current-temperature coefficient, G_n being the nominal solar irradiance (1 kW/m^2), and T_n being the nominal cell temperature (25°C). These values can be found in the datasheets for commercial PV cells.

The diode current i_d can be expressed as an exponential function of the diode voltage as well as the solar cell temperature,

$$i_d = i_{rs}\left[\exp\left(\frac{v_d}{nk_BT/q}\right) - 1\right] = i_{rs}\left[\exp\left(\frac{v_{pv} + i_{pv}R_s}{nk_BT/q}\right) - 1\right] \tag{9.3}$$

where i_{rs} is the diode reverse saturation current, v_d is the diode voltage, v_{pv} is the solar cell voltage, n is the diode ideality constant (i.e., the ideality factor of the p-n junction, typically, $1 \le n \le 2$), k_B is the Boltzmann constant ($k_B = 1.3806503 \times 10^{-23}$ J/K), and q is the electron charge ($q = 1.60217646 \times 10^{-19}$ C). The diode saturation current i_{rs} is mainly affected by the cell temperature as

$$i_{rs} = \frac{I_{scn} + k_i(T - T_n)}{\exp\left[\frac{V_{ocn} + k_v(T - T_n)}{nk_BT/q}\right] - 1} \tag{9.4}$$

in which V_{ocn} is the nominal open-circuit voltage and k_v is the voltage-temperature coefficient.

Substituting Eq. (9.4) into Eq. (9.3), and then Eqs. (9.3) and (9.2) into Eq. (9.1), result in

$$i_{pv} = [I_{scn} + k_i(T - T_n)]\frac{G}{G_n} - \frac{I_{scn} + k_i(T - T_n)}{e^{\frac{V_{ocn} + k_v(T - T_n)}{nk_BT/q}} - 1} \cdot \left(e^{\frac{v_{pv} + i_{pv}R_s}{nk_BT/q}} - 1\right) - \frac{v_{pv} + i_{pv}R_s}{R_p} \tag{9.5}$$

which is an equation that can be solved using the Newton-Raphson method [12]. To simplify, the ideal circuit model shown in Fig. 9.5 ($R_p = \infty\ \Omega$ and $R_s = 0\ \Omega$) is considered. In that case, the mathematical model for a solar PV cell can be obtained as

$$i_{pv} = [I_{scn} + k_i(T - T_n)]\frac{G}{G_n} - \frac{I_{scn} + k_i(T - T_n)}{e^{\frac{V_{ocn} + k_v(T - T_n)}{nk_BT/q}} - 1} \cdot \left(e^{\frac{v_{pv}}{nk_BT/q}} - 1\right) \tag{9.6}$$

It is implied in Eq. (9.6) that, under a given operating condition, when the PV cell output voltage is low, the PV cell will behave like a current source; conversely, when the output current is low, it will behave like a voltage source. In practice, a number of solar cells interconnected in series are encapsulated into a single, long-lasting, and stable unit, called a PV module. By taking the number of cells into account, the output current for a PV module is obtained as

$$i_{\mathrm{pv}} = [I_{\mathrm{scn}} + k_i(T - T_n)]\frac{G}{G_n} - \frac{I_{\mathrm{scn}} + k_i(T - T_n)}{e^{\frac{V_{\mathrm{ocn}} + k_v(T - T_n)}{nk_{\mathrm{B}}TN_s/q}} - 1} \cdot \left(e^{\frac{v_{\mathrm{pv}}}{nk_{\mathrm{B}}TN_s/q}} - 1\right) \quad (9.7)$$

with N_s being the number of cells in series in a PV module.

According to Eq. (9.7), the PV current can be obtained based on a given voltage. A simple solar PV model can then be implemented in MATLAB® using Simulink® Blocks or programmed models. Fig. 9.6 shows the block diagram of the solar PV model in MATLAB®, where the inputs are the solar irradiance G, the cell temperature T, and the PV module operating voltage v_{pv}, correspondingly, and the output is the PV module current i_{pv}. The mathematical model of (9.7) is programmed in C, and the audience of this book are encouraged to model the PV panels in C as an exercise. Based on the model, characteristics of a PV module can be examined.

Fig. 9.7 presents the general power-voltage (*P*-*V*) and current voltage (*I*-*V*) curves of a PV module, where the red dot indicates the maximum power point (MPP) of the *P*-*V* curve—the PV will produce the maximum power (P_{mpp}) at the MPP. Additionally, the corresponding voltage (V_{mpp}) and the current (I_{mpp}) at the MPP are also shown. It can be observed from the *I*-*V* curve in Fig. 9.7 that the PV current is relatively constant as the PV voltage changes at voltages below V_{mpp}—the PV module acts like a current source. In contrast, at voltages above V_{mpp}, the PV module voltage does not change significantly as the PV current changes—the PV module is like a voltage source—when the PV voltage is within V_{mpp} and the open circuit voltage (V_{oc}). Nevertheless, the curves in Fig. 9.7 reflect the nonlinear characteristics of PV modules. Thus, in practice, MPPT control should be applied for optimal power production, which will be discussed in Section 9.3.1. Notably, the *P*-*V* and *I*-*V* curves are for a specific

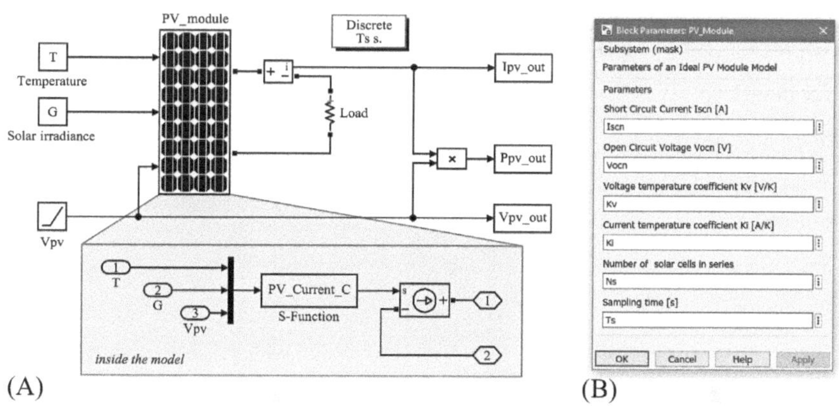

FIG. 9.6 Model of PV modules built up in MATLAB® using the user-definable block (i.e., S-Function) according to the mathematical model of (9.7): (A) Simulink® model and (B) parameter dialog box for the model. A resistor (Load) is connected.

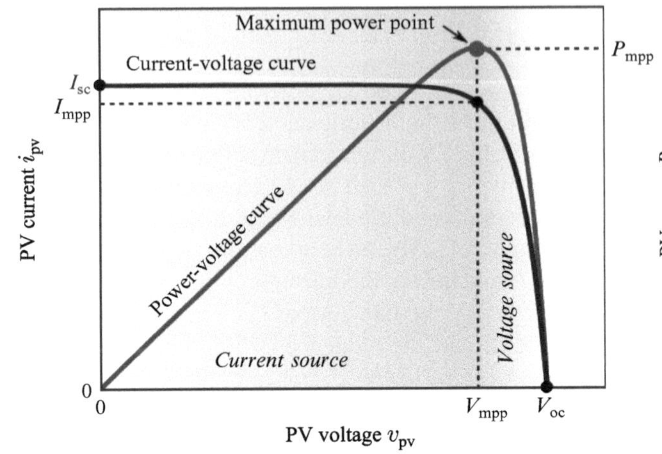

FIG. 9.7 Characteristics of PV modules (MPP: maximum power point, I_{sc}: short circuit current, V_{oc}: open circuit voltage, P_{mpp}: power at the MPP, I_{mpp}: current at the MPP, and V_{mpp}: voltage at the MPP).

condition (e.g., solar irradiance level: 1000 W/m^2 and temperature: 25°C). Over a day, both the solar irradiance level and temperature will change. Often, the changes are gradual as sunlight angles or shading patterns change such that the PV characteristics also change graduate. Sometimes quick changes can occur, e.g., due to passing clouds or birds [13], such that the curves can change drastically for just a short period of time.

As discussed previously, the solar irradiance and temperature will affect the power output from PV modules. To further investigate the impacts and also to validate the effectiveness of the model in Fig. 9.6, a commercial PV module has been investigated. Table 9.1 lists the main parameters of the PV

TABLE 9.1 Parameters of a Commercial Solar PV Module [14].

Parameter	Symbol	Value	Unit
Maximum power	P_{mpp}	65	W
Voltage at P_{mpp}	V_{mpp}	17.6	V
Current at P_{mpp}	I_{mpp}	3.69	A
Short-circuit current	I_{scn}	3.99	A
Open-circuit voltage	V_{ocn}	22.1	V
Current-temperature coefficient	k_i	0.065 ± 0.015	%/°C
Voltage-temperature coefficient	k_v	-80 ± 10	mV/°C
Number of cells	N_s	36	–

module [14], which are then programmed into the Simulink® model. Subsequently, the characteristics of the PV module under various operating conditions are obtained as shown in Fig. 9.8. It can be observed in Fig. 9.8 that the irradiance level is directly proportional to the PV current. This is in agreement with the mathematical model, where it is indicated in Eq. (9.2) that the photocurrent is proportional to the solar irradiance level. In contrast, the temperature is inversely proportional to the PV voltage. That is, as the PV cell temperature increases, the open-circuit voltage and the MPP voltage will decrease. According to these trends, a solar PV module will produce the most power when the sunlight intensity is high (i.e., the solar irradiance is strong) but the cell temperature is relatively low. However, such environmental conditions are not common in practice, since high-intensity sunlight hitting an object tends to increase its temperature at the same time. The nonlinearity of solar PV modules is also shown in Fig. 9.8.

Notably, the previously described ideal model is a relatively straightforward mathematical model. Any parasitic components are not considered. Thus, the Simulink® model may not be able to reflect the characteristics of a real PV module very accurately. For instance, in the datasheet of the adopted PV module, the power at the MPP is 65 W under standard test conditions, while the Simulink® model gives the maximum power of 70 W under the same conditions. Parasitic components in PV cells or modules, introduced and shown in Fig. 9.5, can be added to increase the model accuracy at the cost of higher model complexity. For example, if the series and shunt resistance components are considered in the model, as in Eq. (9.5), the PV characteristics become very nonlinear. In this case, much more efforts should be devoted to estimating the series and shunt resistance values according to datasheets. Further, a nonlinear solver, like the Newton-Raphson method, should be employed to solve the model numerically [9,12], as aforementioned. Nonetheless, in a similar way, a relatively accurate model for PV cells/modules can be built up in Simulink®, which is left as an exercise for the readers.

9.2.3 Configuration of PV Cells

In most PV applications, many solar PV cells are combined to produce a larger amount of power. When added together, the solar PV cells can be connected in series, parallel, or a combination of both, which form PV modules, PV panels, and PV arrays. The connection type and number of PV cells used is dependent on the voltage and current needs of the application. In most of today's PV applications, like roof-top PV installations (e.g., Fig. 9.3) and satellites, a higher voltage than a single cell is required; thus, many solar PV cells are connected in series. Typically, a standard commercial PV module is made up of 60–96 cells connected in series so that the sum at the terminals achieves an overall voltage in the range of 30–60 V.

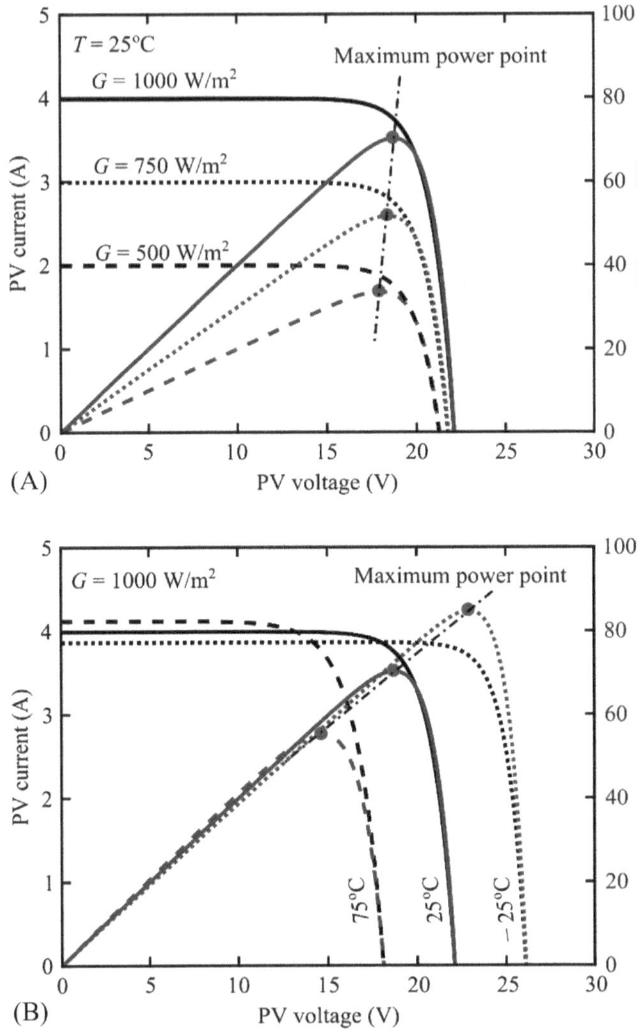

FIG. 9.8 Characteristics of a commercial PV module under different conditions: (A) fixed temperature (i.e., $T = 25°C$) but varying solar irradiance level and (B) fixed solar irradiance level (i.e., $G = 1000\ W/m^2$) but varying temperature.

The physical connection structure of a PV module/panel consisting of 60 solar PV cells is shown in Fig. 9.9. To cover the area of a panel, solar cells are typically connected down one row and back up the next, as shown in Fig. 9.9. Each series connection that goes down and back up is called a substring. It can be observed that the PV module consists of three substrings with a diode that is connected across the ends of each substring. This diode is called a bypass diode [15], which allows for an additional current path around the

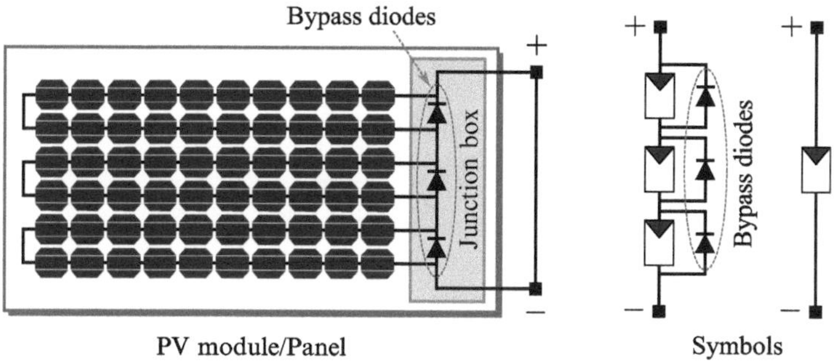

FIG. 9.9 Physical structure of a PV module consisting of 60 solar PV cells (6 × 10) and its electrical symbols.

substring, if solar PV cells in that substring experience a problem, such as partial shading, degradation, or a connection break in the substring. Bypass diodes help to reduce the voltage losses through the module, and the degree that these kinds of problems affect the total power production. However, various types of faults can occur in a solar PV panel that can lead to reduced power production [16].

Many PV systems are installed on rooftops or remote areas that are not easy to service if problems arise. Thus, both the panels and the associated power electronics must be able to work properly with minimal maintenance. Solar PV panels are typically rated to work continuously for 25 years or even longer, with expected degradation, generally, around 0.5%–1% per year for Si PV cells [17,18]. However, in certain conditions, PV panels tend to degrade more quickly, specifically in high heat (e.g., in the desert) or high humidity environments (e.g., tropical or seaside areas). These extreme conditions can degrade and corrode the materials that protect the cells and also degrade the PV cells themselves. Once the cells are directly exposed to the outside environment, effects of degradation occur more rapidly [19–21]. When working with PV systems, it is important that the PV cells are properly configured to meet the voltage and current needs of the application and that the cells are protected from harsh environmental conditions. Today, there are ongoing research and development efforts on fault detection and protection to improve the longevity of PV panels and the entire PV system [22–25].

9.3 CONTROL OF PV SYSTEMS

In solar PV systems, power electronic converters are normally heavily involved in transferring PV power to the loads, as illustrated in Fig. 9.10. These power electronic converters make the outputs of the power source (i.e., the PV modules) and the load compatible in terms of voltage level and frequency. For instance, in the case of grid-connected PV systems, the DC power from PV modules must be

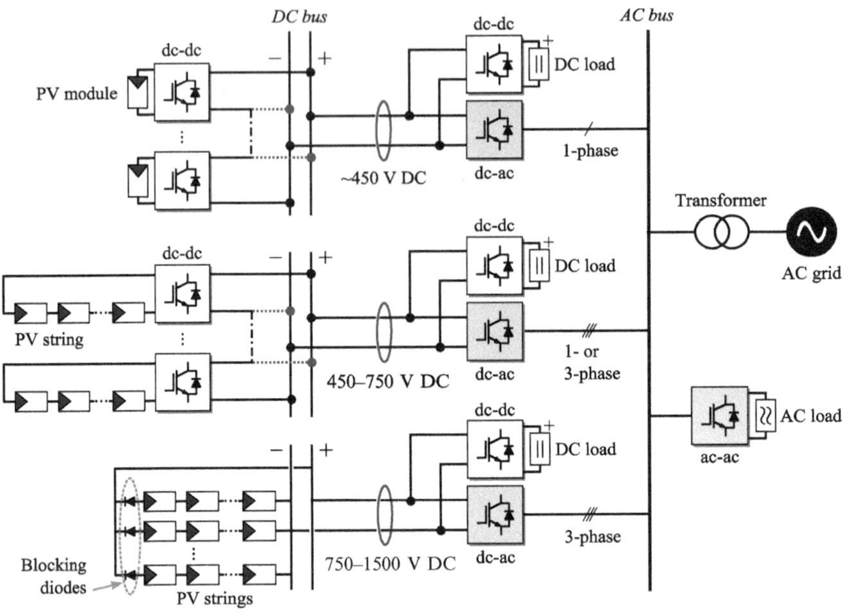

FIG. 9.10 Photovoltaic power conversion systems. As indicated, power electronic converters (e.g., DC-DC, DC-AC, and AC-AC converters) are widely used in the conversion of solar PV energy.

converted into AC power (e.g., 400 V, 50 Hz). Additionally, an important role of the power electronic converters is in that it enables the optimization of the PV power production even under varying environmental conditions.

9.3.1 Maximum Power Point Tracking

As discussed previously, the output power of PV modules is time variant, dependent on the solar irradiance intensity and solar cell temperature (indirectly, the ambient temperature). To maximize the power production (i.e., optimize the energy capture), MPPT control is necessary for PV systems. Hence, there have been numerous MPPT algorithms reported in the literature [26–30]. The major MPPT control algorithms are summarized in Table 9.2. Among those, the Perturb and Observe (P&O) method is the most commonly used MPPT algorithm due to its simplicity and accuracy. Actually, most of the reported MPPT control schemes are realized in accordance to the *P-V* characteristic of PV modules, as shown in Fig. 9.7, which is a hill-like curve. More specifically, on the top of the hill, the PV modules produce the maximum power, called the MPP. On the left side of the MPP, the output power has a nearly linear relationship with the PV operating voltage; while above the MPP voltage, the power drops with the increase of the PV voltage. Fig. 9.11 redepicts the general *P-V* curve of solar PV modules for illustrating the MPPT control algorithms. Ideally, the

TABLE 9.2 Pros and Cons of Major Maximum Power Point Tracking (MPPT) Methods

MPPT Methods	Advantages	Disadvantages
Perturb & Observe (P&O)/Incremental Conductance	■ Simple ■ Low computation ■ Generic	■ Trade-off between speed and accuracy ■ Inaccurate under fast changing conditions
Constant Voltage (CV)	■ Much simple ■ No perturbation ripple	■ Energy is wasted during V_{oc} measurement ■ Inaccuracy
Short-Current Pulse (SCP, i.e., constant current)	■ Simple ■ No perturbation ripple	■ Extra switch needed for short-circuit ■ Inaccuracy
Ripple Correction Control	■ Ripple amplitude offers the MPP information ■ No need for perturbation	■ Trade-off between efficiency loss due to the MPPT and due to the ripple

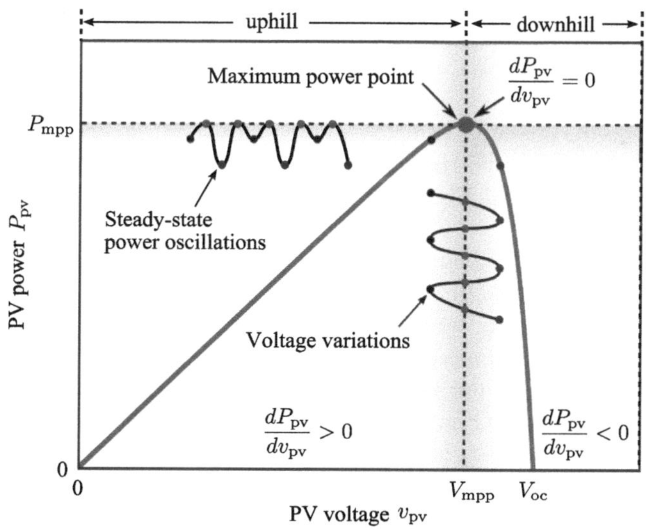

FIG. 9.11 Power-voltage (*P-V*) characteristic of PV modules.

MPPT control should maintain the PV module operating point at the MPP to be robust and act quickly in the face of environmental variations (e.g., passing clouds inducing a sudden drop in the solar irradiance level). Those become the basic demands for the MPPT controller design.

According to Fig. 9.11, a simple P&O MPPT controller can be developed, which enables the operating point to continuously and repeatedly climb up to the top of the hill. Based on its name, the operating voltage of PV modules is perturbed (changed incrementally) in one direction first. Then, the power of the PV modules is calculated and compared with the previously-sampled value. The maximum power extraction is achieved by continuously adjusting the reference signal that is given to the controller of the power electronic converters (e.g., a boost DC-DC converter), which then adjusts the operating point of the PV module. The reference signal of the MPPT control block can be a voltage v_{pv}^*, a current i_{pv}^*, a power P_{pv}^*, or a duty cycle D^*. Taking the voltage as an example, the reference signal can be expressed as

$$v_{\mathrm{pv}}^*(k) = v_{\mathrm{pv}}^*(k-1) + \mathrm{sgn}\left(\frac{\Delta P_{\mathrm{pv}}(k)}{\Delta v_{\mathrm{pv}}(k)}\right) \cdot v_{\mathrm{step}} \tag{9.8}$$

where k indicates the time step, v_{step} is the perturbation step size, Δ presents the difference between the kth sampled data and the $(k-1)$th sampled data, and sgn $(\cdot)$ is a function that determines the perturbation direction. The function sgn$(\cdot)$ is defined as

$$\mathrm{sgn}(x) = \begin{cases} 1 & x > 0 \\ 0 & x = 0 \\ -1 & x < 0 \end{cases} \tag{9.9}$$

which reflects the operating principle of the P&O MPPT control scheme. It also indicates that the power will be oscillating around the MPP in steady state due to the repeated perturbation. The flowchart of the P&O MPPT control algorithm is shown in Fig. 9.12.

According to Fig. 9.12 and Eq. (9.8), it is known that the design of the P&O MPPT involves the selection of the step-size and the perturbation frequency (i.e., the MPPT sampling frequency). A large perturbation step-size results in fast tracking of the maximum power in response to operating condition changes. The operating point of the P&O MPPT will inevitably oscillate around the MPP, as demonstrated in Fig. 9.11. In that case, a large step-size, in return, leads to more power losses (i.e., lower accuracy), and thus low MPPT control efficiency. A rule-of-thumb for designing the step-size is to set it as around 0.1%–1% of the nominal MPP voltageV_{mpp} at STC [31]. In regard to the sampling frequency of the MPPT algorithm, in general, it should ensure that before applying the next perturbation, the entire system is already in steady-state operation. Typically, the sampling frequency can be around a few Hz and up to hundreds of Hz [26,31–34]. However, in recent studies, it has been found that the MPPT perturbation may induce interharmonics in the systems [35,36]. This may also be considered when designing the MPPT controller.

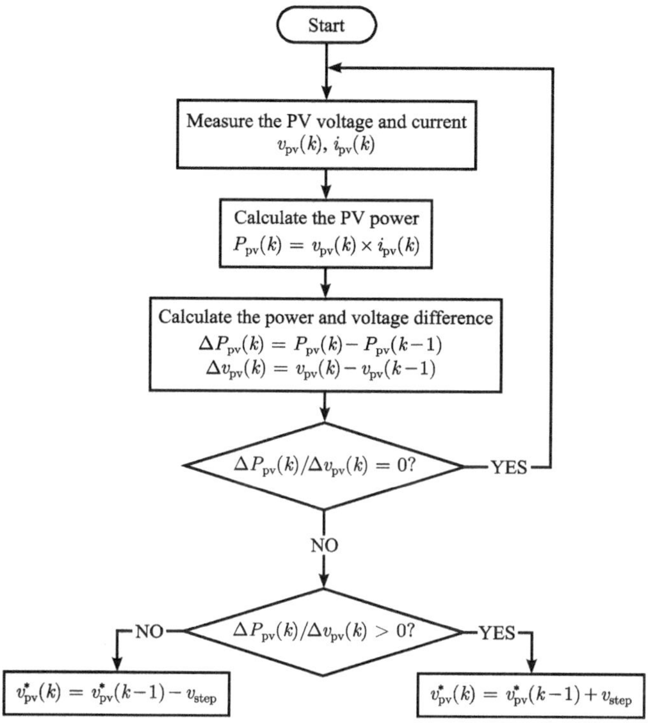

FIG. 9.12 Flowchart of the Perturb and Observe (P&O) MPPT control algorithm.

9.3.2 PV-Fed DC-DC Conversion Systems

PV modules can be connected to different loads, as demonstrated in Fig. 9.10. When powering a DC load (e.g., batteries) or a DC microgrid, DC-DC converters should be adopted either to boost the PV output voltage or to step it down, according to the load demands. In certain cases, high-voltage conversion gains are required (e.g., in microinverter applications, 30–60 V DC inputs should be converted to high DC voltages for the grid-connection to a 50-Hz 220-V AC network). PV modules employing advanced DC-DC converters (e.g., quasi-Z-source converters) are gaining more popularity in today's PV markets [3,37]. Nevertheless, the boost DC-DC converter is the most common one, which will be detailed in Section 9.4. Fig. 9.13 shows the schematic of a PV-fed DC-DC conversion system, where a resistive load is adopted.

As mentioned previously, the output of the MPPT control block can be the reference voltage, current, and power for the PV modules and also the duty cycle for the DC-DC converters. The main control objective of the DC-DC conversion systems is to optimize the power of the PV modules even under nonuniform solar irradiance and ambient temperature profiles. Normally, measuring

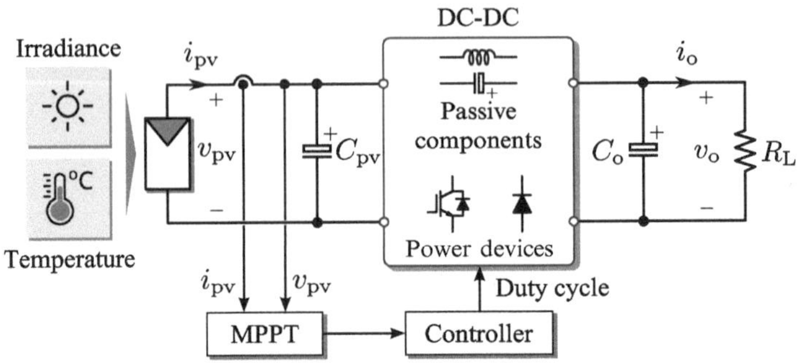

FIG. 9.13 Structure of a DC-DC conversion system fed by PV modules, where the load is a resistive load denoted as R_L, C_{pv} and C_o are the input filter and the output filter capacitor, respectively. The MPPT control algorithm shown in Fig. 9.12 can be implemented in the "MPPT" block in this figure.

the PV voltage and current is sufficient, while in some advanced control algorithms, other variables may be measured, which incurs more cost and complexity. Nonetheless, the criteria to select the MPPT algorithms include (1) implementation complexity, (2) tracking accuracy, and (3) robustness to fast changing conditions. For the controllers of the DC-DC converters, Chapter 3 has presented a detailed overview, and the discussed controllers (e.g., a proportional integral controller) can be employed. In addition, considering the dynamics of the PV modules, Eq. (9.7), the models of DC-DC converters (e.g., the small-signal models) together with the model of the PV modules presented in this chapter are beneficial to parameter tuning. Thus, the controller design of PV-fed DC-DC converters is not covered here; interested readers are directed to Chapter 3 of this book.

9.3.3 DC-AC Inverters for PV Systems

In many applications, voltage source converters are adopted as the interface for solar PV modules and the load (e.g., in grid-connected applications). There are mainly two ways to arrange the systems: single-stage and double-stage (i.e., DC-DC with DC-AC), as demonstrated in Fig. 9.10. For double-stage PV systems, the control is more straightforward, where the MPPT control is commonly implemented in the first stage (i.e., by controlling the DC-DC converters) and the control of the PV inverters (i.e., the DC-AC converters) is mainly to ensure proper power injection (by keeping the DC-link voltage constant). That is, the DC-link capacitor decouples the control of the DC-DC converter and the DC-AC inverter. Therefore, the MPPT control strategies can be designed according to the discussions in Section 9.3.2; while the control schemes for DC-AC converters discussed in Chapter 6 are applicable to the PV inverters.

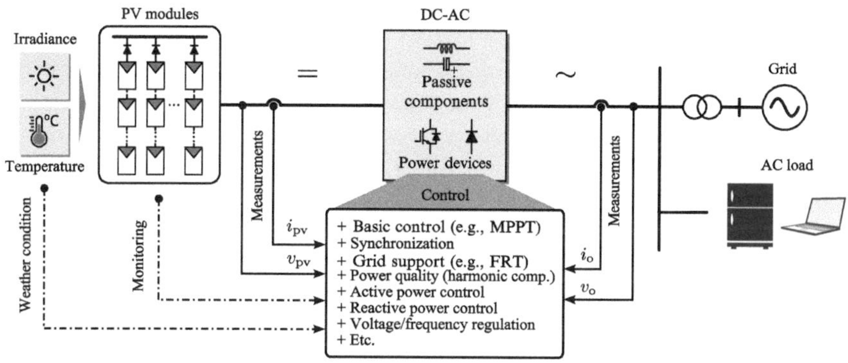

FIG. 9.14 General control structure for PV systems connected to the grid and AC loads, where the DC-AC conversion stage can also be a double-stage system (i.e., a DC-DC converter and a DC-AC converter). FRT—Fault Ride-Through.

In contrast, if a single-stage configuration is adopted, both control objectives (i.e., PV power optimization and power injection) should be accomplished in the inversion stage. Nevertheless, the major control objectives of a PV inversion system include maximum power extraction, synchronization, and power quality. There are also advanced control functionalities for PV systems, which have been illustrated in Fig. 9.14.

As shown in Fig. 9.14, the control of the PV-fed DC-AC converters is complicated in contrast to the DC-DC converters. In certain cases (e.g., the grid-connected PV systems), the entire control should be implemented at a system level and also coordinated with other controllers. For instance, in utility-scale grid-connected PV systems, it may be mandatory for PV systems to follow frequency and voltage regulations at the distribution level [38,39]. Generally, the control can be separated into two cascaded loops [4,40]: an inner current control loop and an outer loop.The outer loop is to regulate the power or the DC-link voltage, which then generates the current references for the inner control loop. Compared with the current control loop, the outer loop is slow, and a proportional or a proportional integral controller can be adopted. For the inner control loop, it should be much faster than the outer loop, and it can be implemented in different reference frames, as detailed in Chapter 6. It should be pointed out that the nonlinear hysteresis controller (also known as a bang-bang or an on-off controller) can be adopted as the inner-loop controller. The design considerations for these controllers can also be found in that chapter. In addition, the inner current should be synchronized with the AC voltage. For grid-connected systems, the synchronization can be achieved by means of phase-locked loops; while in standalone applications, an arbitrary phase may be employed depending on the load requirements. Nevertheless, details of the modeling and associated controller design of DC-AC converters can be referred to Chapter 6.

9.4 OPERATION EXAMPLES

In this section, the P&O MPPT control scheme is demonstrated first, where the PV modules are feeding power to a resistive load via a boost converter, as shown in Fig. 9.15. Three PV modules are connected in series, forming a system with the rated power of around 900 W (3 × 300 W). The parameters for the adopted PV module are shown in Table 9.3, where the other system parameters

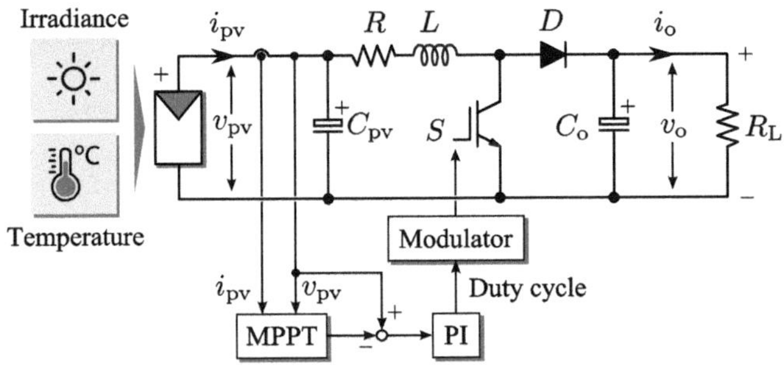

FIG. 9.15 Electrical circuit diagram of a boost conversion system fed by PV modules (PI—Proportional Integral), where the load is a resistive load R_L.

TABLE 9.3 Parameters of a PV-Fed Boost Conversion System With Three Commercial Solar PV Modules in Series [41]

Parameter	Symbol	Value	Unit
PV module			
Maximum power	P_{mpp}	300	W
Voltage at P_{mpp}	V_{mpp}	35.8	V
Current at P_{mpp}	I_{mpp}	8.37	A
Short-circuit current	I_{scn}	8.86	A
Open-circuit voltage	V_{ocn}	45.2	V
Current-temperature coefficient	k_i	0.0044	A/°C
Voltage-temperature coefficient	k_v	−0.1446	V/°C
Number of cells	N_s	72	–
Boost converter			
Boost inductor	L	1.5	mH
Resistance of the boost inductor	R	1	mΩ
PV output capacitor	C_{pv}	220	μF
Boost output capacitor	C_o	1100	μF
Resistive load	R_L	40	Ω

are also listed. The model presented in Section 9.2 is implemented (see Fig. 9.6). The switching frequency for the boost converter is $f_{sw} = 20$ kHz. A PI controller is adopted as shown in Fig. 9.15, and it can be expressed as

$$G_{PI}(s) = k_p + \frac{k_i}{s} \tag{9.10}$$

with k_p and k_i being the proportional and the integral gain, respectively. The gains are designed as $k_p = 0.0025$ and $k_i = 0.58$. As stated, the P&O MPPT algorithm has been employed to track the maximum power of the PV modules, as shown in Fig. 9.12. The MPPT sampling rate is 20 Hz, and the perturbation step-size is set as 0.33% of the voltage at the nominal MPP. First, the conversion system was simulated under constant conditions (solar irradiance level: 1000 W/m^2, ambient temperature: 25°C). Then, a trapezoidal solar irradiance profile has been applied to validate the robustness of the MPPT algorithm, where the ambient temperature was fixed at 25°C. In this profile, the solar irradiance was changed from 500 to 1000 W/m^2 in 5 seconds and then experienced a fast drop to 800 W/m^2 after an operation of 8 seconds (at STC). Simulation results are shown in Figs. 9.16 and 9.17.

As shown in Fig. 9.16, the P&O MPPT algorithm with the designed parameters can effectively optimize the power production of the system. Furthermore,

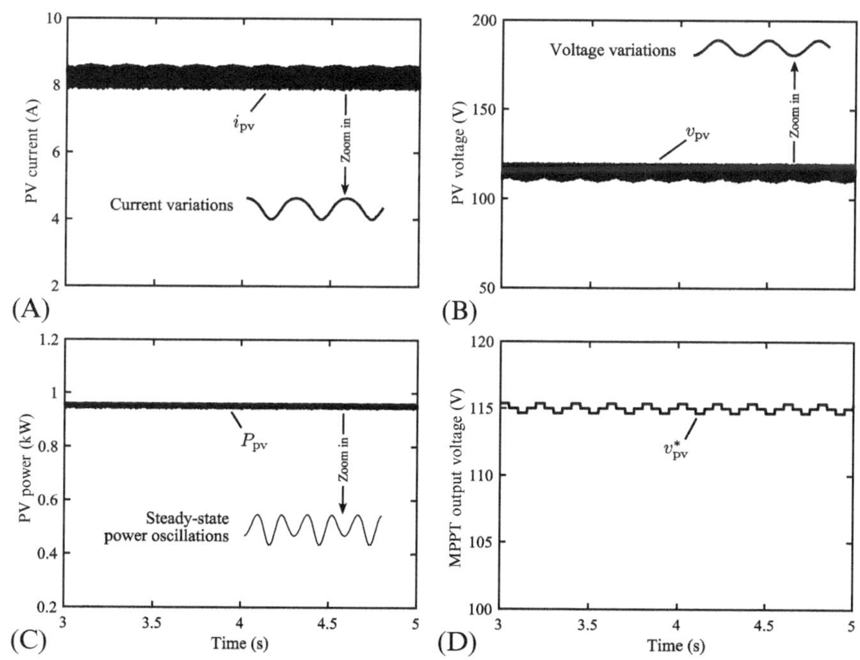

FIG. 9.16 Performance of the PV-fed boost conversion system: (A) PV current, (B) PV voltage, (C) PV power, and (D) MPPT output voltage (i.e., PV voltage reference), where the solar irradiance level is 1000 W/m^2 and the ambient temperature is 25°C.

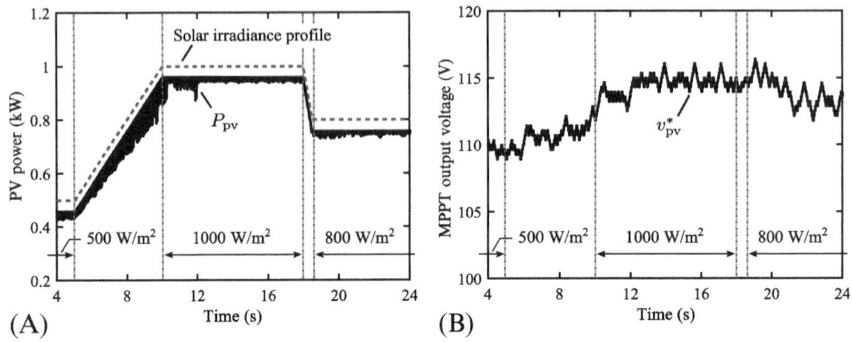

FIG. 9.17 Performance of the PV system under a trapezoidal solar irradiance profile: (A) PV power and solar irradiance profile, and (B) MPPT output voltage (i.e., PV voltage reference). In this case, the ambient temperature is constant at 25°C.

it is observed in Fig. 9.16D that the MPPT output voltage reference v^*_{pv} has three levels, which is considered as an optimal and stable operation for the MPPT controller. Additionally, this also indicates that the steady-state power will oscillate around the MPP. The resultant power losses are related to the perturbation step size. Readers are advised to explore the impact of perturbation step size on the efficiency of the MPPT control by referring to Refs. [30,31].

Following, a trapezoidal solar irradiance profile was applied to the system, as one input of the PV module model. The simulation results are shown in Fig. 9.17. As noted, under the changing solar irradiance condition the P&O MPPT algorithm may get confused, leading to loss of power production. The phenomenon has been observed in Fig. 9.17 when the solar irradiance slowly changed from 500 to 1000 W/m^2—the power variations are large. This situation (i.e., the P&O algorithm gets confused) may become severe when the solar irradiance changes fast during cloudy days. In that case, advanced MPPT controllers should be developed. It should also be mentioned that the above cases are presented to show the P&O MPPT control of PV systems and to validate the modeling of PV modules in Section 9.2. More details about the parameter tuning and design are directed to Chapter 3, [26,30] in order to address the challenges under fast changing conditions.

Since many PV systems are grid-connected, a two-stage grid-connected single-phase PV system is simulated to demonstrate the basic operation and general control. Fig. 9.18 shows the hardware schematic of the exemplified system and its control structure. The PV module model presented in Section 9.2.2 has been adopted, and the entire system has been modeled in MATLAB® Simscape/Power Systems. The electrical and control parameters for the PV modules and the boost converter are the same as those for the system in Fig. 9.15, except for the PV output capacitor (C_{pv} = 470 μF). The P&O MPPT algorithm is again employed in this case study. As shown in Fig. 9.18B, the inverter is controlled

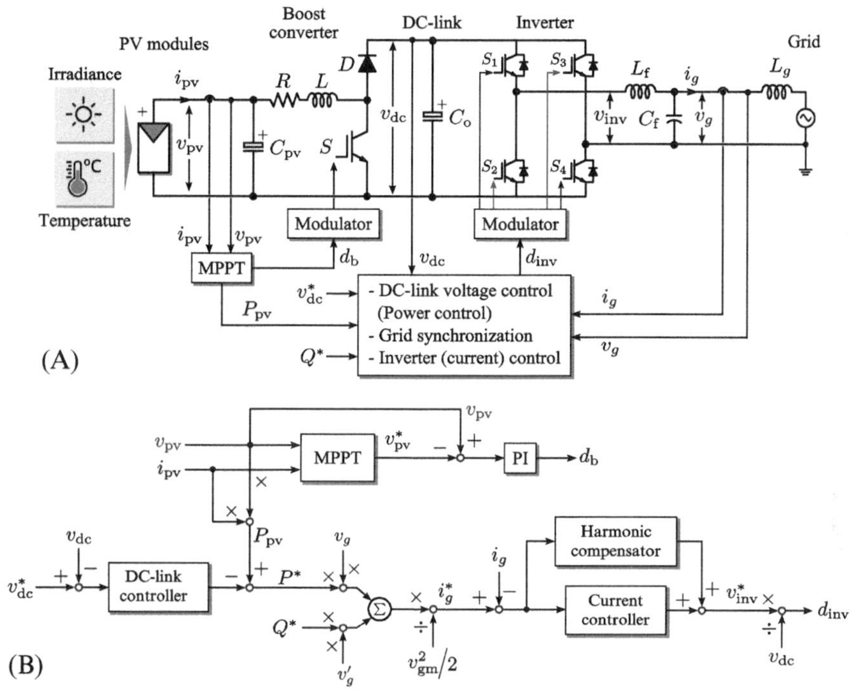

FIG. 9.18 A single-phase two-stage grid-connected PV system: (A) system hardware schematic and its general control structure and (B) detailed control block diagrams, where v_g' is the voltage in-quadrature with the grid voltage v_g, v_{gm} is the grid voltage amplitude, Q^* is the reactive power reference, and d_b, d_{inv} are the duty-cycle and modulation signal of the boost converter and the inverter respectively.

with a cascaded control system: the inner loop control regulates the grid current with a harmonic compensator, and the output loop controls the DC-link voltage and the power. A PI controller is adopted as the DC-link voltage controller $G_{PI\text{-}dc}(s)$, and a proportional resonant (PR) $G_{PR}(s)$ with multiple resonant controllers $G_{HC}(s)$ is used to regulate the injected grid current. These controllers can be expressed as

$$\begin{aligned} G_{PI-dc}(s) &= k_{p-dc} + \frac{k_{i-dc}}{s} \\ G_{PR}(s) &= k_{pr} + \frac{k_{ir}s}{s^2+\omega_0^2} \\ G_{HC}(s) &= \frac{k_{i3}s}{s^2+(3\omega_0)^2} + \frac{k_{i5}s}{s^2+(5\omega_0)^2} + \frac{k_{i7}s}{s^2+(7\omega_0)^2} \end{aligned} \tag{9.11}$$

where $k_{p\text{-dc}}$ and $k_{i\text{-dc}}$ are the proportional and integral gain of the DC-link voltage controller; k_{pr} and k_{ir} are the proportional and resonant gain of the PR current controller; k_{i3}, k_{i5}, and k_{i7} are the gains for the resonant controllers of the harmonic compensator; and ω_0 is the grid fundamental frequency. It is clear that only the third-, fifth-, and seventh-order harmonics are compensated,as they are dominant in single-phase systems [39, 41]. The controller and the other system parameters are shown in Table 9.4. In addition, as seen in Fig. 9.18, a virtual voltage v'_g in quadrature with the grid voltage should be generated in order to calculate the current reference for the inner loop. To achieve this, a second-order generalized integrator is adopted (see Chapter 4). Simulation results are shown in Fig. 9.19, where the DC-link voltage is controlled at 450 V (i.e., $v^*_{dc} = 450$ V).

Two scenarios are simulated, where an ambient temperature of 25°C was assumed for both. In the first case, the PV system is generating the maximum power and injecting currents at unity power factor, where the solar irradiance has been decreased to 700 W/m^2 from 1000 W/m^2 at $t = t_1$. As it can be seen in Fig. 9.19A that, when there are solar irradiance changes (for instance, in practice, due to passing clouds), the entire two-stage PV system can still optimize the power production from PV modules. Additionally, the inverter is able to deliver the optimized maximum power to the grid with high-quality currents.

Furthermore, in certain applications, the PV systems may be required to provide reactive power to support the grid. Hence, the second case demonstrates the reactive power controllability of the two-stage PV system. As shown in Fig.

TABLE 9.4 Parameters of a Single-Phase Two-Stage Grid-Connected PV System

Parameter	Symbol	Value
PV inverter and grid		
Inverter output filter inductance	L_f	3.6 mH
Inverter output filter capacitor	C_f	2.35 μF
Grid voltage amplitude	V_{gm}	325 V
Grid frequency	ω_0	314 rad/s
Grid inductor	L_g	4 mH
Controllers		
DC-link PI proportional gain	$k_{p\text{-dc}}$	30
DC-link PI integral gain	$k_{i\text{-dc}}$	430
Current controller proportional gain	k_{pr}	12
Current controller resonant gain	k_{ir}	2000
Harmonic compensation resonant gains	k_{i3}, k_{i5}, k_{i7}	1400, 1200, 600

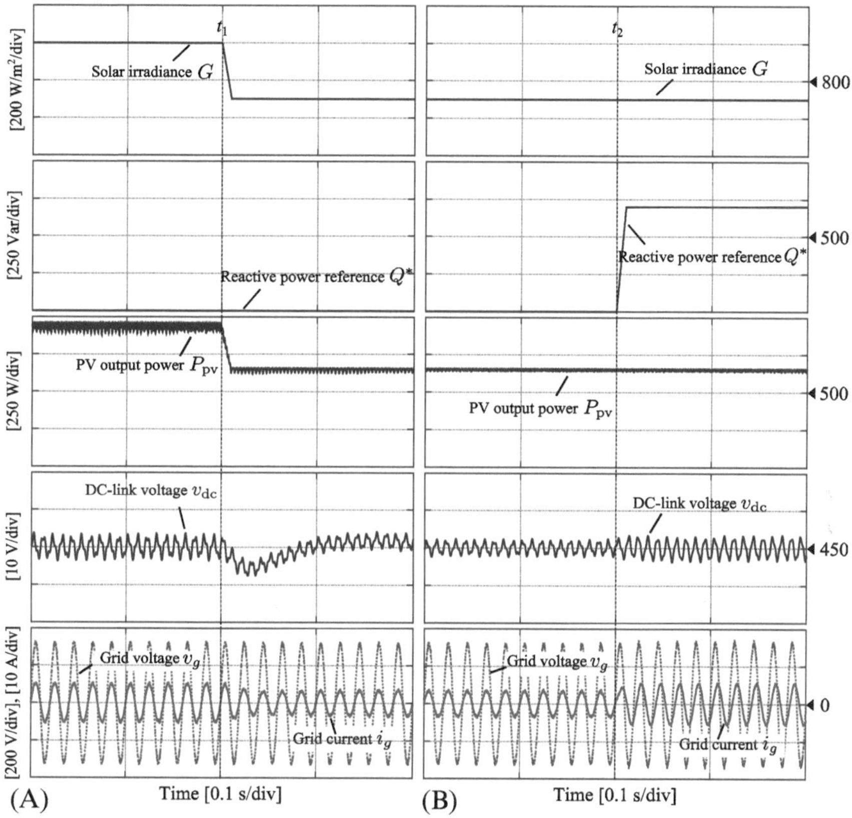

FIG. 9.19 Performance of the single-phase two-stage grid-connected PV system under various operating conditions: (A) the solar irradiance has changed to 700 W/m^2 and (B) the reactive power reference was changed to 700 Var. Ambient temperature: 25°C.

9.19B, the adopted controllers enable a proper reactive power injection to the grid upon demands. The DC-link voltage experienced minor variations during the transients, and the MPPT control of the PV modules is not affected by the control of the PV inverter.This confirms that the DC-link capacitor balances the DC power from the PV modules and the ac power injected to the grid. That is, it "decouples" the control of the entire PV system, and thus the MPPT control and the PV inverter control can be designed separately. The above examples have shown the basic operation and control of PV systems.

9.5 SUMMARY

This chapter first presented the modeling of PV systems, from a solar PV cell to a PV module, which can be configured according to the load requirements. As power electronic converters are the key to an optimal utilization of solar PV energy, the configuration of PV power conversion systems was presented, including general controls for PV power systems. An MPPT control algorithm (i.e., the P&O MPPT algorithm) has been demonstrated on a boost DC-DC converter, with simulations under constant environmental conditions and a changing solar irradiance profile. Results have demonstrated the effectiveness of the P&O MPPT algorithm to optimize the power of PV systems and validated the PV model presented in this chapter. Notably, many efforts can be made to enhance the tracking performance of MPPT algorithms in terms of accuracy and dynamics. Additionally, a case of a two-stage single-phase grid-connected PV system was also exemplified to illustrate the basic operation and control of PV systems. The model of the PV modules was adopted and simulated to show its operation.

REFERENCES

[1] IRENA, "Renewable Capacity Statistics 2017," International Renewable Energy Agency (IRENA), Abu Dhabi, 2017. ISBN: 978-92-9260-018-1. Available from: http://www.irena.org/DocumentDownloads/Publications/IRENA_RE_Capacity_Statistics_2017.pdf (accessed 27.04.17).

[2] J. Nelson, The Physics of Solar Cells, Imperial College Press, London, UK, 2003.

[3] Y. Yang, F. Blaabjerg, Overview of single-phase grid-connected photovoltaic systems, Electr. Power Compon. Syst. 43 (12) (2015) 1352–1363.

[4] F. Blaabjerg, R. Teodorescu, M. Liserre, A.V. Timbus, Overview of control and grid synchronization for distributed power generation systems, IEEE Trans. Ind. Electron. 53 (5) (2006) 1398–1409.

[5] M.A. Green, Silicon photovoltaic modules: a brief history of the first 50 years, Prog. Photovolt. Res. Appl. 13 (5) (2005) 447–455.

[6] W.T. Jewell, R. Ramakumar, The history of utility-interactive photovoltaic generation, IEEE Trans. Energy Convers. 3 (3) (1988) 583–588.

[7] NREL National Center for Photovoltaics, Best Research-Cell Efficiencies for Solar PV Cells From NREL. Available from: http://www.nrel.gov/ncpv/ (accessed 27.04.17).

[8] M.A. Green, K. Emery, Y. Hishikawa, W. Warta, E.D. Dunlop, Solar cell efficiency tables (version 47), Prog. Photovolt. 24 (1) (2015) 3–11.

[9] M.G. Villalva, J.R. Gazoli, E. Filho, Comprehensive approach to modeling and simulation of photovoltaic arrays, IEEE Trans. Power Electron. 24 (5) (2009) 1198–1208.

[10] F. Spertino, J. Sumaili, H. Andrei, G. Chicco, PV module parameter characterization from the transient charge of an external capacitor, IEEE J. Photovolt. 3 (4) (2013) 1325–1333.

[11] K.A. Kim, C. Xu, L. Jin, P.T. Krein, Dynamic photovoltaic model incorporating capacitive and reverse-bias characteristics, IEEE J. Photovolt. 3 (4) (2013) 1334–1341.

[12] Y. Yang, K. Zhou, Photovoltaic cell modeling and MPPT control strategies, Trans. China Electrotechn. Soc. 26 (Sup.-1) (2011) 229–234 (in Chinese).

[13] R.J. Serna, B.J. Pierquet, J. Santiago, R.C.N. Pilawa-Podgurski, in: Field measurements of transient effects in photovoltaic panels and its importance in the design of maximum power point trackers, Proc. of IEEE APEC, 2015, , pp. 3005–3010.

[14] bp solar. "BP 365 Datasheet—65 W Photovoltaic Module," Available from: http://www.solarpanelstore.com/pdf/bp-365.pdf (accessed 12.05.17).

[15] Texas Instruments "SM74611 Datasheet—Smart Bypass Diode (Rev. B)," 2016. Available from: http://www.ti.com/lit/ds/symlink/sm74611.pdf (accessed 13.05.17).
[16] K.A. Kim, P.T. Krein, Reexamination of photovoltaic hot spotting to show inadequacy of the bypass diode, IEEE J. Photovolt. 5 (5) (2015) 1435–1441.
[17] A. Sangwongwanich, Y. Yang, D. Sera, F. Blaabjerg, Lifetime evaluation of grid-connected pv inverters considering panel degradation rates and installation sites, IEEE Trans. Power Electron. 99 (2017) 1–11.
[18] D.C. Jordan, S.R. Kurtz, Photovoltaic degradation rates—an analytical review, Prog. Photovolt. Res. Appl. 21 (1) (2013) 12–29.
[19] E.L. Meyer, E.E. van Dyk, Assessing the reliability and degradation of photovoltaic module performance parameters, IEEE Trans. Reliab. 53 (1) (2004) 83–92.
[20] M. Vazquez, I. Rey-Stolle, Photovoltaic module reliability model based on field degradation studies, Prog. Photovolt. Res. Appl. 16 (5) (2008) 419–433.
[21] D.C. Jordan, S.R. Kurtz, Photovoltaic degradation rates—an analytical review, Prog. Photovolt. Res. Appl. 21 (1) (2011) 12–29.
[22] J. Johnson, K. Armijo, in: Parametric study of PV arc-fault generation methods and analysis of conducted DC spectrum, Proc. of IEEE PVSC, 2014, , pp. 3543–3548.
[23] C.B. Jones, J.S. Stein, S. Gonzalez, B.H. King, Photovoltaic system fault detection and diagnostics using Laterally Primed Adaptive Resonance Theory neural network, Proc. of IEEE PVSC, 2015, pp. 1–6.
[24] M.K. Alam, F. Khan, J. Johnson, J. Flicker, A comprehensive review of catastrophic faults in PV arrays: types, detection, and mitigation techniques, IEEE J. Photovolt. 5 (3) (2015) 982–997.
[25] K.A. Kim, G.-S. Seo, B.-H. Cho, P.T. Krein, Photovoltaic hot-spot detection for solar panel substrings using AC parameter characterization, IEEE Trans. Power Electron. 31 (2) (2016) 1121–1130.
[26] N. Femia, G. Petrone, G. Spagnuolo, M. Vitelli, Optimization of perturb and observe maximum power point tracking method, IEEE Trans. Power Electron. 20 (4) (2005) 963–973.
[27] T. Esram, P.L. Chapman, Comparison of photovoltaic array maximum power point tracking techniques, IEEE Trans. Energy Convers. 22 (2) (2007) 439–449.
[28] S. Lyden, M.E. Haque, Maximum power point tracking techniques for photovoltaic systems: a comprehensive review and comparative analysis, Renew. Sust. Energ. Rev. 52 (2015) 1504–1518.
[29] E. Koutroulis, K. Kalaitzakis, N.C. Voulgaris, Development of a microcontroller-based, photovoltaic maximum power point tracking control system, IEEE Trans. Power Electron. 16 (1) (2001) 46–54.
[30] D. Sera, L. Mathe, T. Kerekes, S.V. Spataru, R. Teodorescu, On the perturb-and-observe and incremental conductance MPPT methods for PV systems, IEEE J. Photovolt. 3 (3) (2013) 1070–1078.
[31] H. Schmidt, B. Burger, U. Bussemas, S. Elies, How fast does an MPP tracker really need to be?, Proc. of EU PVSEC, 21–25 Sept., 2009, pp. 3273–3276.
[32] N. Femia, G. Petrone, G. Spagnuolo, M. Vitelli, Optimizing sampling rate of P&O MPPT technique, Proc. of PESC, vol. 3, Jun. 2004, pp. 1945–1949.
[33] M. Rosu-Hamzescu, S. Oprea, Practical guide to implementing solar panel MPPT algorithms, in: Application Note (No.: AN1521), Microchip Technology Inc, 2013.
[34] J. Kivimaki, S. Kolesnik, M. Sitbon, T. Suntio, A. Kuperman, Revisited perturbation frequency design guideline for direct fixed-step maximum power point tracking algorithms, IEEE Trans. Ind. Electron. 64 (6) (2017) 4601–4609.

[35] M. Aiello, A. Cataliotti, S. Favuzza, G. Graditi, Theoretical and experimental comparison of total harmonic distortion factors for the evaluation of harmonic and interharmonic pollution of grid-connected photovoltaic systems, IEEE Trans. Power Delivery 21 (3) (2006) 1390–1397.

[36] A. Sangwongwanich, Y. Yang, D. Sera, F. Blaabjerg, in: Interharmonics from grid-connected PV systems: Mechanism and mitigation, Proc. of IFEEC—ECCE Asia, 2017, pp. 1–6.

[37] Y. Liu, H. Abu-Rub, B. Ge, F. Blaabjerg, O. Ellabban, P.C. Loh, Impedance Source Power Electronic Converters, Wiley—IEEE, West Sussex, UK, 2016.

[38] F. Blaabjerg, Y. Yang, D. Yang, X. Wang, Distributed power-generation systems and protection, Proc. IEEE 105 (7) (2017) 1311–1331.

[39] Y. Yang, F. Blaabjerg, H. Wang, M.G. Simões, Power control flexibilities for grid-connected multi-functional photovoltaic inverters, IET Renew. Power Gener. 10 (4) (2016) 504–513.

[40] F. Blaabjerg, D.M. Ionel (Eds.), Renewable Energy Devices and Systems With Simulations in MATLAB® and ANSYS®, CRC Press, Boca Raton, FL, 2017.

[41] Yingli Green Energy Holding Co. Ltd., "YGE-U 72 Cell Series—YL300P-35b PV Module," Product Datasheet, Available from: http://d9no22y7yqre8.cloudfront.net/assets/uploads/products/downloads/YGEU72CellSeries_US_201602_V02.pdf, (accessed 22.07.17).

Chapter 10

Control of Wind Turbine System

Dao Zhou, Yipeng Song and Frede Blaabjerg
Aalborg University, Aalborg, Denmark

10.1 OVERVIEW OF WIND TURBINE SYSTEMS

Over the last decade, the wind power industry has expanded greatly with governmental support all around the world. According to the statistics provided by Global Wind Energy Council (GWEC), the cumulative wind power has increased from 17.4GW in 2000 to 318.1GW at the end of 2013, almost 18.2 times higher [1]. In this section, the mainstream configurations of the wind power generation system and their state-of-art control strategies will be investigated and evaluated.

10.1.1 Mainstream Configurations

The steady growth of wind power capacity has a consequence to the wind turbine system—lower cost per kWh, increased power density, and higher reliability [2]. According to the popular products of the top 10 wind turbine manufacturer, the constant-speed system is fading out pushed by stricter grid connection requirements, and variable-speed system in the wind turbine becomes the most popular system [3]. Although the wind turbine system can be categorized into several concepts in terms of the generator type, with and without gearbox, or the rating of power electronics converter, it is common to divide the wind turbine system into a partial-scale power converter equipped with a doubly fed induction generator (DFIG) and a full-scale power converter with either a synchronous generator (SG) or an induction generator (IG).

The first concept of wind turbine configuration, which is based on the DFIG, employs a power converter rated to approximately 30% of the nominal generator power and it is shown in Fig. 10.1A. The power converter is connected to the rotor through slip ring and controls the rotor current as well as rotor speed, while the stator is linked to the grid without any decoupling. The fraction of slip power through the converter makes this concept attractive from an economical point of view. However, the main drawbacks lie in the use

Control of Power Electronic Converters and Systems. https://doi.org/10.1016/B978-0-12-805245-7.00010-X

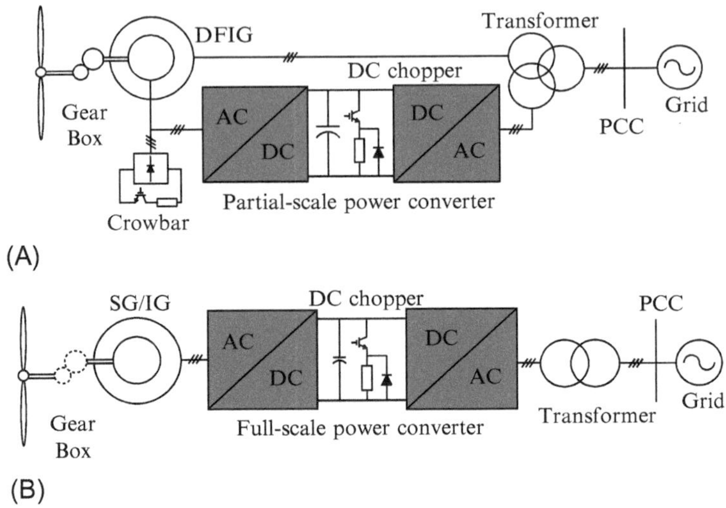

FIG. 10.1 Mainstream wind turbine system with (A) partial-scale power converter; and (B) full-scale power converter.

of slip rings, and also an additional crowbar might be needed to protect the rotor-side converter under grid faults [2,4].

As shown in Fig. 10.1B, a full-scale power converter configuration is considered as an alternative promising technology for multi-MW wind turbine system. The generator stator winding is connected to the grid through the full-scale power converter, which performs the reactive power compensation and a smooth grid connection for the entire speed range of the generator. Some variable speed wind turbine systems are gearless by introducing a multipole generator. The elimination of the slip rings, a simpler gearbox, and full power controllability during the grid faults are the main advantages of such topologies [2,4]. However, in order to satisfy the full power rating in the low-voltage level (below 1 kV), a common approach is to use several power modules or power converters in parallel. Furthermore, the voltage level goes to medium voltage (1–35 kV) is another possibility, which also gives the challenge of voltage stress to the existing power semiconductor technologies.

10.1.2 State-of-Art Control Solutions

Although different wind turbine manufacturers may have their own favorite generators—DFIG, permanent-magnet synchronous generator (PMSG), electrically excited SG/IG, the control structure for the wind turbine system is almost the same. As shown in Fig. 10.2, the general control scheme for modern wind turbine system includes typical three control units [4,5]—power converter control strategy (Level I), wind turbine control strategy (Level II), and grid integration control strategy (Level III). This chapter only focuses on

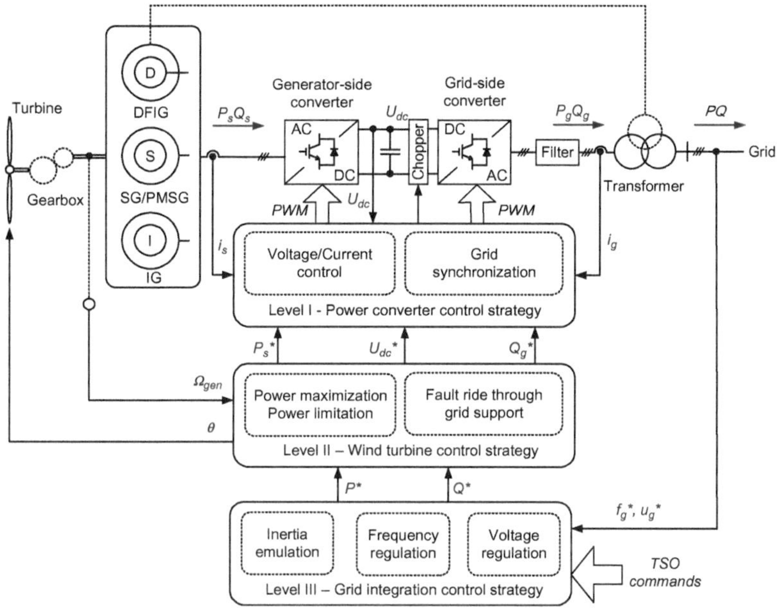

FIG. 10.2 General control strategy for modern wind turbine systems.

Level I—power converter control strategy in the configuration of the DFIG system, where the vector control and the direct control are most commonly used and representing techniques for both the generator-side converter and the grid-side converter [6–8].

The vector control can be further divided into the stator flux oriented control (FOC) and the stator voltage-oriented control (VOC) by different orientations of the synchronous rotating frame. In the case of the FOC, the vector of the stator flux is aligned to the *d*-axis of the rotating frame for the decouple control of the electromagnetic torque and the excitation current. In the case of the VOC, the vector of the stator voltage is aligned to the *d*-axis of the rotating frame for the decouple control of the active and the reactive power [9–11]. Similarly, the direct control can be further categorized into the direct torque control (DTC) and the direct power control (DPC), whose control objectives are the electromagnetic torque together with rotor flux, and the active and the reactive power, respectively [12,13]. Compared with the vector control, although the direct control has the advantages of the easy-implementing algorithm, fast response, and better robust, its switching frequency is not fixed, which results in the fluctuation of the electromagnetic torque, generator current and flux, and the reduced reliability of the mechanical gearbox [13]. As a consequence, the vector control is more popular to be adopted in the wind turbine system. Both the FOC and the VOC enable the good steady-state performance of the DFIG system, but they behave different in respect to the stability of the control.

As mentioned in [6,13], in the case of the stator FOC, the damping of the system is reduced with the increasing excitation current. Moreover, due to the existence of the DC component of the stator flux during the grid abnormal conditions, the reference angle can be distorted remarkably, which induces the instability of the stator FOC. As the rotating speed of the grid voltage is still constant under the abnormal grid conditions, the stability of the VOC can be maintained. Above all, the stator VOC is widely adopted in modern DFIG wind turbine system [14–16].

The remainder of the chapter is organized as follows. It starts with control of the DFIG-based wind turbine system equipped with the partial-scale power converter, where the vector control and the direct control are both in focus. Afterward, the vector control of the PMSG system with full-scale power converter is investigated and verified by the simulation. The summary and the conclusion are drawn at the last section.

10.2 CONTROL OF DFIG-BASED WIND TURBINE SYSTEM

Owing to the fact that the partial-scale power converter is normally realized by a DFIG, while the full-scale power converter normally matches a direct-drive permanent-magnet synchronous generator (PMSG), for simplicity, these two configurations are named as the DFIG system and the PMSG system, respectively. In order to evaluate and compare the performance of these two systems, the mathematical modeling and control of the power converter will be described.

10.2.1 Modeling of DFIG

Fig. 10.3 shows a DFIG system that consists of a generator, a partial-scale power converter, a filter, and a transformer. Since the stator is directly connected to power grid, the DC-link voltage is set as low as possible seen from the power device switching loss. Moreover, the filter inductor is designed to limit the current ripple within 0.25 pu [17].

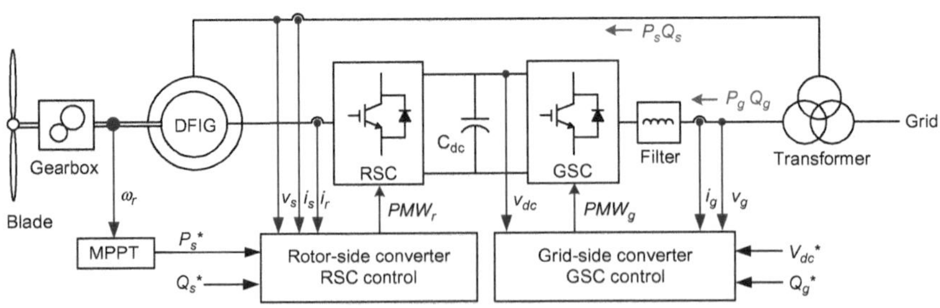

FIG. 10.3 DFIG configuration equipped with back-to-back power converter.

The grid-side converter contains a simple inductor between the voltage source converter and the power grid. However, the rotor-side converter employs a DFIG in between, where the rotor of the DFIG is connected to the converter, and the stator of the DFIG is linked to the power grid. As the DFIG is much more complicated than the inductor, its modeling is in focus in this section.

With the transformation under the synchronous rotating frame and rotor variables referred to the stator-side, the mathematical model of the DFIG can be simplified with reduced order [6,14]. The voltage equation can be expressed as,

$$\begin{cases} u_{sd} = R_s i_{sd} + \dfrac{d}{dt}\psi_{sd} - \omega_1 \psi_{sq} \\ u_{sq} = R_s i_{sq} + \dfrac{d}{dt}\psi_{sq} + \omega_1 \psi_{sd} \\ u'_{rd} = R_r i'_{rd} + \dfrac{d}{dt}\psi'_{rd} - \omega_s \psi'_{rq} \\ u'_{rq} = R_r i'_{rq} + \dfrac{d}{dt}\psi'_{rq} + \omega_s \psi'_{rd} \end{cases} \tag{10.1}$$

where u_s and u_r denote the stator and rotor voltage, i_s and i_r denote the stator and rotor current, R_s and R_r denote the stator and rotor resistance, and ψ_s and ψ_r denote the stator and rotor flux. ω_1 and ω_s represent the stator and slip angular frequency. The subscripts d and q are the variables under d-axis and q-axis, and superscript ´ means the rotor variables referred to the stator-side.

The flux equation can be expressed as,

$$\begin{cases} \psi_{sd} = L_s i_{sd} + L_m i'_{rd} \\ \psi_{sq} = L_s i_{sq} + L_m i'_{rq} \\ \psi'_{rd} = L_m i_{sd} + L_r i'_{rd} \\ \psi'_{rq} = L_m i_{sq} + L_r i'_{rq} \end{cases} \tag{10.2}$$

where L_s, L_r, and L_m represent the stator inductance, the rotor inductance, and magnetizing inductance, respectively.

Based on Eqs. (10.1) and (10.2), the equivalent circuit of the DFIG can be depicted in Fig. 10.4. It can be seen that the d-axis circuit and q-axis circuit are strongly coupled with each other. Besides, many variables are included, like the

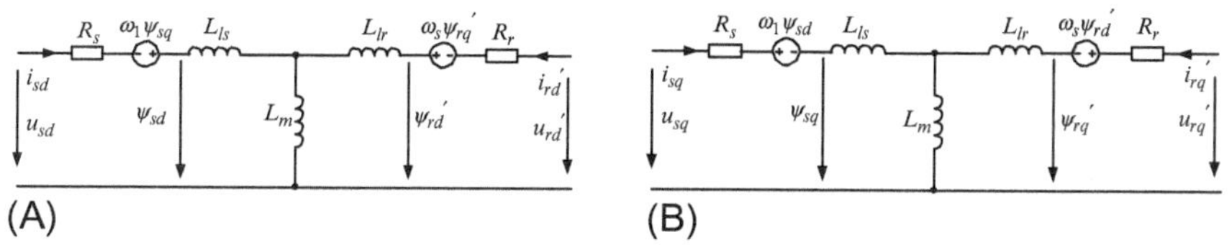

FIG. 10.4 Equivalent circuit of DFIG under synchronous rotating frame. (A) d-axis circuit; (B) q-axis circuit.

voltage, the flux, and the current of the stator-side and rotor-side, which makes the control more difficult.

Different with the rotor or stator flux oriented control (FOC) used in electric motor drives, the stator voltage-oriented control (VOC) is normally applied in the DFIG due to its constant voltage and frequency, as the stator of the generator is directly connected to the grid. The stator VOC is achieved by aligning the d-axis of the synchronous reference frame with the stator voltage vector. The resultant stator voltage under d-axis and q-axis are,

$$\begin{cases} u_{sd} = U_s \\ u_{sq} = 0 \end{cases} \tag{10.3}$$

where U_s is the peak phase voltage of the stator.

During the steady-state operation, the stator flux normally stays constant. Meanwhile, neglecting the stator resistance, usually very low for the high-power induction generator, the relationship between the stator voltage and stator flux can be simplified on basis of Eq. (10.1),

$$\begin{cases} \psi_{sd} = 0 \\ \psi_{sq} = -\dfrac{U_s}{\omega_1} \end{cases} \tag{10.4}$$

Substituting Eq. (10.4) into Eq. (10.2), the stator current can be expressed in term of the rotor current,

$$\begin{cases} i_{sd} = -\dfrac{L_m}{L_s} i'_{rd} \\ i_{sq} = -\dfrac{U_s}{\omega_1 L_s} - \dfrac{L_m}{L_s} i'_{rq} \end{cases} \tag{10.5}$$

As a result, the rotor flux can be expressed by using rotor current,

$$\begin{cases} \psi'_{rd} = \sigma L_r i'_{rd} \\ \psi'_{rq} = -\dfrac{U_s L_m}{\omega_1 L_s} + \sigma L_r i'_{rq} \end{cases} \tag{10.6}$$

where σ denotes the leakage coefficient, and equals $\left(L_s L_r - L_m^2\right)/L_s L_r$.

Afterward, substituting Eq. (10.6) into Eq. (10.3), the relationship between the rotor voltage and the rotor current can be found,

$$\begin{cases} u'_{rd} = R_r i'_{rd} + \sigma L_r \dfrac{d}{dt} i'_{rd} + s_l \dfrac{L_m}{L_s} U_s - \sigma L_r \omega_s i'_{rq} \\ u'_{rq} = R_r i'_{rq} + \sigma L_r \dfrac{d}{dt} i'_{rq} + \sigma L_r \omega_s i'_{rd} \end{cases} \tag{10.7}$$

where s_l denotes the slip of the induction generator, and equals ω_s/ω_1.

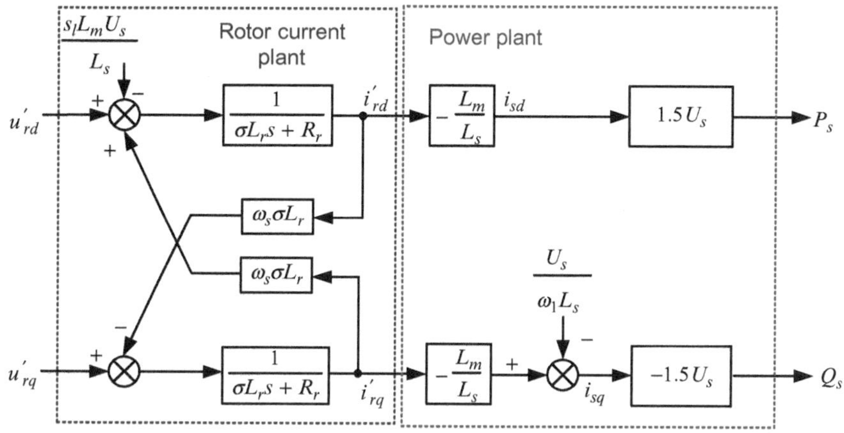

FIG. 10.5 Block diagram of the DFIG mathematical model.

Meanwhile, the stator-side active power P_s and reactive power Q_s can be expressed by,

$$\begin{cases} P_s = \frac{3}{2}U_s i_{sd} = -\frac{3}{2}\frac{L_m U_s}{L_s} i'_{rd} \\ Q_s = -\frac{3}{2}U_s i_{sq} = \frac{3}{2}\left(\frac{U_s^2}{\omega_1 L_s} + \frac{L_m U_s}{L_s} i'_{rq}\right) \end{cases} \tag{10.8}$$

Eventually, the block diagram of the DFIG mathematical model is depicted in Fig. 10.5. It is noted that the equivalent model is much more simplified compared with Fig. 10.4, but the variables under d-axis and q-axis are still coupled with each other.

10.2.2 Vector Control of Back-to-Back Power Converters

The power converter can be divided into the rotor-side converter and the grid-side converter due to their positions. Owing to the DC capacitor decoupling, each of the control schemes can be designed separately.

10.2.2.1 Control of Rotor-Side Converter

Although the generator may operate in the super-synchronous or sub-synchronous mode, one of the control objectives focuses on transferring the produced active power to the grid. The other purpose is to provide the excited-current for the DFIG. The active and reactive power in the stator can be described by rotor *d*-axis and *q*-axis current, respectively, when the stator voltage-oriented control is applied. Therefore, the reference of *d*-axis current is determined by the produced active power, while the reference of *q*-axis is related to desired reactive power in the stator of the DFIG. The cascade control is implemented in the rotor-side converter. Rotor current is controlled as the inner loop, while the stator-side power is regulated as the outer loop (Fig. 10.6).

Design of Inner-Current Controller

Although the control plant from the rotor voltage to the rotor current is illustrated in Fig. 10.5, it is established with rotor-side variables referred to the stator-side. In the real control system, with the relationship of the rotor voltage and current between the stator-side and the rotor-side,

$$\begin{cases} u_r = \dfrac{1}{k_{sr}} u_r' \\ i_r = k_{sr} i_r' \end{cases} \tag{10.9}$$

The control plant referred to the rotor-side can be obtained by using the winding ratio between the stator-side and rotor-side k_{sr} as shown in Fig. 10.7.

The feedforward control is used in the current controller to cancel out the coupling components, but a few proportional changes are required to transfer the variables back to the rotor-side. An inertia unit $G_{PMW}(s)$ is introduced by the digital control [18]. The transfer functions of the rotor current controller $G_{PI\_ir}(s)$, the DFIG model referred to the rotor-side $G_{pl\_ir}(s)$ can be expressed as,

$$G_{PI\_ir}(s) = K_{p\_ir} + \frac{K_{i\_ir}}{s} \tag{10.10}$$

$$G_{pl\_ir}(s) = \frac{k_{sr}^2}{R_r + \sigma L_r s} \tag{10.11}$$

where K_{p_ir} and K_{i_ir} denote the proportional and integral coefficients of the rotor current controller.

Neglecting the DC component and the decoupling effects, the transfer function of the open-loop rotor current $G_{ol\_ir}(s)$ can be achieved as shown in Fig. 10.8, and it yields,

$$G_{ol\_ir}(s) = G_{PI\_ir}(s) G_{PWM}(s) G_{pl\_ir}(s) \tag{10.12}$$

According to the parameters of the 2 MW DFIG, the switching period and the current controller are listed in Table 10.1. The Bode plots of the control plant, current controller, and open-loop rotor current can be obtained as shown in Fig. 10.9.

The amplitude curve of the control plant starts with the constant magnitude, but it bends to −20 dB/dec due to its 2.4 Hz pole. Besides, a 212.3 Hz pole of the delay unit causes another −20 dB/dec bending. In order to realize the amplitude curve of the open-loop rotor current to crossover 0 dB with the slope −20 dB/dec, its bandwidth is regulated at 100 Hz, 1/20 of the 2 kHz switching frequency.

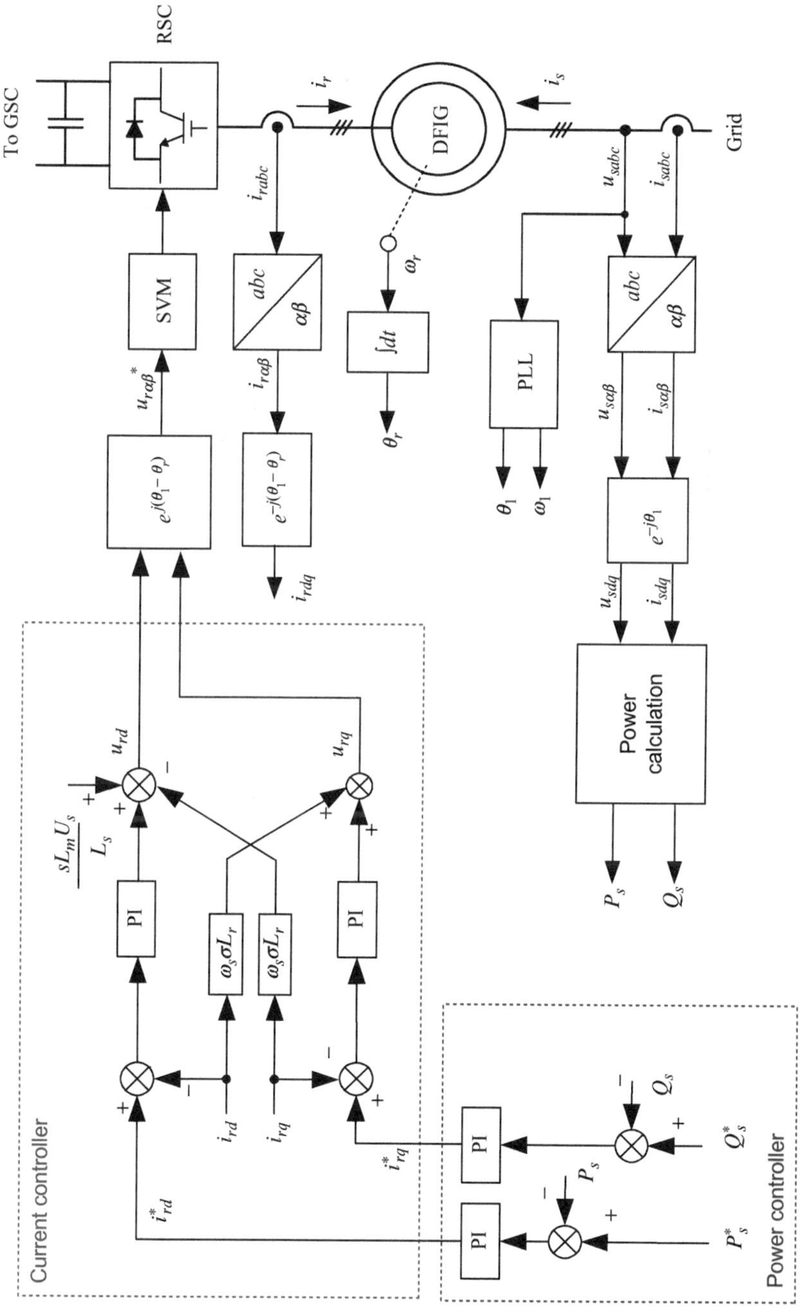

FIG. 10.6 Control block diagram of the rotor-side converter for the DFIG system.

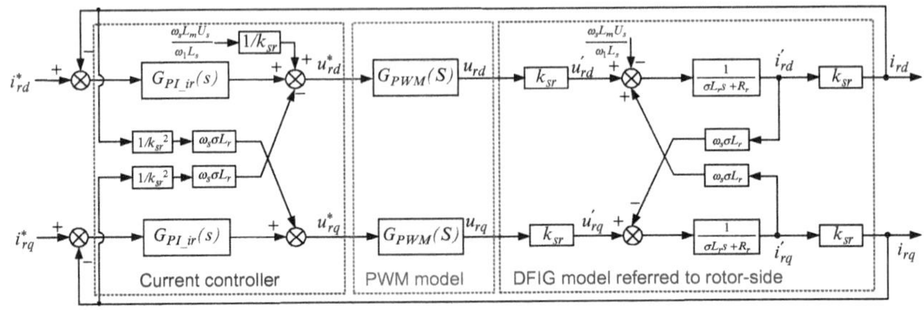

FIG. 10.7 Control block diagram of the rotor current loop.

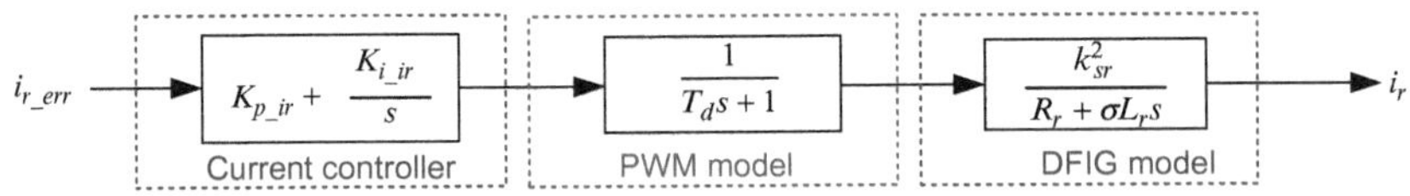

FIG. 10.8 Transfer function of open-loop rotor current.

TABLE 10.1 Parameters of Rotor-side Converter and Generators in 2 MW DFIG System (Referred to Stator-Side)

	2 MW
Stator peak phase voltage, U_s (V)	563
Stator resistance, R_s (mΩ)	1.69
Stator inductance, L_s (mH)	2.95
Rotor resistance, R_r (mΩ)	1.52
Rotor inductance, L_r (mH)	2.97
Magnetizing inductance, L_m (mH)	2.91
Leakage coefficient, σ	0.03
Winding ratio between stator and rotor, k_{sr}	0.369
Switching frequency, f_{sw} (kHz)	2
Delay time introduced by PWM, T_d (μs)	625
Proportional coefficient of current controller, K_{p_ir}	0.5
Integral coefficient of current controller, K_{i_ir}	7.5

Meanwhile, the zero of the PI controller is designed to counteract the pole of the control plant for the sake of the constant −20 dB/dec before the crossover frequency. Viewed from the mitigation of the switching harmonics, a phase margin of 64.8° can be achieved.

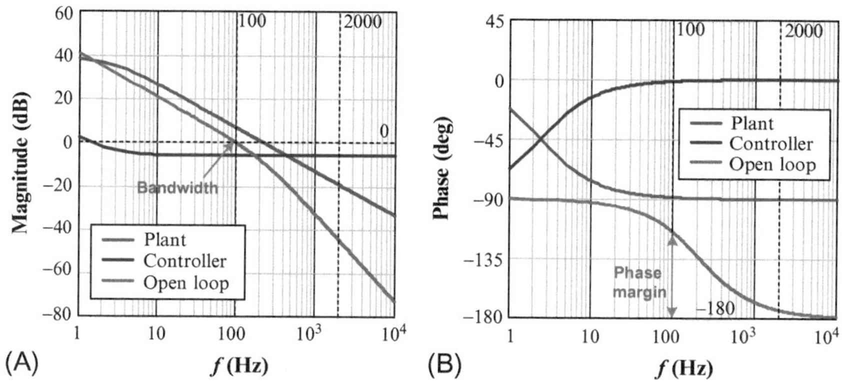

FIG. 10.9 Bode plots of open-loop rotor current in terms of (A) amplitude characteristics; and (B) phase characteristics.

Design of Outer-Power Controller

According to Eq. (10.8), as the leakage inductance is much smaller than the magnetizing inductance, the rotor current in *d*-axis is only determined by the stator active power, which results in the open-loop control of the active power. However, the rotor current in q-axis is not only related to the reactive power, but also coupled to the stator inductance. Due to the reduced magnetizing inductance during the normal operation, the rotor reference in *q*-axis cannot be determined by the defined reactive power. As a consequence, a closed-loop control of the reactive power is required due to the sensitivity of rotor reference in *q*-axis with parameters of the induction generator. The control of the active power is open-loop, while the control of the reactive power is closed-loop. As a result, this section only deals with the controller design of the reactive power.

According to Eq. (10.5), considering $U_s/(\omega_1 L_s)$ as a DC bias, the stator current can be expressed by the rotor current. It is worth mentioning that a proportion unit is required to transfer the rotor current back to the stator-side. Together with Eq. (10.8), the control plant of the reactive power $G_{pl\_Q}(s)$ can be established as shown in Fig. 10.10. A PI controller $G_{PI\_Q}(s)$ is applied to realize the reactive power closed-loop. In order to cancel out the DC bias from the control plant, a feedforward control is used. Meanwhile, the control of the rotor current

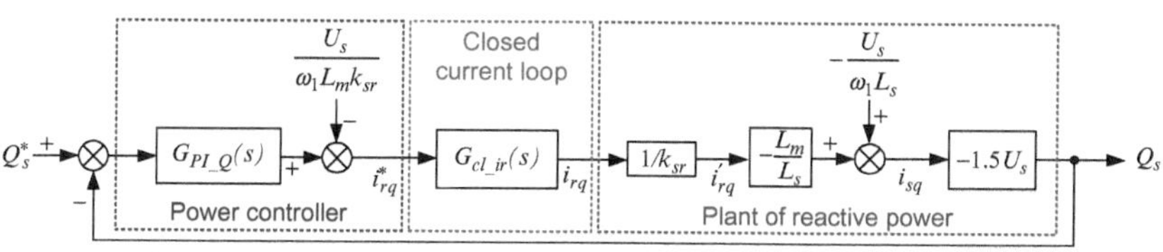

FIG. 10.10 Control block diagram of the reactive power loop.

can be considered as the closed-loop unit $G_{cl\_ir}(s)$. It is noted that the transfer function of the power controller, the closed rotor current loop, and the control plant can be expressed by,

$$G_{PI\_Q}(s) = K_{p\_Q} + \frac{K_{i\_Q}}{s} \tag{10.13}$$

$$G_{cl\_ir}(s) = \frac{G_{ol\_ir}(s)}{1 + G_{ol\_ir}(s)} \tag{10.14}$$

$$G_{pl\_Q}(s) = \frac{1.5U_s L_m}{k_{sr} L_s} \tag{10.15}$$

where K_{p_Q} and K_{i_Q} denote the proportional and integral coefficients of the power controller, respectively.

By neglecting the disturbances, the transfer function of the open-loop reactive power $G_{pl\_Q}(s)$ can be found as shown in Fig. 10.11, and it yields,

$$G_{ol\_Q}(s) = G_{PI\_Q}(s) G_{cl\_ir}(s) G_{pl\_Q}(s) \tag{10.16}$$

To independently design the inner loop and the outer loop, the design principle of the reactive power PI controller is to regulate the bandwidth of outer loop between 1/20 and 1/10 of the inner loop. At the same time, the amplitude curve is required to cross over the bandwidth with the slope rate −20 dB/dec seen from the stability point of view.

With the PI parameters of the power controller (Table 10.2), the Bode plots of the open-loop reactive power are depicted in Fig. 10.12. It is evident that the amplitude curve and the phase curve of the plant keeps constant, as it is a proportion unit. Viewed from the controller perspective, a pole at zero frequency is required for the zero error at the steady-state operation. At the same time, the turning point of the PI controller is set at 24 Hz. The bandwidth of

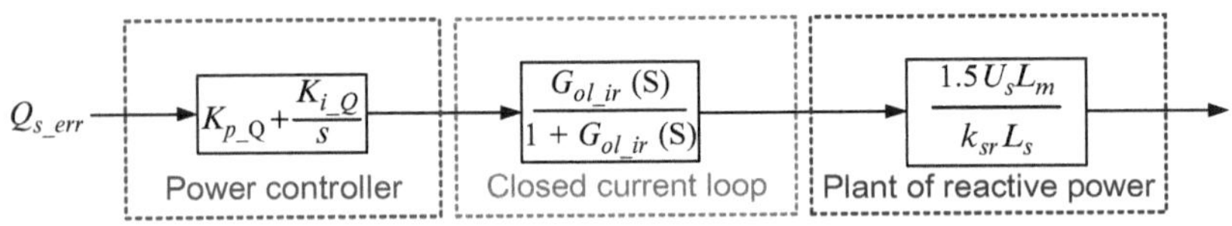

FIG. 10.11 Transfer function of open-loop reactive power.

TABLE 10.2 PI Parameters of Power Controller in 2 MW DFIG Systems

	2 MW
Proportional coefficient, K_{p_Q}	0.00009
Integral coefficient, K_{i_Q}	0.0135

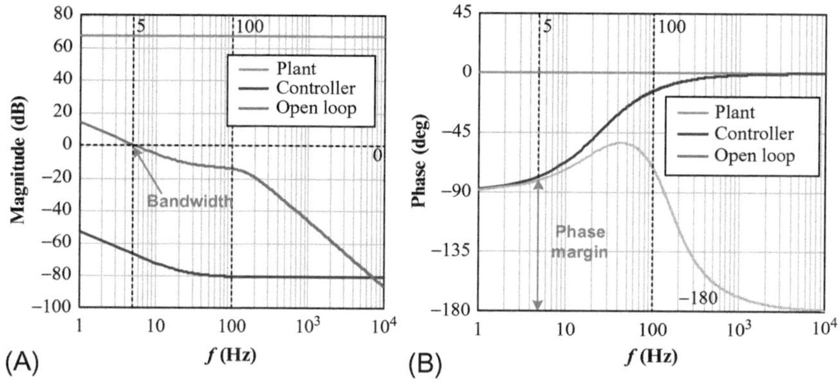

FIG. 10.12 Bode plots of open-loop reactive power in terms of (A) amplitude characteristics; and (B) phase characteristics.

the reactive power loop is set at 5 Hz with phase margin of 99.1°. Additionally, if the frequency is higher than the inner-current loop, it can be seen that the amplitude curve bends with the slope −40 dB/dec, and the phase angle approaches −180° due to two poles introduced by the transfer function of the closed-loop rotor current.

10.2.2.2 Control of Grid-Side Converter

The grid-side converter keeps the DC-link voltage fixed and meets the reactive power demand according to the grid codes. As shown in Fig. 10.13, the active and reactive power can simply be controlled by *d*-axis and *q*-axis current using the grid voltage-oriented control. This control strategy contains two cascaded loops. The inner loop takes care of the grid current; the outer loop regulates the DC-link voltage and reactive power for the grid-side converter. The DC-link voltage is closely dependent on the active power, thus the adjustment can be regarded as the reference of the *d*-axis current. Normally, a unity power factor is required, in other words, the reference of *q*-axis is set to zero unless the grid operator needs reactive power compensation. Furthermore, in order to efficiently utilize the DC voltage, the space vector modulation (SVM) is used to generate the switching signals.

If a single inductance is used as filter as shown in Fig. 10.3, defining that the *d*-axis is aligned with the supply voltage angle, the voltage and current relationship between the grid and the converter output are,

$$\begin{cases} u_{cd} = u_{gd} - R_g i_{gd} - L_g \dfrac{d}{dt} i_{gd} + \omega_1 L_g i_{gq} \\ u_{cq} = u_{gq} - R_g i_{gq} - L_g \dfrac{d}{dt} i_{gq} - \omega_1 L_g i_{gd} \end{cases} \tag{10.17}$$

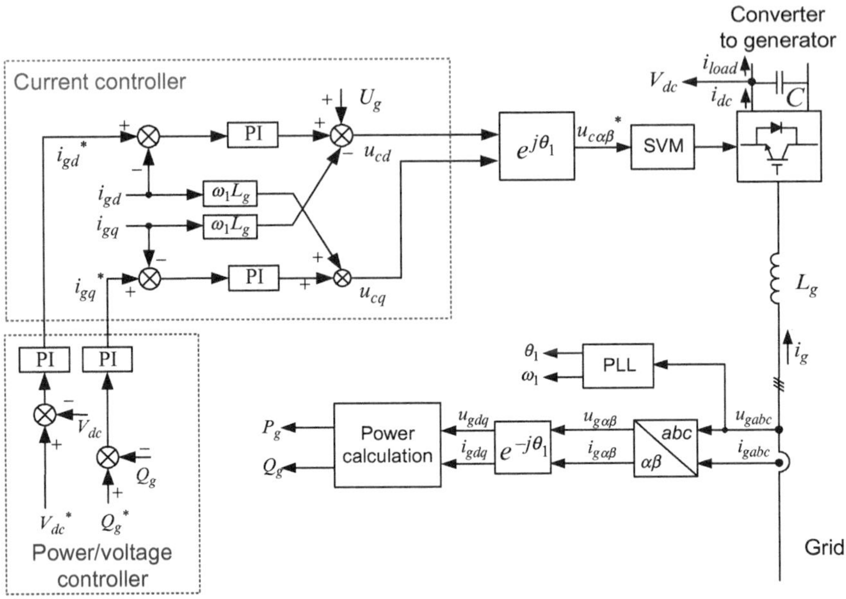

FIG. 10.13 Control block diagram of the grid-side converter for the DFIG system.

where u_g and u_c denote the grid and converter voltage, i_g denotes the grid current, R_g denotes the equivalent resistance of the grid filter, and L_g denotes inductance of the grid filter. The subscripts d and q are the variables under d-axis and q-axis, respectively. As a result, the equivalent circuit of the grid-side converter under the synchronous rotating reframe is shown in Fig. 10.14.

Based on Eq. (10.17), the plant of the grid filter can be described in Fig. 10.15. Meanwhile, the active and reactive power is solely related to the d-axis and q-axis current through the grid filter

$$\begin{cases} P_g = \dfrac{3}{2} U_g i_{gd} = V_{dc} i_{dc} \\ Q_g = -\dfrac{3}{2} U_g i_{gq} \end{cases} \tag{10.18}$$

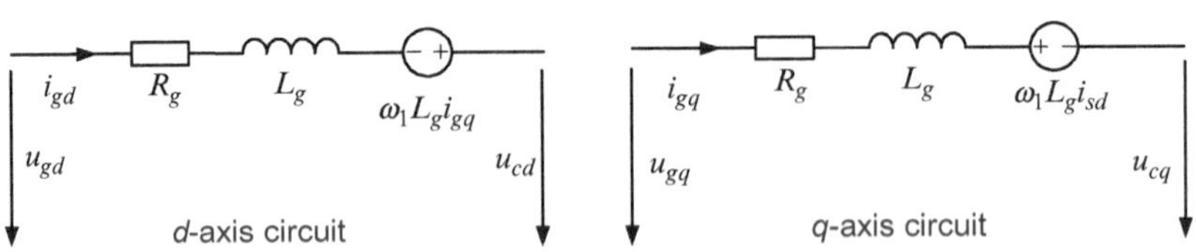

FIG. 10.14 Equivalent circuit of the grid-side converter for the DFIG system under synchronous rotating frame.

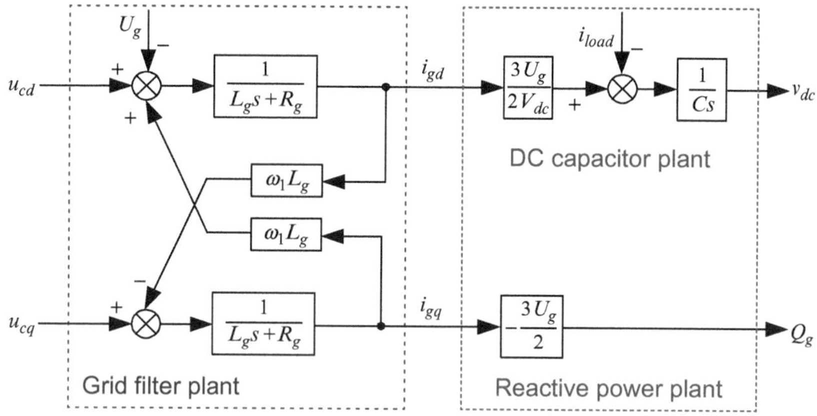

FIG. 10.15 Block diagram of grid-side converter mathematical model.

Neglecting the power loss of the power converter, the power balance can be achieved between the converter AC side and DC side. Thus, the DC voltage can be calculated by the integration of the current through the DC-link capacitor as shown in Fig. 10.15. As a consequence, the control plant of the DC capacitor and reactive power can be derived as well. In respect to the inner-current loop and outer power/voltage loop, the design procedure of the PI controller is the same with the rotor-side converter.

10.2.3 Direct Power Control of Back-to-Back Power Converters

Besides the popular VOC control of the DFIG-based wind turbine system, the direct power control (DPC) strategy is also well investigated for decades. The DPC technique has been proved to be prominent for the DFIG control, such as simple implementation, fast dynamic response, robustness against parameter variations, and grid disturbance. Conventional DPC adopts the look-up table to determine the on-off state of the IGBT switches, however this method unfortunately results in the variable switching frequency, which places challenges to the filter design. Therefore, in order to overcome the traditional DPC drawback of variable switching frequency, the DPC integrated with space vector modulation (DPC-SVM) will be introduced in this section.

10.2.3.1 RSC Mathematical Modeling Adopting DPC

The DFIG stator flux $\boldsymbol{\psi}_{sdq}$ and rotor flux $\boldsymbol{\psi}_{rdq}$ can be presented respectively as,

$$\boldsymbol{\psi}_{sdq} = L_s \mathbf{I}_{sdq} + L_m \mathbf{I}_{rdq} \tag{10.19}$$

$$\boldsymbol{\psi}_{rdq} = L_m \mathbf{I}_{sdq} + L_r \mathbf{I}_{rdq} \tag{10.20}$$

where $\boldsymbol{\psi}$ is flux; $\mathbf{I}$ is current; subscripts d and q represent components at the d and q axes rotating at the grid angular speed of ω_1; subscripts s and r represent stator and rotor components of DFIG, respectively. $L_s = L_{\delta s} + L_m$ and $L_r = L_{\delta r} + L_m$ are total self-inductances of stator and rotor winding, $L_{\delta s}$, $L_{\delta r}$, and L_m are stator and rotor leakage inductances and mutual inductance, respectively.

Based on Eqs. (10.19) and (10.20), the stator current and rotor current can be written as,

$$\mathbf{I}_{sdq} = \frac{L_m}{L_r L_s - L_m^2}\left(\frac{L_r}{L_m}\boldsymbol{\psi}_{sdq} - \boldsymbol{\psi}_{rdq}\right) \tag{10.21}$$

$$\mathbf{I}_{rdq} = \frac{L_m}{L_r L_s - L_m^2}\left(\frac{L_s}{L_m}\boldsymbol{\psi}_{rdq} - \boldsymbol{\psi}_{sdq}\right) \tag{10.22}$$

The stator and rotor voltages $\mathbf{U}_{sdq}$ and $\mathbf{U}_{rdq}$ in the dq reference frame can be expressed as,

$$\mathbf{U}_{sdq} = R_s \mathbf{I}_{sdq} + d\boldsymbol{\psi}_{sdq}/dt + j\omega_1 \boldsymbol{\psi}_{sdq} \tag{10.23}$$

$$\mathbf{U}_{rdq} = R_r \mathbf{I}_{rdq} + d\boldsymbol{\psi}_{rdq}/dt + j\omega_s \boldsymbol{\psi}_{rdq} \tag{10.24}$$

where $\mathbf{U}$ is the voltage, R_s and R_r are stator and rotor resistances, ω_r is rotor angular speed, and $\omega_s = \omega_1 - \omega_r$ is slip angular speed.

The DFIG stator instantaneous active and reactive powers can be expressed as,

$$P_s + jQ_s = \frac{3}{2}\mathbf{U}_{sdq} \cdot \hat{\mathbf{I}}_{sdq} \tag{10.25}$$

where $\hat{\mathbf{I}}_{sdq}$ is the conjugated space vector of $\mathbf{I}_{sdq}$, P_s and Q_s are stator active and reactive power.

When the d-axis of the synchronous reference frame is aligned with the stator voltage vector, the differential of stator flux $\boldsymbol{\psi}_{sdq}$ is zero. Assuming that the stator resistance is ignored, Eq. (10.23) can be written as,

$$\mathbf{U}_{sdq} = j\omega_1 \boldsymbol{\psi}_{sdq} = -\omega_1 \psi_{sq} = U_{sd} \tag{10.26}$$

Substituting Eqs. (10.21) and (10.26) into (10.25), the stator active and reactive power can be yielded as,

$$\begin{aligned} P_s + jQ_s &= \frac{3}{2}U_{sd}\frac{L_m}{L_r L_s - L_m^2}\left(\frac{L_r}{L_m}\hat{\boldsymbol{\psi}}_{sdq} - \hat{\boldsymbol{\psi}}_{rdq}\right) \\ &= k_\sigma U_{sd}\left(\frac{L_r}{L_m}(\psi_{sd} - j\psi_{sq}) - (\psi_{rd} - j\psi_{rq})\right) \\ &= -k_\sigma U_{sd}\psi_{rd} + jk_\sigma U_{sd}\left(\frac{L_r}{L_m}\frac{U_{sd}}{\omega_1} + \psi_{rq}\right) \end{aligned} \tag{10.27}$$

where $k_\sigma = \frac{3}{2}\frac{L_m}{L_s L_r - L_m^2}$

Therefore, the stator active and reactive power can be written respectively as,

$$P_s = -k_\sigma U_{sd} \psi_{rd} \tag{10.28}$$

$$Q_s = k_\sigma U_{sd}\left(\psi_{rq} + \frac{L_r}{L_m}\frac{U_{sd}}{\omega_1}\right) \tag{10.29}$$

Based on Eqs. (10.28) and (10.29), the rotor flux can be shown as,

$$\psi_{rd} = -\frac{1}{k_\sigma U_{sd}} P_s \tag{10.30}$$

$$\psi_{rq} = \frac{Q_s}{k_\sigma U_{sd}} - \frac{L_r}{L_m}\frac{U_{sd}}{\omega_1} \tag{10.31}$$

Eq. (10.24) can be separated into d-axis and q-axis component and rewritten as,

$$U_{rd} = R_r I_{rd} + d\psi_{rd}/dt - \omega_s \psi_{rq} \tag{10.32}$$

$$U_{rq} = R_r I_{rq} + d\psi_{rq}/dt + \omega_s \psi_{rd} \tag{10.33}$$

Besides, based on Eqs. (10.30) and (10.31), during a constant sampling time period T_s, the differential of rotor flux can be calculated as,

$$\frac{d\psi_{rd}}{dt} = -\frac{1}{k_\sigma U_{sd}}\frac{dP_s}{dt} = -\frac{1}{k_\sigma U_{sd}}\frac{P_s^* - P_s}{T_s} \tag{10.34}$$

$$\frac{d\psi_{rq}}{dt} = \frac{1}{k_\sigma U_{sd}}\frac{dQ_s}{dt} = \frac{1}{k_\sigma U_{sd}}\frac{Q_s^* - Q_s}{T_s} \tag{10.35}$$

where P^*_s and Q^*_s are the stator active and reactive power reference, respectively.

Substituting Eqs. (10.30) and (10.31), (10.34) and (10.35) into (10.32) and (10.33), and neglecting the rotor resistance, the rotor control reference voltage $\mathbf{U}_{rdqdpc}$ for the DFIG control based on DPC can be written as,

$$\begin{aligned}\mathbf{U}_{rddpc} &= \mathbf{V}_{rddpc} + \mathbf{E}_{rddpc} = -C_{power}\left(P_s^* - P_s\right) - \omega_s\left(\frac{Q_s}{k_\sigma U_{sd}} - \frac{L_r U_{sd}}{L_m \omega_1}\right) \\ &= -C_{PI}\left(P_s^* - P_s\right) - \omega_s\left(\frac{Q_s}{k_\sigma U_{sd}} - \frac{L_r U_{sd}}{L_m \omega_1}\right)\end{aligned} \tag{10.36}$$

$$\begin{aligned}\mathbf{U}_{rqdpc} &= \mathbf{V}_{rqdpc} + \mathbf{E}_{rqdpc} = C_{power}\left(Q_s^* - Q_s\right) - \omega_s \frac{P_s}{k_\sigma U_{sd}} \\ &= C_{PI}\left(Q_s^* - Q_s\right) - \omega_s \frac{P_s}{k_\sigma U_{sd}}\end{aligned} \tag{10.37}$$

where $\mathbf{U}_{rdqdpc}$ is the rotor control reference voltage, $\mathbf{V}_{rdqdpc}$ is the output voltage of the power controller, $\mathbf{E}_{rdqdpc}$ is the decoupling compensation voltage, and

C_{power} is the proper stator power regulator to restrain the power regulation error. It would be essential to choose proper regulator C_{power} to achieve zero tracking error, and a PI controller is normally adopted.

10.2.3.2 GSC Mathematical Modeling Adopting DPC

The relationship between the AC side voltage of the GSC (which is also the control output voltage of the GSC) and its flux can be presented as,

$$\mathbf{V}_{cdq} = \frac{d\boldsymbol{\Psi}_{cdq}}{dt} + j\omega_1 \boldsymbol{\Psi}_{cdq} \tag{10.38}$$

On the other hand, the relationship between the three-phase AC grid voltage and its flux can be similarly presented as,

$$\mathbf{U}_{gdq} = \frac{d\boldsymbol{\Psi}_{gdq}}{dt} + j\omega_1 \boldsymbol{\Psi}_{gdq} \tag{10.39}$$

According to the mathematical modeling of the GSC, the three-phase AC grid voltage can be expressed as,

$$\mathbf{U}_{gdq} = R_g \mathbf{I}_{gdq} + L_g \frac{d\mathbf{I}_{gdq}}{dt} + j\omega_1 \left(\boldsymbol{\Psi}_{gdq} - \boldsymbol{\Psi}_{cdq}\right) + \mathbf{V}_{cdq} \tag{10.40}$$

Substituting Eqs. (10.38) and (10.39) into Eq. (10.40) gives out the following,

$$\boldsymbol{\Psi}_{gdq} = L_g \mathbf{I}_{gdq} + \boldsymbol{\Psi}_{cdq} \tag{10.41}$$

Based on Eq. (10.41), the output active power and reactive power of the GSC can be presented as,

$$P_g + jQ_g = \frac{3}{2}\mathbf{U}_{gdq}\hat{\mathbf{I}}_{gdq} = \frac{3}{2}\mathbf{U}_{gdq}\frac{1}{L_g}\left(\hat{\boldsymbol{\Psi}}_{gdq} - \hat{\boldsymbol{\Psi}}_{cdq}\right) \tag{10.42}$$

According to Eq. (10.42), the GSC output active and reactive power can be respectively expressed as,

$$P_g = -\frac{3}{2}\frac{u_{gd}}{L_g}\psi_{cd} \tag{10.43}$$

$$Q_g = \frac{3}{2}\frac{u_{gd}}{L_g}\left(\psi_{cq} + \frac{u_{gd}}{\omega_1}\right) \tag{10.44}$$

Then, based on Eqs. (10.38) and (10.43), the d-axis component of the DPC control voltage in the GSC can be deduced as,

$$\begin{aligned} U_{cddpc} &= V_{gddpc} + E_{gddpc} = -G_{PQ}(s)\left(P_g^* - P_g\right) - \frac{2}{3}\frac{\omega_1 L_g}{u_{gd}}Q_g + u_{gd} \\ &= -C_{PI}(s)\left(P_g^* - P_g\right) - \frac{2}{3}\frac{\omega_1 L_g}{u_{gd}}Q_g + u_{gd} \end{aligned} \tag{10.45}$$

Similarly, based on Eqs. (10.38) and (10.44), the q-axis component of the DPC control voltage in the GSC can be deduced as,

$$\begin{aligned} U_{cqdpc} &= V_{gqdpc} + E_{gqdpc} = G_{PQ}(s)\left(Q_g^* - Q_g\right) - \frac{2\omega_1 L_g}{3\ u_{gd}} P_g \\ &= C_{PI}(s)\left(Q_g^* - Q_g\right) - \frac{2\omega_1 L_g}{3\ u_{gd}} P_g \end{aligned} \tag{10.46}$$

where $\mathbf{U}_{cdqdpc}$ is the grid-side converter control reference voltage, $\mathbf{V}_{gdqdpc}$ is the output voltage of the grid-side power controller, and $\mathbf{E}_{gdqdpc}$ is the decoupling compensation voltage.

Since the GSC is responsible for maintaining constant DC-link voltage V_{dc}, the active power reference can be obtained as the following,

$$P_g^* = V_{dc}\left(k_{vp} + k_{vi}/s\right)\left(V_{dc}^* - V_{dc}\right) \tag{10.47}$$

where V_{dc}^* is the DC-link voltage reference, and k_{vp} and k_{vi} are proportional and integral parameters for the DC-link voltage PI controller, respectively.

In the practical situation, the GSC is normally required to operate with unity power factor, thus its reactive power is set at zero as the following,

$$Q_g^* = 0 \tag{10.48}$$

10.2.3.3 Control Block Diagram of the DPC-Based DFIG System

The proposed DPC control strategy for the DFIG system including both the RSC and GSC is shown in Fig. 10.16. For the RSC control block diagram shown in Fig. 10.16A, firstly, the grid voltage phase angle is obtained through the PLL, and the rotor position and speed are achieved by the encoder. The stator active and reactive power can be calculated by the sampled three-phase stator voltage and current. The stator active and reactive power control error, which is the input of PI regulator, can be calculated according to the actual signal and reference signal. The output of PI regulator, together with the compensation back electromagnetic force in Eqs. (10.36) and (10.37), is sent to the SVM to generate the IGBT switching signals.

Similarly, for the GSC control block diagram shown in Fig. 10.16B, the grid voltage phase angle is also obtained through the PLL. The grid active and reactive power can be calculated by the sampled three-phase grid voltage and current. The active power reference can be obtained through the closed-loop control of the DC-link voltage, while the reactive power reference is set at zero due to the unity power factor. Then, the power control error can be regulated by the PI controller. The output of PI regulator, together with the compensation back electromagnetic force in Eqs. (10.45) and (10.46), is sent to the SVM to generate the IGBT switching signals.

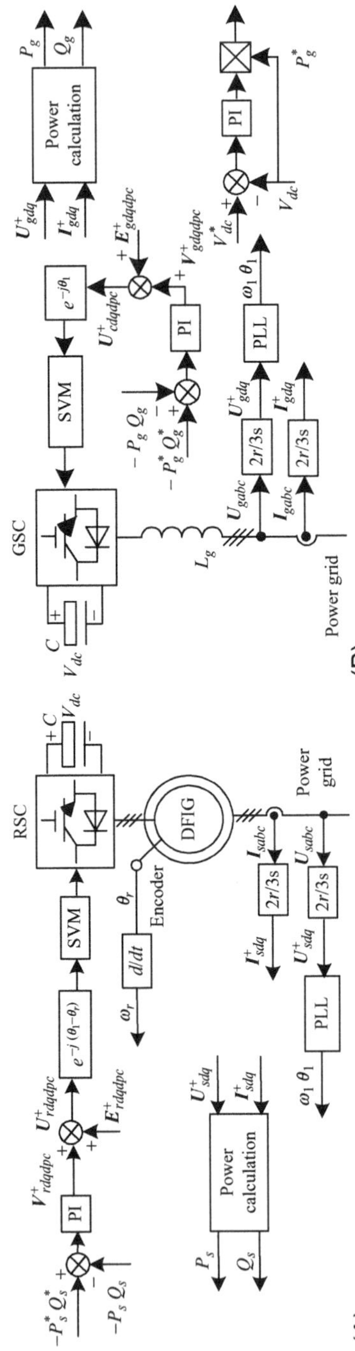

FIG. 10.16 Control block diagram of the DPC-based DFIG system. (A) RSC control strategy; (B) GSC control strategy.

10.2.4 Simulation Validation

In order to validate the vector and direct control for the DFIG system, the simulation result using MATLAB/Simulink is provided. The rated power of the simulated DFIG system is 2 MW.

10.2.4.1 Vector Control

The simulation conditions of the RSC exactly follow the aforementioned parameters, while the parameters of the GSC are listed in Table 10.3. Then, the step changes can be performed at the various loops in order to verify the bandwidth design.

For the rotor-side converter as shown in Fig. 10.17A, if a rotor current step of 0.2 pu and a reactive power step of 0.2 pu are injected at time instant of 2.5 and 3.2 s, their rise time of 3.0 and 70 ms can be observed, respectively. Both of them are consistent with the bandwidth design of 100 Hz inner-current loop and 5 Hz outer power loop.

As the d-axis and the q-axis current loop of the grid-side converter have the same transfer function, the bandwidth of the current loop can be simply verified by using a step response at the q-axis reference. As shown in Fig. 10.17B, a 0.1 pu overexcited reactive current is set to zero, and the rise time is around 3.0 ms. Similarly, a step change of the DC-link voltage from 1200 to 1050 V is set, and the rise time around 70 ms can be observed.

10.2.4.2 Direct Power Control

With the controller parameters listed in Table 10.4, Fig. 10.18 shows the simulation waveforms of the DFIG system using the DPC strategy. It can be seen that the DPC strategy ensures a smooth and stable operation of the DFIG system

TABLE 10.3 Parameters of Grid-Side Converter in 2 MW DFIG System

Power Rating	2 MW
Grid peak phase voltage, U_g (V)	563
DC-link voltage, V_{dc} (V)	1050
Filter inductance, L_g (mH)	0.5
Switching frequency, f_{sw} (kHz)	2
Proportional coefficient of current controller, K_{p_ig}	0.3
Integral coefficient of current controller, K_{i_ig}	15
Proportional coefficient of voltage controller, K_{p_vdc}	2
Integral coefficient of voltage controller, K_{i_vdc}	10

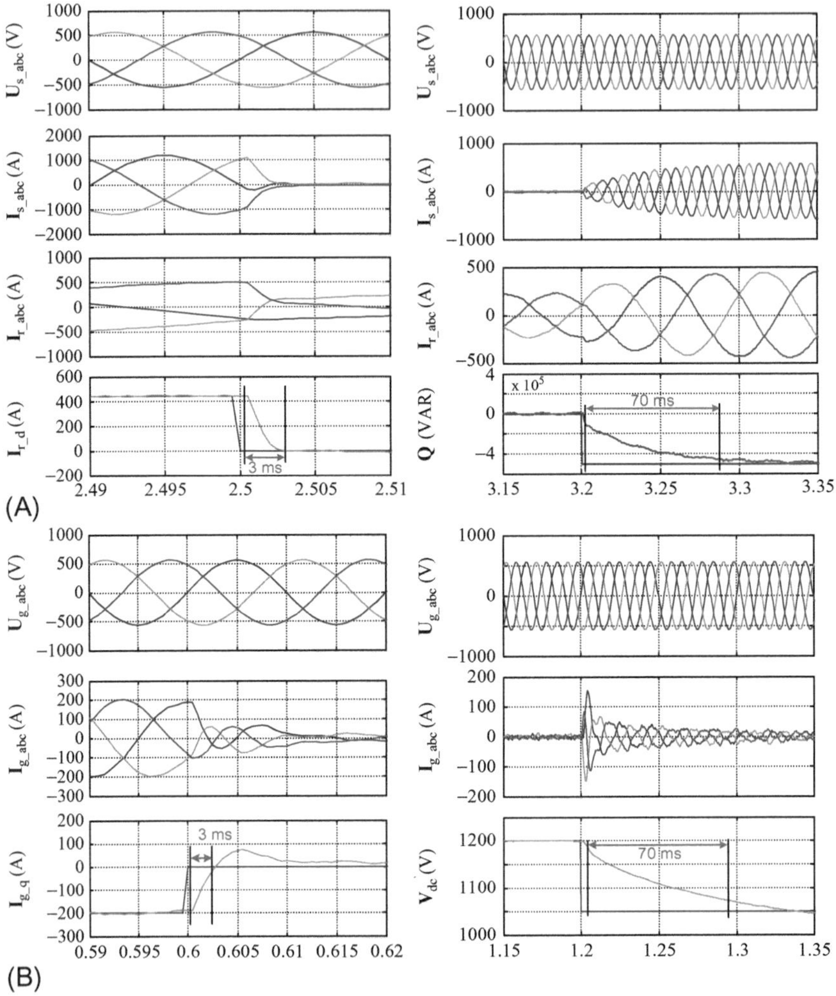

FIG. 10.17 Step response simulation of the inner loop and outer loop of the DFIG system by using the vector control. (A) Rotor-side converter; (B) grid-side converter.

with the output active power of 1.0 p.u. and reactive power of 0.0 p.u. Besides, the stator current and rotor current as well as the grid-side current are able to maintain sinusoidal. The DC-link voltage is also able to be kept of 1200 V.

10.3 CONTROL OF PMSG-BASED WIND TURBINE SYSTEM

After the transition from the constant-speed squirrel-cage induction generator to the variable-speed generator, a number of generator types are adopted by the wind turbine manufacturers and the most optimum concept is still under discussion [2,4,19–21]. Initially, the wind turbine system equipped with the DFIG

TABLE 10.4 Controller Parameters of Direct Power Control in 2 MW DFIG System

Grid/stator peak phase voltage (sV)	563
Rotor-side converter	
Proportional coefficient of power controller, K_{p_PsQs}	1.2
Integral coefficient of power controller, K_{i_PsQs}	15
Grid-side converter	
Proportional coefficient of power controller, K_{p_PgQg}	2.0
Integral coefficient of power controller, K_{i_PgQg}	15
Proportional coefficient of voltage controller, K_{p_vdc}	12
Integral coefficient of voltage controller, K_{i_vdc}	75

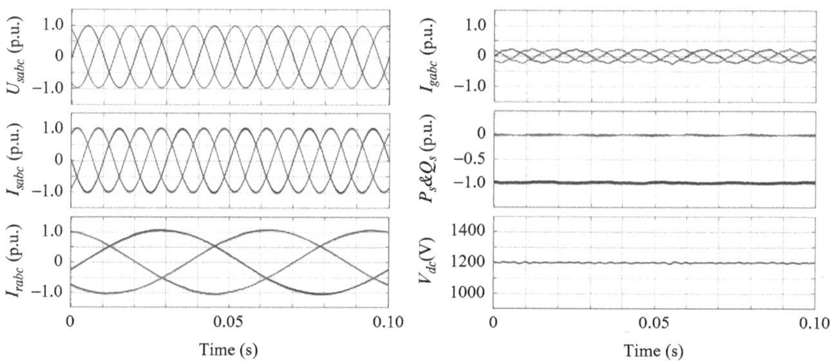

FIG. 10.18 Simulation waveforms of the DFIG system using direct power control.

became attractive due to its traditional generator technology, having an affordable power converter as well as the fully controllability of the active and reactive power [22]. However, with the steady increase of the wind power penetration, grid codes are updated regularly and have become stricter [23,24], which prevents an overwhelming use of this partial-scale power converter-based configuration, because of its poorer low-voltage ride-through capability as discussed in refs. [25,26]. Correspondingly, more and more manufacturers turn to the solution based on the full-scale power converter, whereas the generator type is still uncertain. Synchronous generators (SGs) have been widely used, including permanent-magnet and wound rotor generators, salient and nonsalient pole generators, and the generators with external and internal rotors [27]. The SG-based wind turbine is normally controlled by full-scale power converters. With this structure, it is able to meet various grid codes, including the reactive power compensation and fault ride-through operation, without the need for additional equipment. In the case of the PMSG application, the elimination of slip rings, a simpler gearbox, and better grid support ability are the main advantages compared with the DFIG concept. Nevertheless, it causes more expensive power electronic converters and higher loss dissipation [28].

10.3.1 Modeling of PMSG

As shown in Fig. 10.19, the block diagram of a typical PMSG-based wind turbine system consists of a wind turbine, a gearbox (optional), a synchronous generator, power converters, and a transformer for grid connection. As discussed in ref. [14], a gearbox with a high gear ratio is often required to match the low turbine speed to the high generator speed, especially for the induction generator. However, the gearbox can become one-stage or can even be eliminated, in which a low-speed generator is used to match the turbine speed. The elimination of the gearbox requires use of a generator with high number of poles. Such a generator is more expensive and heavier than one with a small number of poles. The direct-drive wind turbine is competitive for the offshore application, where the system maintenance is costly and inconvenient.

The full-scale power converter consists of a machine-side converter and a grid-side converter. Compared with the DFIG system, the current through both converters is much higher under same power rating of the wind turbines, which means the selection of power devices in the two configurations become quite different in order to realize similar loading. Nevertheless, the design method of the DC-link voltage and the filter inductor in the PMSG system can be referred to the DFIG system.

In order to achieve an independent control of the active and reactive power, d-axis and q-axis equivalent circuits are widely used in the modern drive system. It is modeled as shown in Fig. 10.20, and the stator voltage at the d-axis u_{sd} and at the q-axis u_{sq} can be expressed as,

$$u_{sd} = R_s i_{sd} + L_s \frac{di_{sd}}{dt} - \omega_e L_s i_{sq} \tag{10.49}$$

$$u_{sq} = R_s i_{sq} + L_s \frac{di_{sq}}{dt} + \omega_e L_s i_{sd} + \omega_e \psi_m \tag{10.50}$$

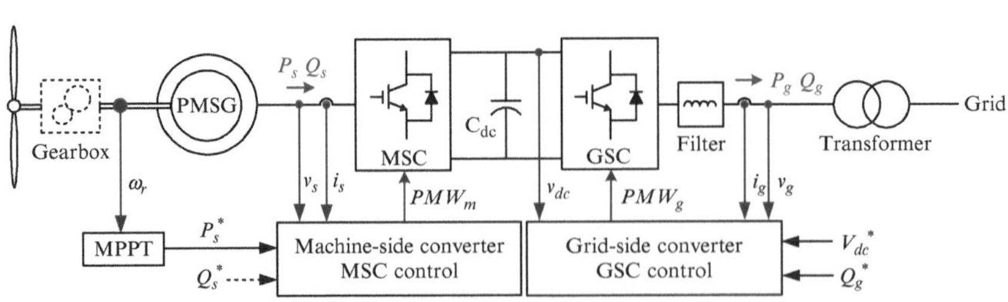

FIG. 10.19 PMSG system with back-to-back power converter.

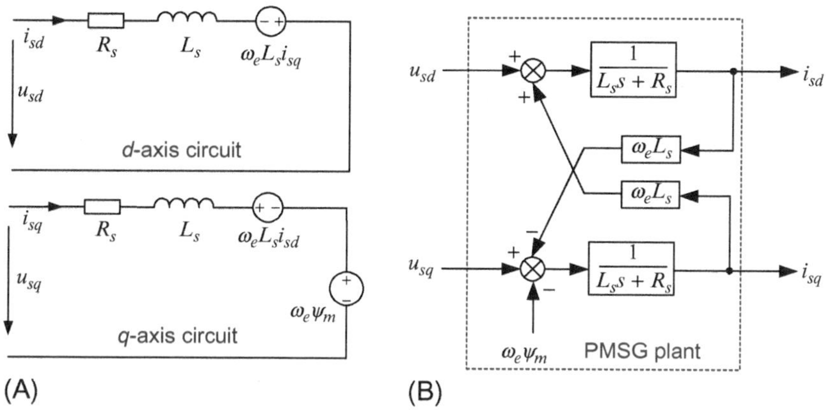

FIG. 10.20 Modeling of the permanent-magnet synchronous generator. (A) Steady-state equivalent circuit; (B) block diagram.

where i_{sd} and i_{sq} denote the stator current in d-axis and q-axis, R_s and L_s denote the stator winding resistance and stator inductance, ω_e denotes the angular frequency of the stator current, and ψ_m denotes the rotor flux linkage.

10.3.2 Vector Control of Back-to-Back Power Converters

The control of a PMSG wind turbine system includes machine-side active power control with the maximum power point tracking (MPPT), the grid-side reactive power control, and the DC-link voltage control. The detailed control structure of the machine-side converter with the aid of the stator field-oriented control (FOC) is shown in Fig. 10.21.

Considering the control scheme of the machine-side converter, the current through the stator of the generator should be regulated to adjust the rotating speed for maximum power. By using the common zero d-axis current control [14], once the rotor speed of the PMSG is detected, the electromagnetic torque is confirmed with the MPPT strategy, which is only determined by the q-axis of the stator current. The cascaded loop structure of the machine-side converter is realized by three PI controllers: one speed controller and two current controllers. The outer PI controller regulates the mechanical speed of the generator and produces the q-axis current reference for inner-current controller. On the contrary, the reference of the d-axis current is set to zero for minimum power loss [29]. Consistent with Eqs. (10.49) and (10.50), the compensation terms are supplied to improve the dynamic response.

For the grid-side converter, the outer DC-link voltage and inner-current loops are used to perform a fast active power response and to control the injected or absorbed reactive power, which are almost the same as the grid-side converter in the DFIG system.

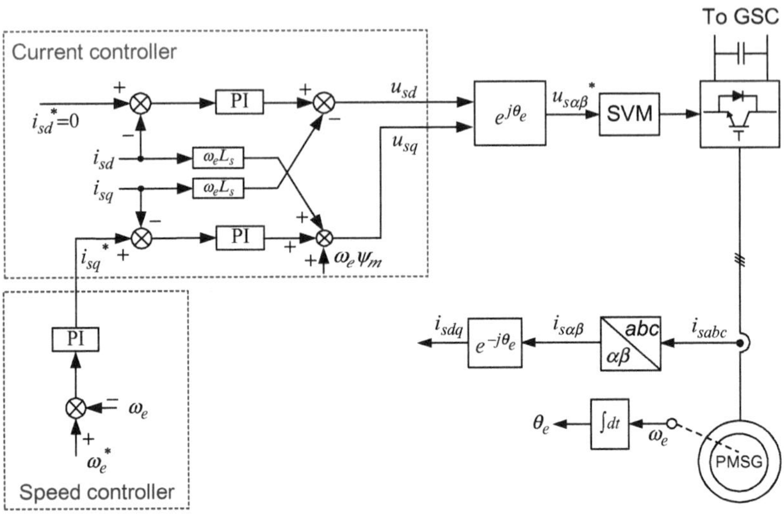

FIG. 10.21 Control block diagram of the machine-side converter for the PMSG system.

TABLE 10.5 Generator and MSC parameters for 2 MW PMSG [30,31]

Rated wind speed, v_{w_rate} (m/s)	12
Rated turbine speed, n_{rot_rate} (rpm)	19
Number of pole pairs, p	102
Rated shaft speed, n_s (rpm)	19
Magnetizing inductance, L_m (mH)	0.276
Rated fundamental frequency, f_e (Hz)	32.3
Rated output voltage (V_{rms})	554
Rated current (A_{rms})	2085
Proportional coefficient of speed controller, K_{p_ω}	50
Integral coefficient of speed controller, K_{i_ω}	1000
Proportional coefficient of current controller, K_{p_is}	0.5
Integral coefficient of current controller, K_{i_is}	50

10.3.3 Simulation Validation

Typical parameters of the generator and MSC are summarized in Table 10.5, where the PMSG system is a direct-drive with a multipole structure. Meanwhile, the most important parameters of the GSC are listed in Table 10.6.

TABLE 10.6 GSC Parameters for 2 MW PMSG

DC-link voltage, U_{dc} (V_{dc})	1050
Switching frequency, f_s (kHz)	2
Filter inductance (mH)]	0.15
Rated output voltage (V_{rms})	704
Rated current (A_{rms})	1641
Proportional coefficient of voltage controller, K_{p_vdc}	50
Integral coefficient of voltage controller, K_{i_dvc}	1500
Proportional coefficient of current controller, K_{p_ig}	0.3
Integral coefficient of current controller, K_{i_ig}	30

The switching frequency of both converters is set to 2 kHz. Based on the active power, together with the rated converter voltage output, the current in the converter needs to be calculated.

Fig. 10.22 shows the simulation of the back-to-back power converter in the PMSG system. For the grid-side converter, it can be seen that the DC-link voltage can be kept at the desired value, and the grid current in *d*-axis and *q*-axis efficiently tracks the reference value. For the machine-side converter, a 1.0 pu active power is produced at the rated operating condition with the stator current frequency of 32.3 Hz. Besides, the stator current in *d*-axis and *q*-axis well follows the reference, and the rotor speed is regulated at the desired value.

10.4 SUMMARY

This chapter starts with the description of the mainstream wind power generation configurations and their state-of-art control solutions. Two popular configurations, including the doubly fed induction generator (DFIG) system and the permanent-magnet synchronous generator (PMSG) system, are evaluated in detail in terms of the control scheme of the back-to-back power converters. With the control objectives of the stator active and reactive power, the performance of the vector control and direct control implemented in the DFIG power is analyzed and compared. Meanwhile, with the help of the modeling of the PMSG and the power converter, the main control tasks—machine-side active power control, grid-side reactive power control, and the DC-link voltage are analyzed in detail. The operation principle and important concepts are illustrated by case studies.

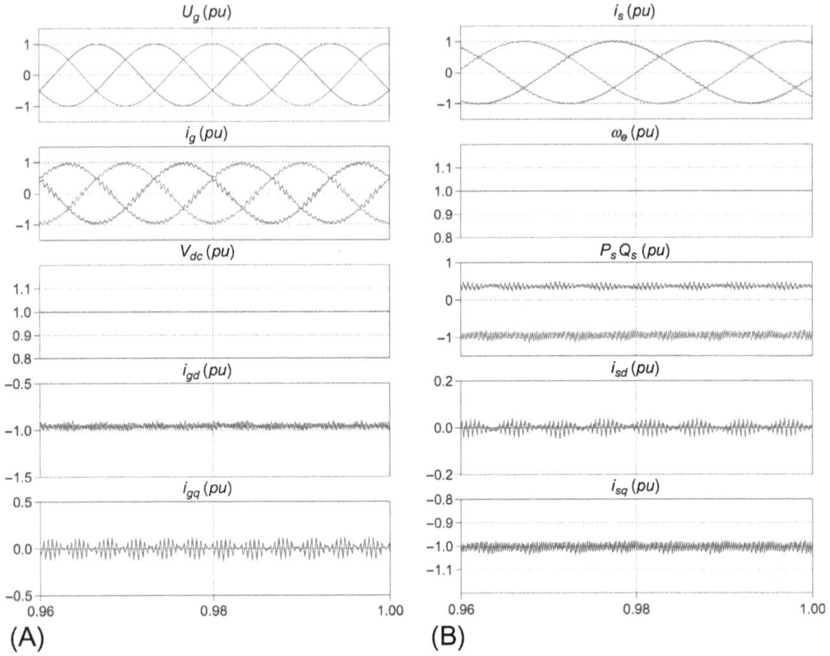

FIG. 10.22 Simulation result of the 2MW PMSG system at the rated power. (A) Grid-side converter; (B) machine-side converter.

REFERENCES

[1] GWEC. Global Wind Report 2013 (website: www.gwec.net).

[2] H. Polinder, J.A. Ferreira, B.B. Jensen, A.B. Abrahamsen, K. Atallah, R.A. McMahon, Trends in wind turbine generator systems, IEEE J. Emerg. Sel. Topics Power Electron. 1 (3) (September 2013) 174–185.

[3] North American wind power. Top 15 Wind turbine suppliers of 2013 revealed. (http://www.nawindpower.com/e107_plugins/content/content.php?content.12710).

[4] F. Blaabjerg, K. Ma, Future on power electronics for wind turbine systems, IEEE J. Emerg. Sel. Topics Power Electron. 1 (3) (September 2013) 139–152.

[5] M. Liserre, R. Cardenas, M. Molinas, J. Rodriguez, Overview of multi-MW wind turbines and wind parks, IEEE Trans. Ind. Electron. 58 (4) (April 2011) 1081–1095.

[6] G. Abad, J. Lopez, M. Rodriguez, et al., Doubly Fed Induction Machine: Modeling and Control for Wind Energy Generation Applications, Wiley-IEEE Press, 2011.

[7] E. Tremblay, S. Atayde, A. Chandra, Comparative study of control strategies for the doubly fed induction generator in wind energy conversion systems a DSP-based implementation approach, IEEE Trans. Sustain. Energy 2 (3) (2011) 288–299.

[8] M.P. Kazmierkowski, R. Krishnan, F. Blaabjerg, et al., Control in Power Electronics: Selected Problems, Academic Press, 2002.

[9] L. Xu, Coordinated control of DFIG's rotor and grid side converters during network unbalance, IEEE Trans. Power Electron. 23 (3) (May 2008) 1041–1049.

[10] J. Hu, Y. He, L. Xu, B.W. Williams, Improved control of DFIG systems during network unbalance using PI–R current regulators, IEEE Trans. Ind. Electron. 56 (2) (February 2009) 439–451.

[11] D. Zhou, F. Blaabjerg, Bandwidth oriented proportional-integral controller design for back-to-back power converters in DFIG wind turbine system, IET Renew. Power Gen. 11 (7) (2017) 941–951.

[12] J. Arbi, M.J.B. Ghorbal, I. Slama-Belkhodja, L. Charaabi, Direct virtual torque control for doubly fed induction generator grid connection, IEEE Trans. Ind. Electron. 56 (10) (October 2009) 4163–4173.

[13] D. Santos-Martin, J.L. Rodriguez-Amenedo, S. Arnalte, Direct power control applied to doubly fed induction generator under unbalanced grid voltage conditions, IEEE Trans. Power Electron. 23 (5) (September 2008) 2328–2336.

[14] B. Wu, Y. Lang, N. Zargari, et al., Power Conversion and Control of Wind Energy Systems, Wiley-IEEE Press, 2011.

[15] R. Cardenas, R. Pena, S. Alepuz, G. Asher, Overview of control systems for the operation of DFIGs in wind energy applications, IEEE Trans. Ind. Electron. 60 (7) (July 2013) 2776–2798.

[16] V. Yaramasu, B. Wu, P.C. Sen, S. Kouro, M. Narimani, High-power wind energy conversion systems: state-of-the-art and emerging technologies, Proc. IEEE 103 (5) (May 2015) 740–788.

[17] M. Liserre, F. Blaabjerg, S. Hansen, Design and control of an LCL-filter-based three-phase active rectifier, IEEE Trans. Ind. Appl. 41 (5) (September 2005) 1281–1291.

[18] D.G. Holmes, T.A. Lipo, Pulse Width Modulation for Power Converters: Principles and Practice, John Wiley & Sons, 2003.

[19] M. Liserre, R. Cardenas, M. Molinas, J. Rodriguez, Overview of multi-MW wind turbines and wind parks, IEEE Trans. Ind. Electron. 58 (4) (April 2011) 1081–1095.

[20] J.M. Guerrero, F. Blaabjerg, T. Zhelev, K. Hemmes, E. Monmasson, S. Jemei, M.P. Comech, R. Granadino, J.I. Frau, Distributed generation: toward a new energy paradigm, IEEE Ind. Electron. Mag. 4 (1) (March 2010) 52–64.

[21] D. Zhou, F. Blaabjerg, T. Franke, M. Tonnes, M. Lau, Reduced cost of reactive power in doubly fed induction generator wind turbine system with optimized grid filter, IEEE Trans. Power Electron. 30 (10) (October 2015) 5581–5590.

[22] R. Cardenas, R. Pena, S. Alepuz, G. Asher, Overview of control systems for the operation of DFIGs in wind energy applications, IEEE Trans. Ind. Electron. 60 (7) (July 2013) 2776–2798.

[23] E.ON-Netz, Requirements for offshore grid connections, April 2008.

[24] M. Tsili, S. Papathanassiou, A review of grid code technical requirements for wind farms, IET Renew. Power Gen. 3 (3) (September 2009) 308–332.

[25] S. Xiao, G. Yang, H. Zhou, H. Geng, An LVRT control strategy based on flux linkage tracking for DFIG-based WECS, IEEE Trans. Ind. Electron. 60 (7) (July 2013) 2820–2832.

[26] J. Lopez, E. Gubia, E. Olea, J. Ruiz, L. Marroyo, Ride through of wind turbines with doubly fed induction generator under symmetrical voltage dips, IEEE Trans. Ind. Electron. 56 (10) (October 2009) 4246–4254.

[27] J. Chivite-Zabalza, C. Girones, A. Carcar, I. Larrazabal, E. Olea, M. Zabaleta, in: Comparison of power conversion topologies for a multi-megawatt off-shore wind turbine, based on commercial power electronic building blocks, Proc. of IECON 2013, 2013, pp. 5242–5247.

[28] D. Zhou, F. Blaabjerg, M. Lau, M. Tonnes, in: Thermal analysis of multi-MW two-level wind power converter, Proc. of IECON 2012, 2012, pp. 5858–5864.

[29] M. Chinchilla, S. Arnaltes, J.C. Burgos, Control of permanent-magnet generators applied to variable-speed wind-energy systems connected to the grid, IEEE Trans. Energy Convers. 21 (1) (March 2006) 130–135.

[30] C. Liu, F. Blaabjerg, W. Chen, D. Xu, Stator current harmonic control with resonant controller for doubly fed induction generator, IEEE Trans. Power Electron. 27 (7) (July 2012) 3207–3220.

[31] H. Li, Z. Chen, H. Polinder, Optimization of multibrid permanent-magnet wind generator systems, IEEE Trans. Energy Convers. 24 (1) (March 2009) 82–92.

Part III

Adjustable Speed Drives

Chapter 11

Basic Control of AC Motor Drives

Kyo-Beum Lee, Hye-Ung Shin and Yeongsu Bak
Ajou University, Suwon, South Korea

11.1 INDUCTION MOTOR

The concept of induction motor was introduced by Arago in 1824 [1,2]. From the Arago's experiment, it is known that when a permanent magnet is rotated around a copper or aluminum disk, the disk is attracted to the magnet and rotates in the same direction as the rotation of the magnet. Subsequently, this phenomenon was explained by Faraday's law of electromagnetic induction. This principle forms the basic working principle of the induction motor.

Current flows in the rotating disk because of the electromotive force induced in the disk and the flux changes in accordance with the rotating magnet. The direction of this current depends on the flux change. A current carrying disk that is placed in a magnetic field experiences a force, whose direction is given by Fleming's left-hand law [3]. If the flux does not change in the rotating disk, the rotating disk would be rotated in the direction of the magnet. Therefore, the direction of force can be easily known. To realize this structure, the rotating magnet becomes the three-phase source that produces the rotating magnetic field in the actual induction motor. As a result, the rotor of the induction motor rotates.

The induction motors have several merits, including solidity, low cost, and a simple structure. However, the motors have demerits such as low efficiency, low output density, and a low power factor compared to the permanent magnet motors. The induction motors have been widely used in industrial applications owing to their low cost and solidity. The motor almost uses 50 or 60 Hz of the constant power for the constant-speed motor. In the last decades the operating the variable speed of the induction motor is made possible by using the more power electronic conversion systems.

Control of Power Electronic Converters and Systems. https://doi.org/10.1016/B978-0-12-805245-7.00011-1

11.1.1 Principle of the Induction Motor

11.1.1.1 Stator

The stator core uses an isolated thin silicon steel plate that consists of oxide and varnish. Therefore, the hysteresis loss and eddy current loss in the inner magnetic material can be minimized. In addition, the isolated coil overlapped is installed in the inner slot of the core. The overlapped coil is connected to the series or parallel type and forms a phase group. The phase group uses Δ or Y connection depending on voltage and current requirements.

11.1.1.2 Rotor

The three-phase induction motor is classified into the squirrel cage type and the wound-rotor type according to the rotor structure. The squirrel cage type is composed of connected rings at both ends and a long conductor inserted in the slot. To reduce the torque ripple, each conductor is arranged askew per slot as shown in Fig. 11.1. Although this motor has the merits of low cost, solidity, and a simple structure, it is difficult to control the speed and to reduce the starting torque. In addition, a spark does not appear in the squirrel cage type because there is no slip ring.

The wound-rotor type consists of a distributed-winding type stator connected to slip rings at the end of the rotor winding. The slip ring needs to increase the number of phases although the speed can be easily controlled; this type of motor is more costly and complex. In addition, sparks may occur in the slip ring during operation.

11.1.1.3 Rotation Principle of the Induction Motor

The rotating magnetic field is produced by the three-phase current of the stator in the actual three-phase induction motor. It can be replaced by permanent magnets in a permanent magnet synchronous motor. The three-phase windings of the inner stator are spaced 120° electrical degrees apart. In addition, the conductor of each winding is distributed such as a sinusoidal wave. Therefore, when the current flows into the three-phase windings, the magneto motive force (MMF) of the sinusoidal waveform is produced by the current. Fig. 11.2 shows the MMF distribution of $\boldsymbol{F}_a$ in the space when the AC current flows into the phase

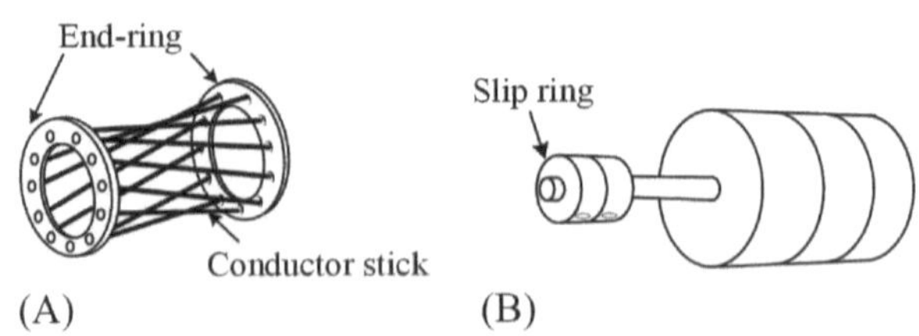

FIG. 11.1 Structure of rotors in induction motor (A) squirrel cage and (B) Wound.

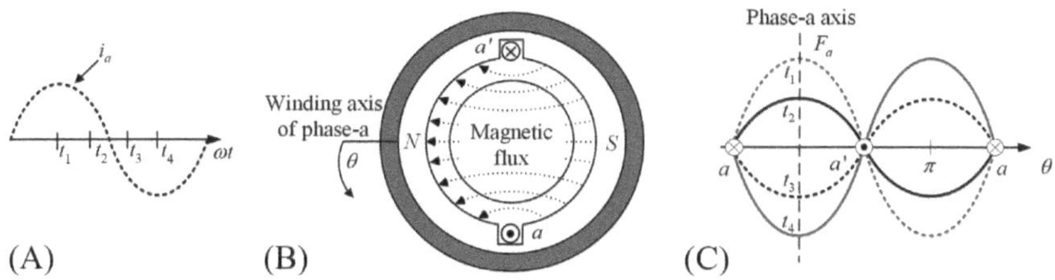

FIG. 11.2 Basics of induction motor. (A) Current of phase-a, (B) winding of phase-a, and (C) magnetic flux distribution of phase-a air gap.

winding. The sinusoidal MMF changes instantaneously according to the input AC source [4].

Sinusoidal MMFs are produced by the windings of phase b and phase c. The windings are located to each other at 120° spatially. The three-phase balanced current of the stator winding can be expressed as:

$$\begin{aligned} i_a &= I_m \cos \omega_s t \\ i_b &= I_m \cos \left(\omega_s t - 120^\circ\right) \\ i_c &= I_m \cos \left(\omega_s t + 120^\circ\right) \end{aligned} \tag{11.1}$$

Fig. 11.3 shows the instantaneous change of the three-phase current. The MMF is determined by the magnitude and direction in accordance with the instantaneous value of current. The MMF in each phase can be expressed as the space vector of the phase axis, and the magnitude is proportional to the instantaneous value of the current. The total MMF can be expressed as the summation of each component of the three-phase MMF.

When $T = t_0$, the current of each phase winding can be expressed as:

$$\begin{aligned} i_a &= I_m \\ i_b &= -I_m/2 \\ i_c &= -I_m/2 \end{aligned} \tag{11.2}$$

Fig. 11.4 shows the position and magnitude of the changed instantaneous total MMF. When the current flows to the phase-a winding in maxima, i.e., $t = t_0$,

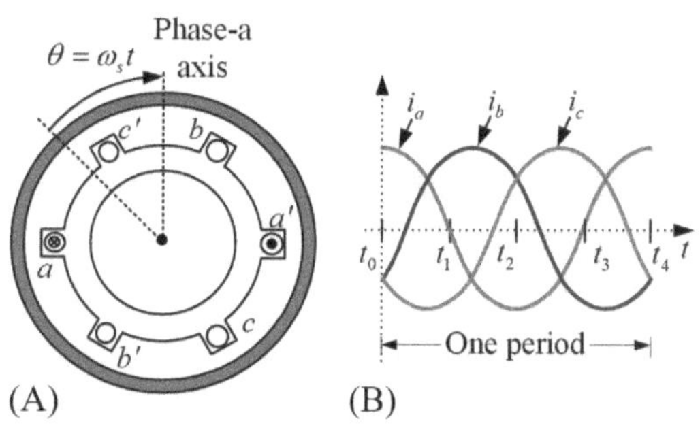

FIG. 11.3 Basics of induction motor. (A) Three-phase winding and (B) three-phase current.

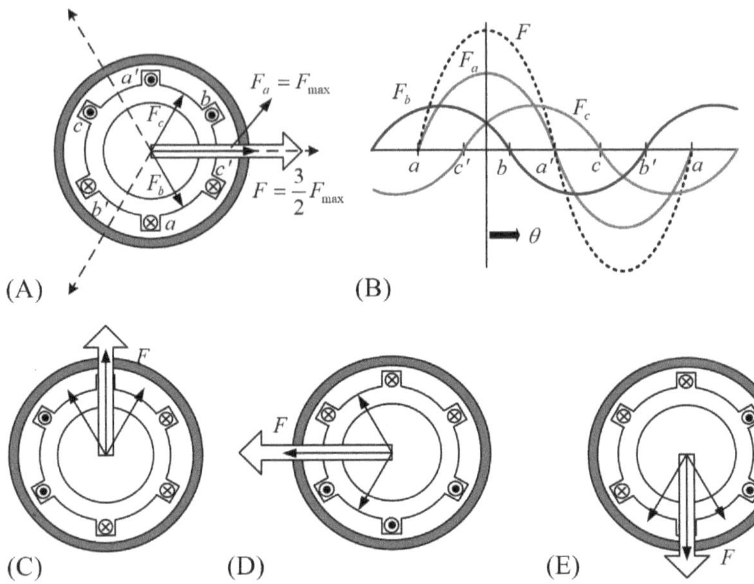

FIG. 11.4 Various instantaneous MMFs in the induction machine at different time instants. (A) $t=t_0=t_4$, (B) $t=t_0$, magneto motive force & synthesis graph, (C) $t=t_1$, (D) $t=t_2$, and (E) $t=t_3$.

the MMF is maximum and $F_a=F_{max}$ into the positive direction of phase-a axis. The MMFs of phase-b and phase-c are expressed as F_b and F_c, respectively. The magnitude is $F_{max}/2$, and the direction is the positive direction of each axis. The vector direction of the MMF summation is equal to the vector direction of the phase-a axis, and its magnitude is $\mathbf{F}=3/2_{max}$. Therefore, this instantaneous summation MMF is distributed by a sinusoidal wave form similar to the MMF of phase-a.

When $t=t_1$, each phase current and MMF can be expressed as:

$$\begin{aligned} i_a &= 0, \quad F_a = 0 \\ i_b &= \frac{\sqrt{3}}{2} I_m, \quad F_b = \frac{\sqrt{3}}{2} F_{max} \\ i_c &= -\frac{\sqrt{3}}{2} I_m, \quad F_c = -\frac{\sqrt{3}}{2} F_{max} \end{aligned} \tag{11.3}$$

Fig. 11.4 shows the vector of the total MMF and the vector components of each phase. The magnitude of the total MMF vector is approximately 1.5 times the MMF of phase-a. However, its direction is rotated 90° in the counter clockwise direction. Fig. 11.4C shows the MMF waveform of each phase and the total wave. If $t=t_2$ and $t=t_3$, Fig. 11.4D and E shows the current and total MMF. The total MMF waveform is a sinusoidal wave of the same magnitude with respect to time. When the current flows into the three-phase winding, expressed as Eq. (11.1), the rotating MMF rotates clockwise and it has the same rotor direction. If the three-phase winding is changed to two phase, the rotor of the motor rotates in the opposite direction. The total MMF wave of the current change during one period returns back to Fig. 11.4A.

$$F(\theta)=F_a(\theta)+F_b(\theta)+F_c(\theta) \tag{11.4}$$

Each phase MMF is proportional to the number of available windings N_s, and the instantaneous current of each phase can be expressed as:

$$\begin{aligned} F_a(\theta)&=N_s i_a \cos\theta \\ F_b(\theta)&=N_s i_b \cos\left(\theta-120^\circ\right) \\ F_c(\theta)&=N_s i_c \cos\left(\theta+120^\circ\right) \end{aligned} \tag{11.5}$$

Eq. (11.6) can be obtained by substituting Eq. (11.4) in Eq. (11.5) as follows:

$$F(\theta)=N_s i_a \cos\theta+N_s i_b \cos\left(\theta-120^\circ\right)+N_s i_c \cos\left(\theta+120^\circ\right) \tag{11.6}$$

$$F(\theta,t)=\frac{3}{2}N_s I_m \cos\left(\omega_s t-\theta\right) \tag{11.7}$$

Eq. (11.7) represents the total MMF in the air gap. This indicates that the total MMF is rotated to a constant angular speed, and it is distributed in the sinusoidal waveform.

11.1.2 Voltage/Frequency Constant Control of Induction Motor

The synchronous speed of the induction motor is controlled by changing the stator frequency because it rotates synchronously with the rotating magnetic field rotating at the frequency of the power source. Therefore, relational expression of the synchronous speed n_s and the stator frequency f_s can be written as:

$$n_s=\frac{120f_s}{P} \tag{11.8}$$

In case the frequency is changed for the speed control, if the stator voltage is constant, the problem that the generated torque is varied as well as the speed of the induction motor is occurred. In addition, the rotor current is affected by the applied frequency at the induction motor. In case the speed of the induction motor is controlled by changing the stator frequency, in order to maintain the generated torque and the rotor current constantly, the stator voltage should be proportionally varied depending on the frequency variation as shown in Fig. 11.5.

This control method is called the voltage/frequency constant control. Fig. 11.6 shows the characteristic of the speed-torque during the voltage/frequency constant control. Although the frequency is changed for the speed control, the generated torque and the rotor current are independent with the stator frequency. Additionally, the synchronous speed of the induction motor is proportional to the stator frequency. However, in the case of increasing the

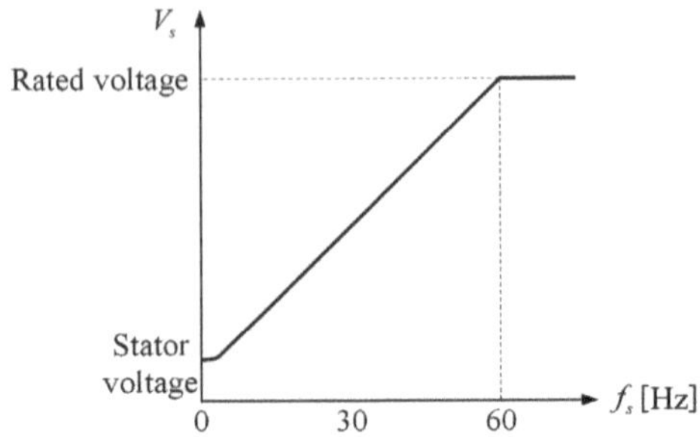

FIG. 11.5 Stator voltage proportional to frequency of the induction motor.

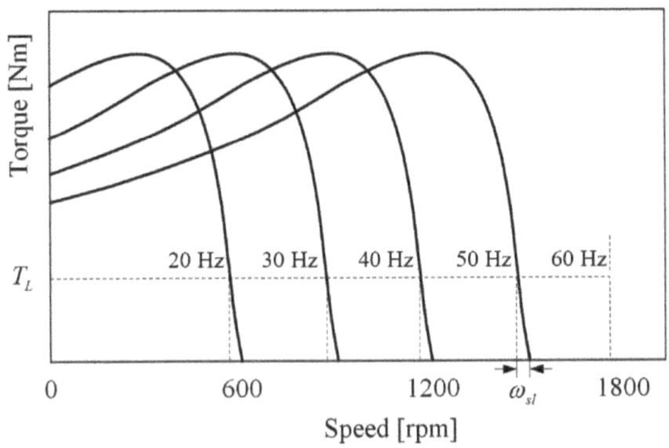

FIG. 11.6 Characteristic of speed-torque during the voltage/frequency constant control.

frequency above the rated frequency, in proportion to this, the stator voltage cannot be applied above the rated voltage. Therefore, the voltage/frequency constant control is used only for the speed control below the rated speed of the induction motor.

11.1.3 Vector Control of Induction Motor

11.1.3.1 Direct Vector Control Scheme

The direct vector scheme of the induction motor controls the d-axis component during the fixed rotor flux. Because this rotor flux is rotated to the synchronous speed ω according to the angular frequency of stator current, the position θ of the rotor flux is changed depending on the time. Therefore, the d-q axis frame of reference through the rotating reference frame rotated at the same speed of the rotor flux is always aligned to the d_e-axis. The rotor flux is equal to d_e-axis component, and q_e-axis component becomes zero. Therefore, the q_e-axis component is controlled to zero, and d_e-axis component is controlled by using the same rotating flux. The required flux angle θ can be calculated for each axis component that is produced by the *d-q* axis reference frame transformation of the rotor flux [5,6].

The relational expression can be written as:

$$\theta = \tan^{-1}\left(\frac{\lambda_{qr}^{s}}{\lambda_{dr}^{s}}\right) \tag{11.9}$$

The rotor-flux vector control scheme is used to calculate the flux angle and to estimate the rotor flux.

Stator *d*-Axis Current and Rotor Flux

Because the q_e-axis is obtained from the rotor voltage equation, and the rotor reference frame of the rotating rotor flux in the synchronous speed ω is zero, d_e-axis component and the stator d_e-axis current are expressed as follows:

$$v_{dr}^{e} = R_r i_{dr}^{e} + p\lambda_{dr}^{e} = 0 \tag{11.10}$$

$$i_{dr}^{e} = -\frac{p\lambda_{dr}^{e}}{R_r} \quad \left(\lambda_{dr}^{e} = L_r i_{dr}^{e} + L_m i_{ds}^{e}\right) \tag{11.11}$$

$$\lambda_{dr}^{e} = -\frac{R_r L_m}{R_r + L_r p} i_{ds}^{e} = \frac{L_m}{1 + \frac{L_r}{R_r} p} i_{ds}^{e} \tag{11.12}$$

where p is a differential operator. The d^e axis component λ_{dr}^{e} of the rotor flux for the stator d_e-axis current i_{ds}^{e} has gain L_m and time constant $T_r(=L_r/R_r)$ has a first-order lag relationship. It implies that the stator d_e-axis current i_{ds}^{e} can linearly control the magnitude of rotor flux λ_{dr}^{e}. Therefore, the stator d_e-axis current i_{ds}^{e} can be called to flux produced by the current. Normally, this flux produced by the current is determined by the rated power, and it is kept to a constant value.

Stator *q*-axis current and torque

When the rotor flux λ_{dr}^{e} is constant, the equation for torque can be expressed as:

$$T_e = K_T i_{qs}^{e} \quad \left(K_T = \frac{3}{2}\frac{P}{2}\frac{L_m}{L_r}\lambda_{dr}^{e}\right) \tag{11.13}$$

Because the torque is proportional to the stator q_e-axis current i_{qs}^{e}, the stator q_e-axis current i_{qs}^{e} can be used to produce the torque as given in Eq. (11.13). Therefore, the induction motor is controlled by the instantaneous torque using control of i_{qs}^{e} magnitude. The instantaneous torque can be controlled by the torque produced by current when keeping the flux in the vector control system constantly. The rotor flux λ_{dr}^{e*} produced by current i_{ds}^{e*} and required torque T_e^* produced by current i_{qs}^{e*} can be expressed as:

$$i_{ds}^{e*} = \frac{\lambda_{dr}^{e*}}{L_m} \tag{11.14}$$

$$i_{qs}^{e*} = \frac{T_e^*}{K_T} \quad \left(K_T = \frac{3}{2}\frac{P}{2}\frac{L_m}{L_r}\lambda_{dr}^{e*}\right) \tag{11.15}$$

As the currents of the flux and the torque are rotating reference frame, Eqs. (11.14), (11.15) should be changed to the reference current (i_a^*, i_b^*, i_c^*) owing to the input of the stator current of the three-phase winding. First, after changing to the stationary reference frame from the rotating reference frame, the three-phase reference current can be calculated by using the inverse transformation. However, control of the induction motor becomes complex. Therefore, the rotating reference frame scheme can be used to control easily as vector control. Vector control of the induction motor uses the rotating reference frame. Fig. 11.7 shows the rotating reference frame current controlled system.

Calculating the flux angle of the rotor instantaneous position for vector control is difficult. A hall sensor and coil for measuring the motor angle may be used to obtain the flux angle of rotor. It is not often used because of the structure and cost. Therefore, flux estimation methods using the voltage equation of the stator and rotor will be introduced in the next paragraph.

Stator Voltage Equation Using Voltage Model

The voltage model is used to calculate the rotor flux obtained from the stator flux using the stator voltage equation as shown in Fig. 11.8. In order to use this method, the information of the stator voltage and current is necessary.

The stator voltage equation of the stationery reference frame can be expressed as:

$$v_{ds}^s = R_s i_{ds}^s + \frac{d\lambda_{ds}^s}{dt} \tag{11.16}$$

$$v_{qs}^s = R_s i_{qs}^s + \frac{d\lambda_{qs}^s}{dt} \tag{11.17}$$

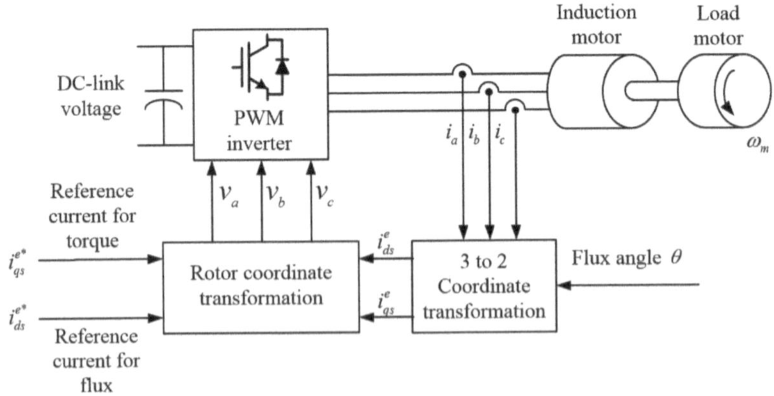

FIG. 11.7 Vector control scheme of the induction motor using current controller.

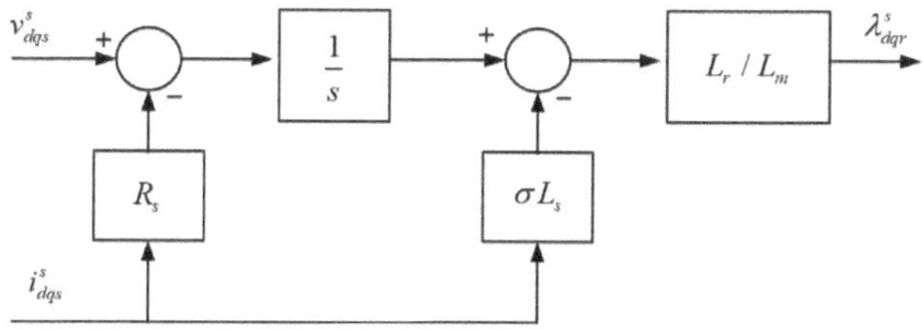

FIG. 11.8 Estimation of the stator flux using voltage model.

The stator flux is calculated by using a pure integrator as follows:

$$\lambda_{ds}^{s} = \int \left(v_{ds}^{s} - R_s i_{ds}^{s}\right) dt \tag{11.18}$$

$$\lambda_{qs}^{s} = \int \left(v_{qs}^{s} - R_s i_{qs}^{s}\right) dt \tag{11.19}$$

The rotor flux can be expressed as:

$$\lambda_{dr}^{s} = L_r i_{dr}^{s} + L_m i_{ds}^{s} = \frac{L_r}{L_m}\left(\lambda_{ds}^{s} - \sigma L_s i_{ds}^{s}\right), \ \left(i_{dr}^{s} = \frac{\lambda_{ds}^{s} - L_s i_{ds}^{s}}{L_m}\right), \ \left(\sigma = 1 - \frac{L_m^2}{L_s L_r}\right) \tag{11.20}$$

$$\lambda_{qr}^{s} = L_r i_{qr}^{s} + L_m i_{qs}^{s} = \frac{L_r}{L_m}\left(\lambda_{qs}^{s} - \sigma L_s i_{qs}^{s}\right), \ \left(i_{qr}^{s} = \frac{\lambda_{qs}^{s} - L_s i_{qs}^{s}}{L_m}\right), \ \left(\sigma = 1 - \frac{L_m^2}{L_s L_r}\right) \tag{11.21}$$

The angle of rotor flux can be calculated by using Eqs. (11.20), (11.21). Because the flux information is obtained by integrating the back EMF, the flux is accurate in the high-speed operation. However, when the back EMF is small in the low-speed operation, the flux is inaccurate due to the voltage drop in the inverter noise caused by stator impedance in the inverter. Therefore, this method is mainly used for high-speed operations.

Rotor Voltage Equation Using Current Model

The current model is used to calculate the rotor flux obtained from rotor speed and stator current using the rotor voltage equation as shown in Fig. 11.9.

The rotating reference frame of the rotor angular velocity is expressed as:

$$v_{dr}^{r} = R_r i_{dr}^{r} + \frac{d\lambda_{dr}^{r}}{dt} \tag{11.22}$$

$$v_{qr}^{r} = R_r i_{qr}^{r} + \frac{d\lambda_{qr}^{r}}{dt} \tag{11.23}$$

Because the rotor current cannot be measured, the stator current can be calculated from the rotor flux equation.

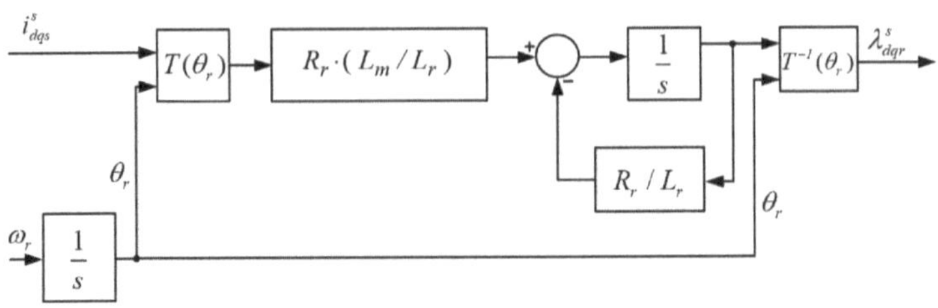

FIG. 11.9 Estimation of the rotor flux using current model.

$$\frac{d\lambda^r_{dr}}{dt} = -\frac{R_r}{L_r}\lambda^r_{dr} + R_r\frac{L_m}{L_r}i^r_{ds}, \quad \left(i^r_{dr} = \frac{\lambda^r_{dr} - L_m i^r_{ds}}{L_r}\right) \tag{11.24}$$

$$\frac{d\lambda^r_{qr}}{dt} = -\frac{R_r}{L_r}\lambda^r_{qr} + R_r\frac{L_m}{L_r}i^r_{qs}, \quad \left(i^r_{qr} = \frac{\lambda^r_{qr} - L_m i^r_{qs}}{L_r}\right) \tag{11.25}$$

The stator current can be calculated by using θ_r as the following:

$$i^r_{ds} = i^s_{ds}\cos\theta_r + i^s_{qs}\sin\theta_r \tag{11.26}$$

$$i^r_{qs} = -i^s_{ds}\sin\theta_r + i^s_{qs}\cos\theta_r \tag{11.27}$$

Finally, the rotor flux of the stationary reference frame is obtained from the rotating reference frame as follows:

$$\lambda^s_{dr} = \lambda^r_{dr}\cos\theta_r - \lambda^r_{qr}\sin\theta_r \tag{11.28}$$

$$\lambda^s_{qr} = \lambda^r_{dr}\sin\theta_r + \lambda^r_{qr}\cos\theta_r \tag{11.29}$$

The information related to the speed, resistance, and inductance is required to estimate the rotor flux using the current model. This method is used for low-speed operations owing to fluctuation of estimated value in high-speed operations.

11.1.3.2 Indirect Vector Control Scheme

In the previous section, the direct vector control method is described to calculate the rotor flux angle directly for the vector control of the induction motor. Determining the accurate magnetic flux is not easy because this method has many problems in the processing of the estimated magnetic flux angle. On the other hand, the indirect vector control method of the induction motor can be easier to implement compared with the direct vector control, as it does not directly use the flux information [7,8].

The stator current of the induction motor is mainly distributed to the magnetic flux current and the torque current by slip *s*. In this regard, the difference between the rotor speed and the synchronous speed is called the slip speed, and the ratio of the slip speed to the synchronous speed is called the slip s, which is expressed as follows:

$$s = \frac{n_s - n}{n_s} \tag{11.30}$$

where n_s is the synchronous speed and n is the rotor speed.

Therefore, the indirect vector control is regulated by controlling the slip, which has an optimal current distribution ratio of the stator current. The equivalent circuit in the steady state of the induction motor is shown in Fig. 11.10.

If the rotor magnetic flux and the d-axis component are the same, then the q-axis component is 0. The slip angular velocity is calculated by applying the q-axis component to the rotor voltage equations.

$$v_{qr}^e = R_r i_{qr}^e + p\lambda_{qr}^e + (\omega_e - \omega_r)\lambda_{dr}^e = R_r i_{qr}^e + (\omega_e - \omega_r)\lambda_{dr}^e = 0 \tag{11.31}$$

The slip angular velocity is expressed as follows:

$$\begin{aligned} \omega_{sl} &= \omega_e - \omega_r = -\frac{R_r i_{qr}^e}{\lambda_{dr}^e} \\ &= \frac{R_r L_m}{L_r}\frac{i_{qs}^e}{\lambda_{dr}^e}, \quad \left(\lambda_{qr}^e = L_r i_{qr}^e + L_m i_{qs}^e = 0\right), \quad \left(i_{qr}^e = -\frac{L_m}{L_r} i_{qs}^e\right) \end{aligned} \tag{11.32}$$

where ω_e is the synchronous angular velocity and ω_r is the angular velocity.

The slip angular velocity can be expressed in terms of d-axis and q-axis components as follows. T_r is a rotor time constant, which is not changed depending on the time. Therefore, pT_r is 0, where p is a differential operator.

$$\omega_{sl} = \frac{1 + pT_r}{T_r}\frac{i_{qs}^e}{i_{ds}^e} = \frac{1}{T_r}\frac{i_{qs}^e}{i_{ds}^e}, \quad \left(\lambda_{dr}^e = \frac{R_r L_m}{R_r + pL_r} i_{ds}^e = \frac{L_m}{1 + p\dfrac{L_r}{R_r}} i_{ds}^e\right), \quad \left(T_r = \frac{L_r}{R_r}\right) \tag{11.33}$$

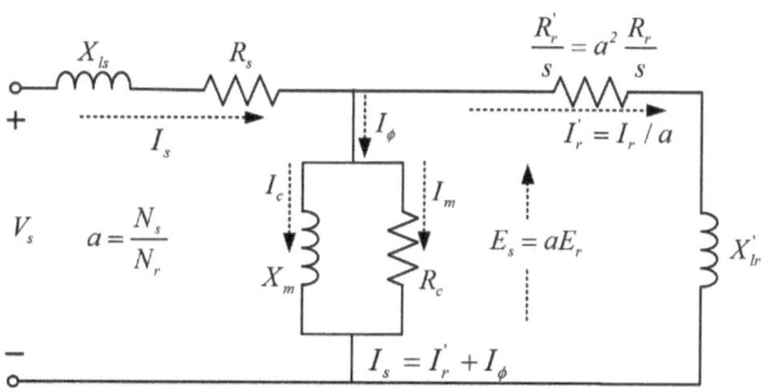

FIG. 11.10 Steady-state equivalent circuit of the induction motor.

The slip angular speed ω_{sl} can be determined by using Eq. (11.33). If i_{ds}^e and i_{qs}^e require the stator flux and the torque to be selected, according to Eq. (11.33), the slip angle speed is calculated and the flux angle θ_e can be calculated by the following formula:

$$\theta_e = \int \omega_e dt = \int (\omega_{sl} + \omega_r) dt \tag{11.34}$$

where the rotor flux angle is not obtained directly; in indirect vector control, the rotor flux angle is obtained indirectly from the slip angle speed ω_{sl} calculated from the current ratio of i_{ds}^e and i_{qs}^e that produces the required rotor flux and torque. This method does not directly obtain the rotor flux data and it is easy to realize, but the rotor angular speed information is needed. Rotor angular speed information can be obtained from the position sensor such as an encoder but during operate, rotor resistor R_r and rotor magnetizing inductance L_r can easily change; hence this method has the disadvantage that the accuracy of the slip angle speed can be low. Therefore, for precise vector control, the errors of those variables need to be compensated. Fig. 11.11 shows a block diagram of the indirect vector control system [9,10].

11.1.4 Simulation of Control of Induction Motor

11.1.4.1 Basic Setting

Fig. 11.12 shows the circuit schematic of the induction motor used for simulation. A PSIM-based tool and Visual C Studio are used to simulate the vector control of the induction motor. The circuit is divided into three parts: power

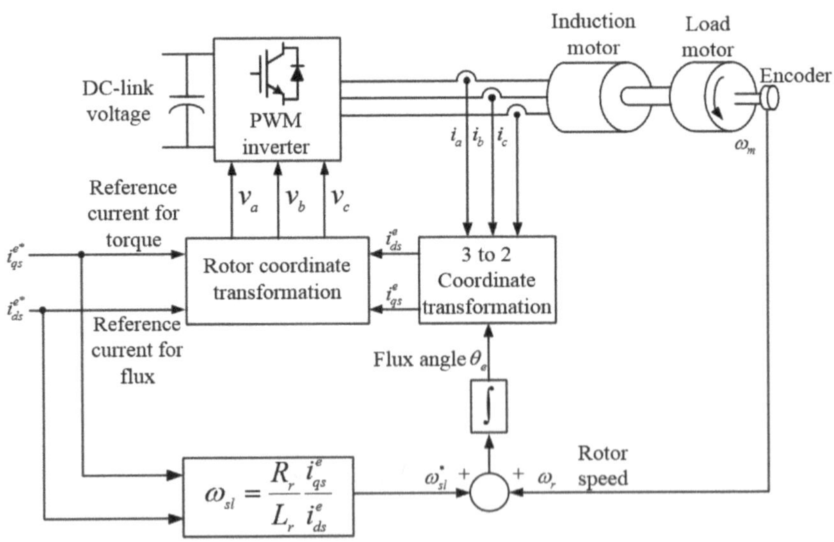

FIG. 11.11 Indirect vector control system of induction motor using a slip estimator.

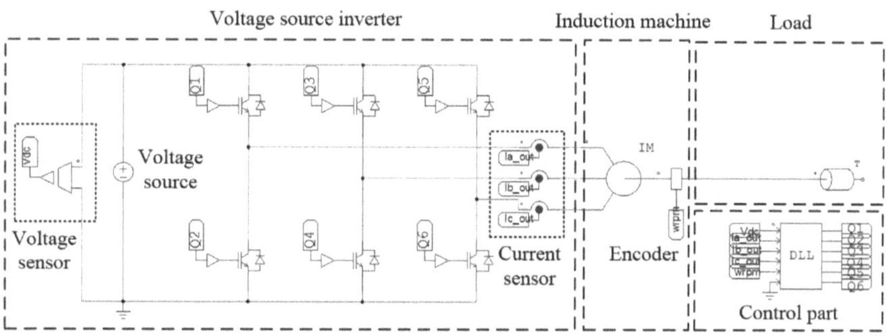

FIG. 11.12 Circuit schematic of induction motor.

TABLE 11.1 Specifications of Simulation Parameters and a 5 HP Induction Motor

Parameter	Value	Unit
Voltage source (V_{DC})	600	V
Sampling time (T_s)	100	μs
Switching frequency (f_{sw})	10	kHz
Rated power (P_{rated})	5	HP
Rated speed (ω_{rated})	1750	rpm
Rated voltage (V_{rated})	220	V
Rated current (I_{rated})	12.8	A
Pole (P)	4	P
Stator resistance (R_s)	0.295	Ω
Rotor resistance (R_r)	0.379	Ω
Magnetizing inductance (L_m)	59	mH
Leakage inductance (L_{ls})	1.794	mH
Stator and rotor inductance (L_s and L_r)	60.794	mH

converter, induction motor, and control. The power conversion part and the induction motor part consist of the motor, switch, and sensor. The control part is used to execute the DLL (dynamic link library) using C code. The specifications of the simulation parameters and the induction motor are listed in Table 11.1.

Fig. 11.13 shows the control block diagram of the motor system. The motor speed can be controlled by controlling the motor torque. It is effective to directly control the motor torque in order to control the motor speed; however, it is

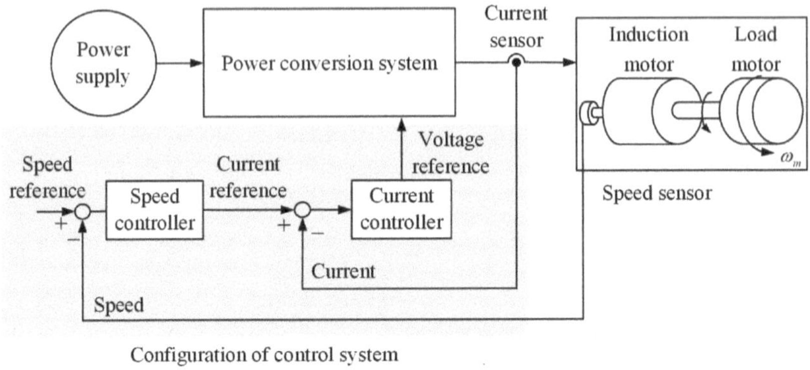

FIG. 11.13 Control block diagram of the motor system.

difficult to apply. The motor torque can be controlled by controlling the armature current of the motor through the principal that the motor torque is proportional to the armature current when the magnetic flux is constant. Therefore, in order to control the motor speed, current control of the motor must be preceded.

11.1.4.2 Direct Vector Control

Fig. 11.14 shows the flux angle estimation of the induction motor using the voltage model and current model in direct vector control as explained in Section 11.1.3.1. The estimated rotor flux shows the *d-q* axis flux in the stationary reference frame as shown in Fig. 11.14C and D. The flux angle ranges from 0 to 2π. When the voltage model is used in the low-speed operation, the performance of voltage model is not good. However, the voltage model is similar to the current model owing to the ideal simulation.

11.1.4.3 Indirect Vector Control

Fig. 11.15 shows the indirect vector control of the induction motor. The indirect vector control does not estimate the rotor flux and uses slip angular velocity ω_{sl} obtained by the current of the flux i_{de} and torque i_{qe} in the rotational reference frame. When the induction motor is controlled by the speed controller from 0.5 to 1 s, the current reference of the torque I_{qe_ref} is produced by the load torque. In addition, the slip angular velocity ω_{sl} becomes almost zero. When the speed reference increases at 1 s, I_{qe_ref} and slip speed increase. This means that rotor flux ω_e is added to the current and voltage references using reference frame transformation.

Finally, when the speed approaches to the speed reference, I_{qe_ref} is reduced and the speed is maintained constant according to the slip speed. Fig. 11.15D–G shows the speed and current in the same situation. The *d*-axis current I_{de_ref} has a constant value to maintain the rated flux of the induction motor. The actual current I_{de} estimates I_{de_ref} through the current controller. The *q*-axis current I_{qe} is

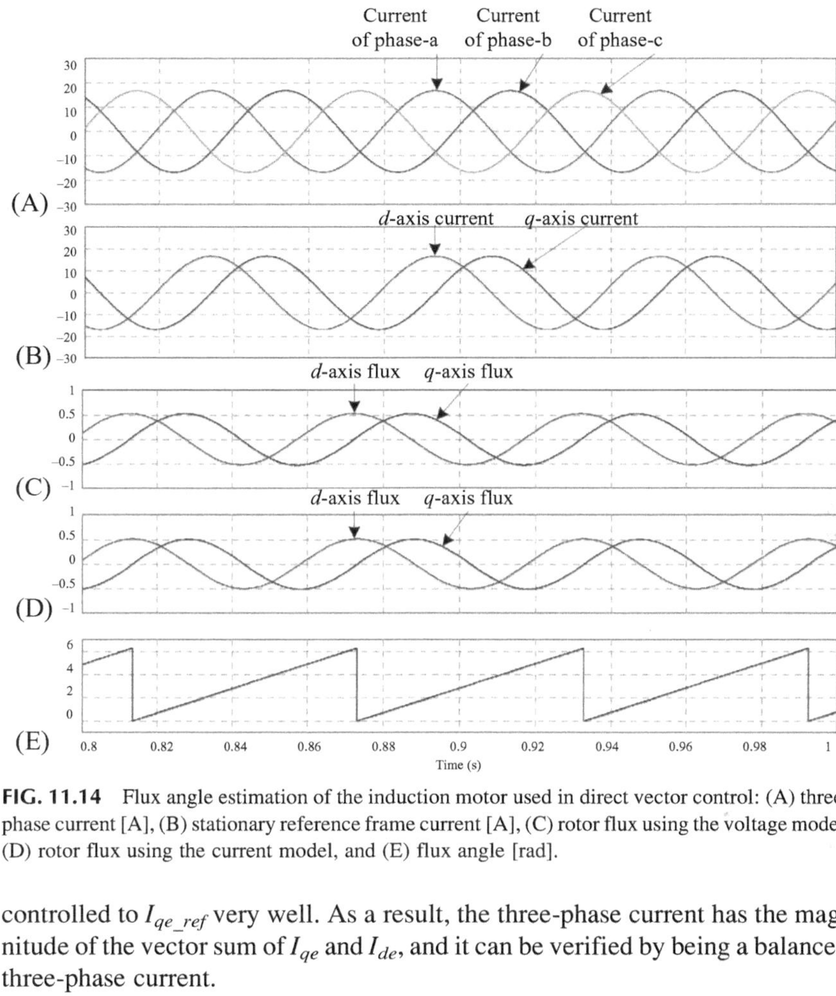

FIG. 11.14 Flux angle estimation of the induction motor used in direct vector control: (A) three-phase current [A], (B) stationary reference frame current [A], (C) rotor flux using the voltage model, (D) rotor flux using the current model, and (E) flux angle [rad].

controlled to I_{qe_ref} very well. As a result, the three-phase current has the magnitude of the vector sum of I_{qe} and I_{de}, and it can be verified by being a balanced three-phase current.

11.2 PERMANENT MAGNET SYNCHRONOUS MOTOR

The permanent magnet synchronous motor rotates at a constant speed regardless of the torque when synchronized to the input voltage frequency irrespective of the alternation of the load and the terminal voltage [11].

From the structure perspective, the permanent magnet synchronous motor minimizes the power used and increases the efficiency of the whole system by producing the rotor flux without using an external source unlike the induction motor because its rotor is composed of a permanent magnet. Moreover, the source is only delivered to the stator; hence, there is no rotor losses. Thus, the permanent magnet synchronous motor has a better cooling characteristic, higher

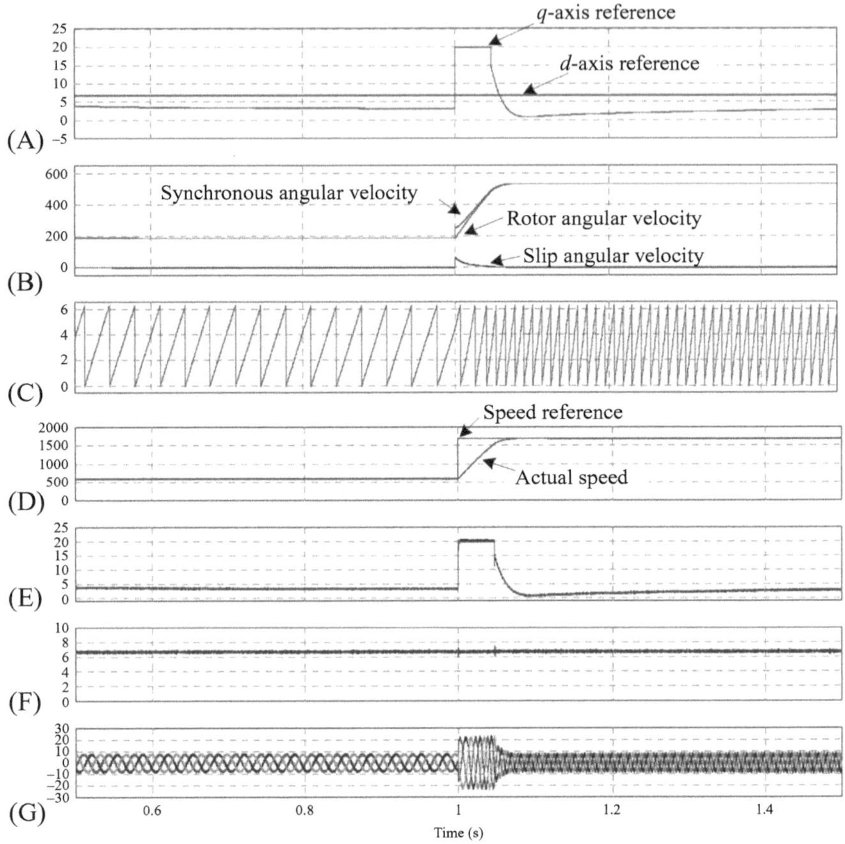

FIG. 11.15 Indirect vector control of induction motor: (A) Current reference [A], (B) angular velocity [rad/s], (C) synchronous phase angle [rad], (D) Speed [rpm], (E) q-axis reference and actual current [A], (F) d-axis reference and actual current [A], and (G) output phase current [A].

efficiency, and higher performance and can be highly developed to advanced application. Therefore, the permanent magnet synchronous motor finds increasing applications in the area of home appliances owing to its high efficiency, high power factor, high torque, and wide speed range motor but also electric system in robotics, wind power generation, etc.

11.2.1 Principle of Permanent Magnet Synchronous Motor

11.2.1.1 Rotation Principle

Magnetic attraction or repulsion acts between a permanent magnet and an electromagnet because of the interaction of the two magnetic fields. If the current flows through a conductor in the magnetic field, the direction of the magnetic force is given by the Fleming's left-hand law and the force direction is the right-handed screw direction given by the right-handed screw direction rule. The flux is generated by the current flowing through the coil thus the torque of the motor is generated by the flux of magnet and coil, and the direction of that torque is the direction of two fluxes that are overlapped. By considering the torque by the permanent magnets, the formula for the torque of the permanent magnet motor can be written as

$$\begin{aligned}\text{Torque} &= \text{magnitude of flux} \times \text{magnitude of current} \\ &\quad \times \sin(\text{phase differential of flux and current}) \\ &= \text{constant} \times \text{machine current} \\ &\quad \times \cos(\text{induced voltage and phase differential of current})\end{aligned} \tag{11.35}$$

The flux is generated by the rotating permanent magnet, and the magnetic flux interlinkage to the stator wire is varied by the rotor position. Therefore, the torque of the motor is influenced by those two factors of phase difference.

If the current flows through a conductor, the flux is generated in the direction of the right-handed screw following the right-handed screw rule, and a magnetic field is generated as if a permanent magnet exists. In the case of a motor, the three-phase AC current that has in time 120° phase difference flows through the three-phase wire in the stator, and the three-phase symmetrical current flows and the magnetic field is generated as if there is a permanent magnet at each stator. As a result, although the stator wire is fixed, a rotating magnetic field is generated by the stator wire. Thus, in a permanent magnet motor, it is possible to generate a rotating field.

Fig. 11.16 shows the working principle of the permanent magnet synchronous motor. The motor rotating is possible by generating continuous interaction force (attraction and repulsion) between the rotating fields and the permanent magnet rotor. The generated torque of the motor is different according to the phase difference of the rotating field of the stator, and the permanent magnet rotor and the maximum torque given in Eq. (11.35) are generated at 90°.

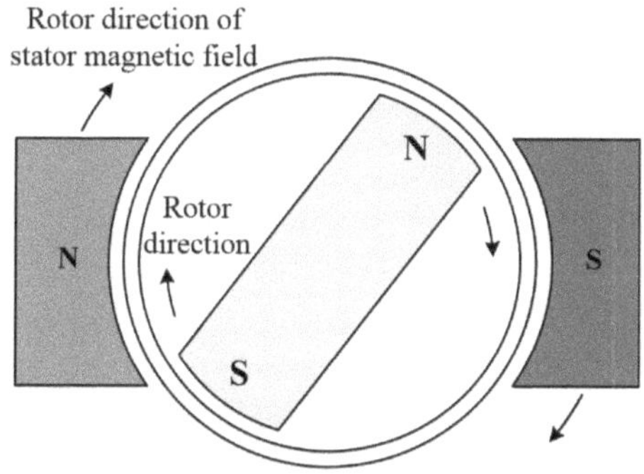

FIG. 11.16 Rotation principle of permanent magnet synchronous motor.

Therefore, the rotor and the rotating field of the stator of the permanent magnet synchronous motor need to be maintained at 90° in order to generate the maximum torque. This can be achieved by obtaining the rotor position by the rotor position sensor and controlling the phase of the stator wire.

Classification According to the Arrangement of the Permanent Magnets

Fig. 11.17 shows different characteristics of the types of the permanent magnet; even when using the same type of permanent magnet, the characteristic of motor can be changed by arranging the permanent magnet and its shape differently. The permanent magnet synchronous motor can be categorized as surface-mounted permanent magnet (SPM) in which the permanent magnet is arranged on the rotor surface and the interior permanent magnet (IPM) in which the permanent magnet is arranged inside of the rotor.

(a) Surface-mounted permanent magnet synchronous motor (SPMSM).
The permanent magnet of the SPMSM is located at the surface of the rotor and the inductance in the motor will be lower. This allows to achieve a high current response. Moreover, because the reluctance torque is not generated, the torque has a strong linearity. Therefore, the SPMSM can be applied in fast respond servo-drive. To get rid of the leakage between the permanent magnet and the stator wire, the valid flux linkage is possible and the flux distribution of the capacitor does not include much harmonics; hence, the noise and ripple are small. Furthermore, the method of locating the magnet at the rotor surface can be simplified by using a ring magnet or glue to ensure low cost and a long life time. Besides, the permanent magnets are located at the surface with the same width; therefore, the axis inductances are the same in the SPMSM. However, as the permanent magnet is located at the surface, the centrifugal force can cause the magnet to

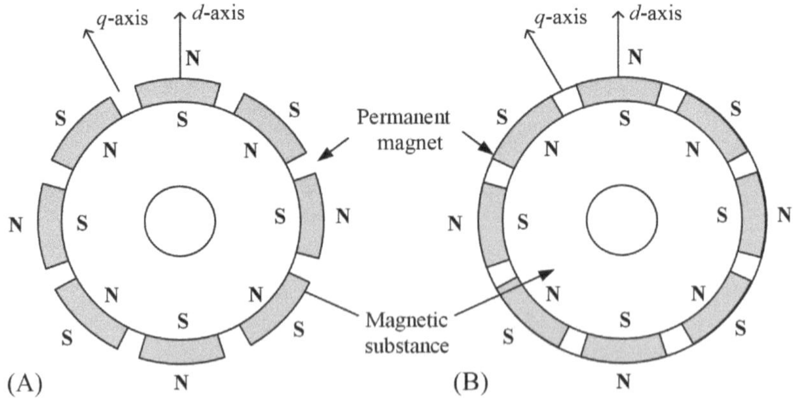

FIG. 11.17 Classification according to rotor arrangement. (A) Surface-mounted permanent magnet synchronous motor (SPMSM) and (B) interior permanent magnet synchronous motor (IPMSM).

be damaged or the glue to be stripped. Therefore, the SPMSM is weak at the very high speed driving. Moreover, the effect of valid air gap is larger than the actual physical air gap of the high magnetic reluctance permanent magnet at the surface so that the flux weakening control becomes more difficult by the lowered inductance. An other drawback is the eddy current loss at the surface of the magnet. Therefore, the SPMSM can be applied when there is little change in load and no flux weakening region needed.

(b) Interior permanent magnet synchronous motor (IPMSM).
The design of the IPMSM's structure addresses the SPMSM's weakness, reduces the valid air gap by putting the permanent magnet inside of the rotor, and improves the robustness of mechanical strength. The surface of the rotor is laminated with a silicon sheet so that the eddy current loss can be lowered and flux weakening can be easy. Moreover, the salient characteristic of the permanent magnet arrangement is that the reluctance torque is generated. The IPMSM is suitable in the cases where the load changes considerably and the flux weakening region exists such as servomotor and motor of electric-hybrid and electric cars. The IPMSM can be categorized by the formation of the inserted permanent magnet shown in Fig. 11.18, and the axis-to-axis inductance differs depending on the location of the inserted permanent magnets.

In the case of a cylindrical structure, because the permanent magnet is located at the aisle of *d*-axis, the inductance of the *d*-axis is lower than that of the *q*-axis as shown in Fig. 11.18B. However, for the cylindrical structure shown in Fig. 11.18C, the *d*-axis inductance is higher than the *q*-axis inductance. The cylindrical structure is usually used in the IPMSM. Unlike the cylindrical structure, the cylindrical structure is more likely to have cogging torque and ripple torque owing to the distortion near the surface of the rotor at the end of the permanent magnet although its structure can focus the flux and increase the air gap flux density and the reluctance torque. Therefore, the cylindrical structure has advantage at the flux weakening region.

The drawback of the IPMSM is the decrease in valid flux. Generally, the structure of the rotor of the IPSM is composed of a laminated silicon sheet

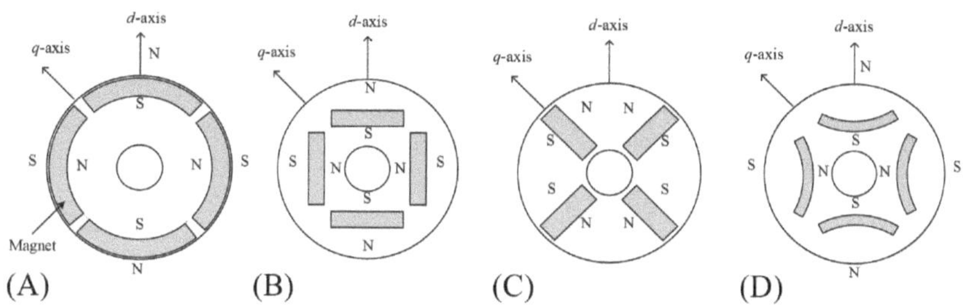

FIG. 11.18 Classification of the interior permanent magnet synchronous motor. (A) Arc type, (B) cylindrical type, (C) radial type, and (D) U type.

inserted in the rotor core along the direction of radius of the permanent magnet as mentioned earlier. As in the IPMSM, the structure is formed by making the salient at rotor core or fixing the permanent magnet by the body of the rotor core to prevent a breakaway of the permanent magnets. Therefore, considerable leakage of flux occurs at the bump at the core, causing the valid flux to decrease. The motor characteristic is related to the effect of decreasing the valid flux, and the unstability can be shown to increase the maximum speed, increase the current rate, and decrease the rated power, thus generating heat at the coil. Another disadvantage is that the flux density of the capacitor is large, and it can be the reason for the oscillations and acoustic noise. Research on the effect of inserting the magnet is ongoing, attempting to decrease this weakness.

Classification According to Rotor Structure

In the synchronous motor, the cylindrical structure indicates that the inductance is the same regardless of the position of the rotor. The air gap is uniform regardless of the position of the rotor, and there is no difference in magnetic reluctance. Generally, the cylindrical structure is used for high-speed driving. In the SPMSM that has a cylindrical structure, the magnetic air gap is uniform owing to form a permanent magnet attached to the surface of the rotor and it is symmetrical with respect not only to the mechanical structure but also to the electrical characteristic.

In the synchronous motor, the salient structure implies that the inductance is changing with the rotor position. For the rotor of the synchronous motor that has a salient structure, the reluctance torque is large owing to the irregular air gap according to the position of the rotor as shown in Fig. 11.19. Therefore, the motor control is difficult, but the speed and output characteristics are improved if the reluctance torque is effectively utilized. The salient structure is mostly applied for low-speed driving. The rotor of the IPMSM has a cylindrical structure, but the salient characteristic is achieved because of the irregular axis-to-axis magnetic air gap generated from the magnet inserted in the rotor.

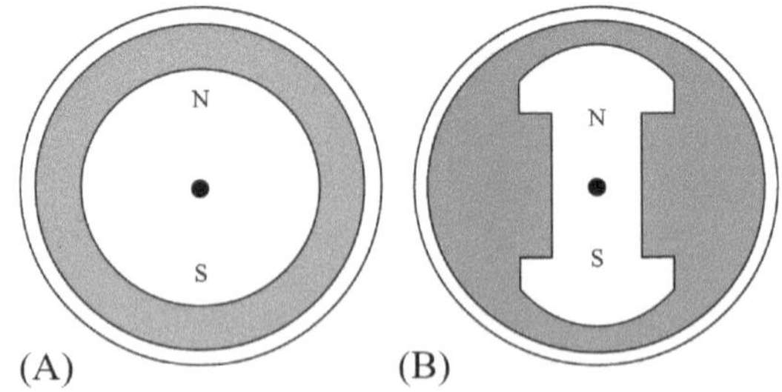

FIG. 11.19 Classification according to the interior of permanent magnet synchronous motor. (A) Cylindrical synchronous machine structure and (B) salient pole synchronous machine structure.

11.2.2 Vector Control of Permanent Magnet Synchronous Motor

11.2.2.1 Modeling of Interior Permanent Magnet Synchronous Motor

The stator circuit of the synchronous motor is similar to that of the induction motor; hence, the *d-q* axis stator voltage equation is derived for the induction motor.

$$\begin{aligned} v_{ds}^r &= R_s i_{ds}^r + \frac{d\lambda_{ds}^r}{dt} - \omega\lambda_{qs}^r \\ v_{qs}^r &= R_s i_{qs}^r + \frac{d\lambda_{qs}^r}{dt} + \omega\lambda_{ds}^r \\ v_{ns}^r &= R_s i_{ns}^r + \frac{d\lambda_{ns}^r}{dt} \end{aligned} \tag{11.36}$$

Unlike the structure of the induction motor, the structure of the permanent magnet synchronous motor has a different linkage flux. The self-inductance part is expressed as the function of θ_r is changed to the $d^\omega - q^\omega$ axis that revolves with rotor angular speed ω_r for the removal of the time-varying inductance such as an important goal of the coordinate conversion, stator inductance L_s is expressed by Eq. (11.35).

$$T(\theta_r)L_sT(\theta_r)^{-1} = \begin{bmatrix} L_{ls} + \frac{3}{2}(L_A - L_B) & 0 & 0 \\ 0 & L_{ls} + \frac{3}{2}(L_A - L_B) & 0 \\ 0 & 0 & L_{ls} \end{bmatrix}$$

$$= \begin{bmatrix} L_{ls} + L_{md} & 0 & 0 \\ 0 & L_{ls} + L_{mq} & 0 \\ 0 & 0 & L_{ls} \end{bmatrix} \tag{11.37}$$

where $L_{md} = (3/2)\cdot(L_A - L_B)$, $L_{mq} = (3/2)\cdot(L_A + L_B)$, and $T(\theta_r)$ is transfer matrix.

Converting the inductance of the permanent magnet L_f to the $d^r - q^r$ axis that revolves with the angular speed of the rotor ω_r can be expressed as follows:

$$T(\theta_r)L_f = L_{sf}\begin{bmatrix} \cos(\theta - \theta_r) \\ -\sin(\theta - \theta_r) \\ 0 \end{bmatrix} = L_{sf}\begin{bmatrix} 1 \\ 0 \\ 0 \end{bmatrix} \tag{11.38}$$

Now, the final *d-q* axis stator linkage flux ($\theta = \theta_r$) in rotor coordinates can be derived.

$$\lambda_{dqns}^r = \begin{bmatrix} L_{ls} + L_{md} & 0 & 0 \\ 0 & L_{ls} + L_{mq} & 0 \\ 0 & 0 & L_{ls} \end{bmatrix}\begin{bmatrix} i_{ds}^r \\ i_{qs}^r \\ i_{ns}^r \end{bmatrix} + L_{sf}\begin{bmatrix} i_f \\ 0 \\ 0 \end{bmatrix} \tag{11.39}$$

This matrix can be expressed as follows:

$$
\begin{aligned}
\lambda_{ds}^{r} &= L_{ls} i_{ds}^{r} + L_{md} i_{ds}^{r} + L_{sf} i_f = L_{ds} i_{ds}^{r} + \phi_f \\
\lambda_{qs}^{r} &= L_{ls} i_{qs}^{r} + L_{mq} i_{qs}^{r} = L_{qs} i_{qs}^{r} \\
\lambda_{ns}^{r} &= L_{ls} i_{ns}^{r}
\end{aligned}
\tag{11.40}
$$

where $L_{ds}=L_{ls}+L_{md}$, $L_{qs}=L_{ls}+L_{mq}$, $\phi_f=L_{sf}i_f$.

Substituting Eq. (11.40) into the stator voltage (Eq. 11.36), it can be expressed as follows:

$$
v_{ds}^{r} = R_s i_{ds}^{r} + \frac{d\lambda_{ds}^{r}}{dt} - \omega_r \lambda_{qs}^{r} = R_s i_{ds}^{r} + L_d \frac{di_{ds}^{r}}{dt} - \omega_r L_q i_{qs}^{r} \tag{11.41}
$$

$$
v_{qs}^{r} = R_s i_{qs}^{r} + \frac{d\lambda_{qs}^{r}}{dt} + \omega_r \lambda_{ds}^{r} = R_s i_{qs}^{r} + L_q \frac{di_{qs}^{r}}{dt} - \omega_r L_d i_{ds}^{r} + \omega_r \phi_f \tag{11.42}
$$

11.2.2.2 Output Torque of Permanent Magnet Synchronous Motor at d-q Axis Reference Frame

The torque of the synchronous permanent magnet motor can be obtained from the mechanical output of the motor. In terms of the *d-q* axis coordinates, the input power of the motor can be expressed as:

$$
P_{in} = \frac{3}{2}\left(v_d^{r} i_d^{r} + v_q^{r} i_q^{r}\right) \tag{11.43}
$$

From the *d*- and *q*-axis coordinates derived early, substituting the stator voltage and linkage flux equation into the power equation, it can be expressed as follows:

$$
\begin{aligned}
P_{in} &= \frac{3}{2}\left\{\left(R_s i_{ds}^{r} + \frac{d\lambda_{ds}^{r}}{dt} - \omega_r \lambda_{qs}^{r}\right) i_{ds}^{r} + \left(R_s i_{qs}^{r} + \frac{d\lambda_{qs}^{r}}{dt} + \omega_r \lambda_{ds}^{r}\right) i_{qs}^{r}\right\} \\
&= \frac{3}{2}\left\{R_s\left(i_{ds}^{r}2 + i_{qs}^{r}2\right) + i_{ds}^{r}\frac{d\lambda_{ds}^{r}}{dt} + i_{qs}^{r}\frac{d\lambda_{qs}^{r}}{dt} + \omega_r \phi_f i_{qs}^{r} + \omega_r\left(L_{ds} - L_{qs}\right) i_{ds}^{r} i_{qs}^{r}\right\}
\end{aligned}
\tag{11.44}
$$

In the above equation, the electrical input is expressed in terms of the stator copper loss, time variation of the integrated magnetic energy, and the mechanical output. The torque is obtained by dividing the mechanical output by motor speed ω_r, and it can be expressed as a *d-q* axis variable as:

$$
T_e = \frac{P}{2}\frac{3}{2}\left[\phi_f i_{qs}^{r} + \left(L_{ds} - L_{qs}\right) i_{ds}^{r} i_{qs}^{r}\right] \tag{11.45}
$$

where P is the number of poles. The generated torque is same to the current as seen from Eq. (11.45). Moreover, there is a reluctance torque coming from the difference of the electromagnetic torque and the inductance generated from flux ϕ_f and current i_{qs}^{r} by the permanent magnet at the torque of the IPMSM.

11.2.2.3 Vector Control of Permanent Magnet Synchronous Motor

As discussed earlier, the permanent magnet synchronous motor is categorized as SPMSM and IPMSM, and the vector control method is slightly different because the output torque characteristics of the two motor are different.

The three conditions for the instantaneous torque control of motor are as follows:

① The flux and torque components always maintained at 90° spatially.
② The flux and torque component current are independent of the counter electromotive force, leakage inductance, and wire resistor, and it can be instantaneously controlled.
③ The torque component current can be immediately controlled.

Let us now consider how to control the permanent magnet synchronous motor satisfying these conditions. First, considering the second condition, the permanent magnet synchronous motor is constructed in such a way in order to do independently control the flux and torque components. The flux is generated from the permanent magnet at the permanent magnet synchronous motor; without relation to this, the torque component current is decided by the stator current. Moreover, the stator current of the torque component can be rapidly and independently controlled independent of the flux by the PWM inverter. Therefore, the second and third conditions are satisfied. Considering the first condition that the flux and torque component always maintain at 90° spatially. To satisfy this condition, the three-phase stator current of the synchronous motor should be separated into two currents of *d*-*q* axis coordinates. In this situation, the *d*-axis is arranged to the position of the permanent magnets that generates the flux. The *q*-axis current is converted to the torque current component that is always maintained at 90° spatially to the rotor flux. The permanent magnet which produces flux of motor is rotating with the rotor angular speed ω_r as shown in Fig. 11.20. Therefore, the *d*-*q* axis should rotate with the same speed as the flux angular speed to maintain the torque component current always at 90° spatially. In other words, the stator current needs to be converted by using the rotating reference frame in the rotor angular speed.

For this coordinate conversion, the data of permanent magnet rotor position θ_r are required. In the synchronous motor, an absolute type rotor position sensor such as a resolver or an encoder is used to obtain the absolute initial position that is along the *d*-axis.

11.2.2.4 Vector Control of Surface-Mounted Permanent Magnet Synchronous Motor

The present position of rotor θ_r is detected and set to be the d^r-axis position and converting stator current coordinates to rotor coordinates. Considering the torque equation of the SPMSM given earlier, the flux component current of

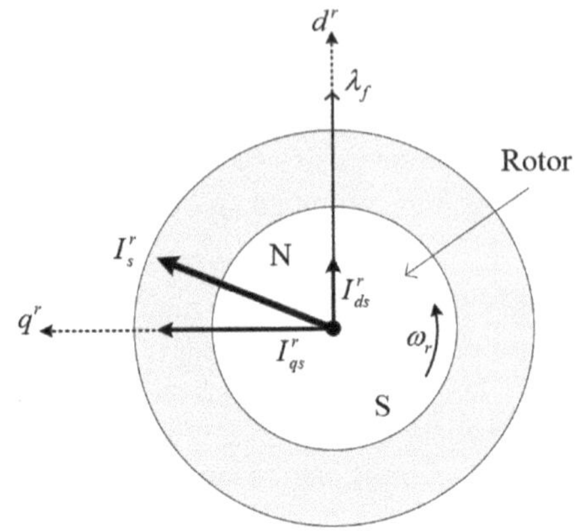

FIG. 11.20 Rotor of permanent magnet synchronous motor and its definitions for control.

the d^r-axis current regardless of the torque component and hence, to utilize maximum stator current, all stator current has to be the torque component current of the q^r-axis current. In other words, d^r-axis current that produces flux ϕ_f is positioned to $d^r_{ds}=0$ and all stator current coordinates are converted to q^r-axis current i^r_{qs}. The reference current can be expressed as:

$$i^{r*}_{ds}=0, \quad i^{r*}_{qs}=\frac{T_e^*}{\frac{P}{2}\frac{3}{2}\phi_f} \tag{11.46}$$

where T_e^* is torque reference and expresses the reference current to the vector diagram that is shown in Fig. 11.21A.

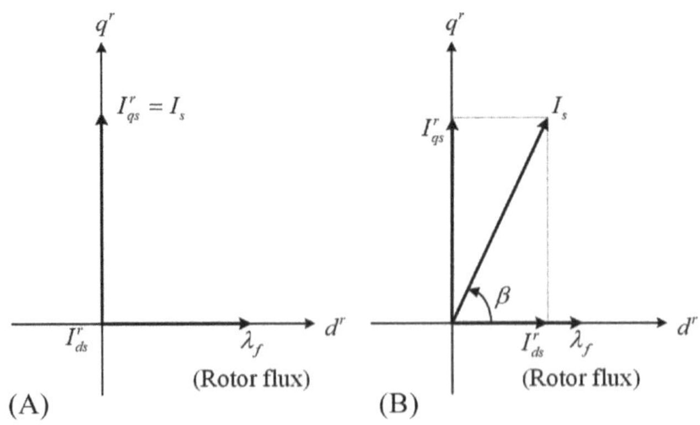

FIG. 11.21 Basic current vector control of permanent magnet synchronous motor. (A) Surface-mounted type and (B) interior type.

11.2.2.5 Vector Control of Interior Permanent Magnet Synchronous Motor

In the torque equation of the IPMSM, reluctance torque exists as well as the electromagnetic torque. Therefore, a slightly larger torque can be achieved by using this reluctance torque component. Similar to the IPMSM vector control method discussed earlier, if the d-axis current is controlled 0, the reluctance torque cannot be utilized and another method should be used. In other words, to generate this torque, d-axis current is needed and the stator current is separated to d^r- and q^r-axis current components. Therefore, the maximum torque can be derived from the current distribution ratio of d^r- and q^r-axis from the given stator current. Fig. 11.21B shows the current vector diagram, and β is the current phase for generating the maximum torque per unit current.

11.2.3 Simulation of Permanent Magnet Synchronous Motor

The basic operation of the permanent magnet synchronous motor is simulated by using the PSIM-based tool. Fig. 11.22 shows the circuit schematic of the permanent magnet synchronous motor for vector control. This circuit is divided into four parts: power converter, permanent magnet synchronous motor, load, and control part. Table 11.2 lists the specifications of the simulation parameters and the permanent magnet synchronous motor for the simulation. When stator inductance L_d is equal to L_q, it is a surfaced permanent magnet synchronous motor, and when the stator inductance L_d is not equal to L_q, it is an interior type permanent magnet synchronous motor. In this paragraph, the surface-mounted permanent magnet synchronous motor is basically used for vector control.

11.2.3.1 Vector Control

Fig. 11.23 shows the reference frame transformation of the three-phase current. When the three-phase stator current is transformed into the stationary reference

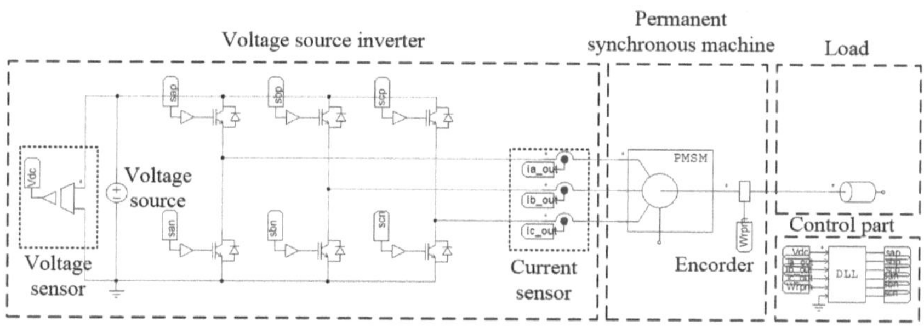

FIG. 11.22 Circuit schematic of permanent magnet synchronous motor.

TABLE 11.2 Specifications of Simulation Parameters and Surface-Mounted Permanent Magnet Synchronous Motor

Parameter	Value	Unit
Voltage source (V_{DC})	600	V
Sampling period (T_s)	100	μs
Switching frequency (f_{sw})	10	kHz
Rated power (P_{rated})	1.5	kW
Rated voltage (V_{rated})	129	V
Rated current (I_{rated})	6.7	A
Rated torque (T_{rated})	9.55	N m
Rated speed (ω_{rated})	1500	rpm
Stator resistance (R_s)	0.123	Ω
Stator inductance (L_d and L_q)	9.52	mH
Moment of inertia (J)	13.1	kg cm^2

frame, the magnitude of the current is the same and the phase difference is 90°. When the three-phase stator current is transformed into the rotating reference frame, the current becomes a DC (direct current). In addition, the magnitude of current is the maximum value of AC (alternative current) as shown in Fig. 11.23D. The current obtained from the reference frame transformation is input to the current controller [12,13].

Fig. 11.24 shows the simulation result of the current controller using PI controllers. The switching frequency is set to 10 kHz, and the bandwidth of the current controller for the stable control characteristic is set to 2000 rad/s. When the current reference is set to be step, the current controller has a fast response without error in steady state. In addition, the mutual coupling of the d- and q-axis can be decoupled through the compensation of the back emf.

Fig. 11.25 shows the speed and torque response result for different speed and torque loads. Although the torque reference is rapidly changed, the actual torque responds immediately as shown in Fig. 11.25A. When the speed reference changes from 800 to −800 rpm, the actual speed responds immediately as shown in Fig. 11.25B.

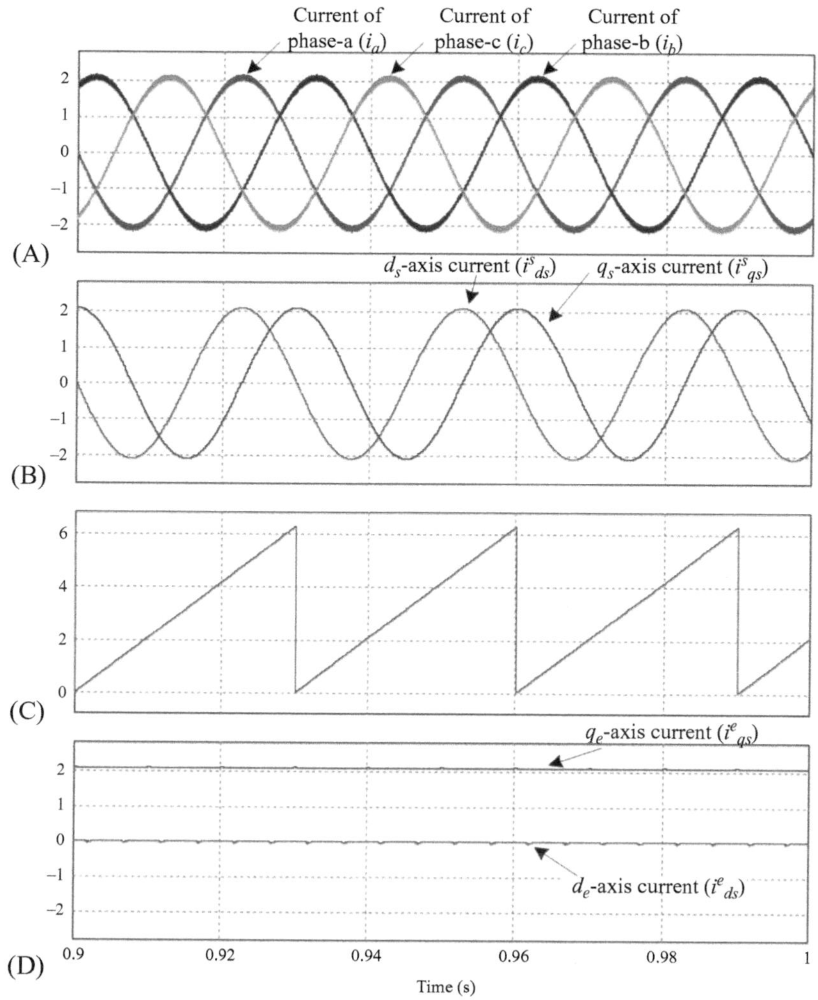

FIG. 11.23 Simulation of a control of surface-mounted permanent magnet motor reference frame transformation of three-phase current: (A) three-phase current [A], (B) stationary reference frame [A], (C) rotor flux angle [rad], and (D) rotating reference frame current [A].

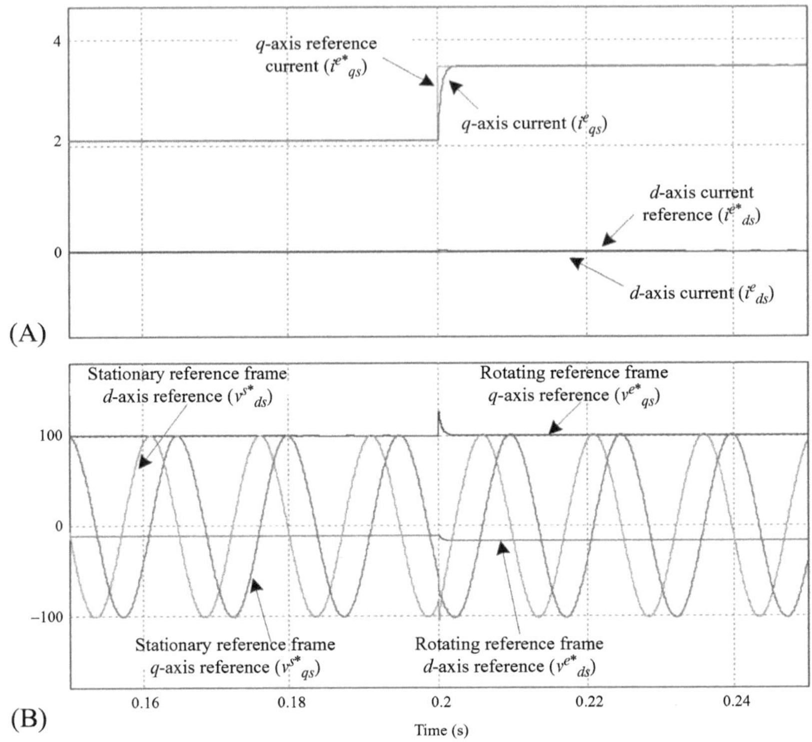

FIG. 11.24 Simulation of a control of surface-mounted permanent magnet motor current control and voltage reference: (A) reference and actual currents [A] and (B) voltage reference [V].

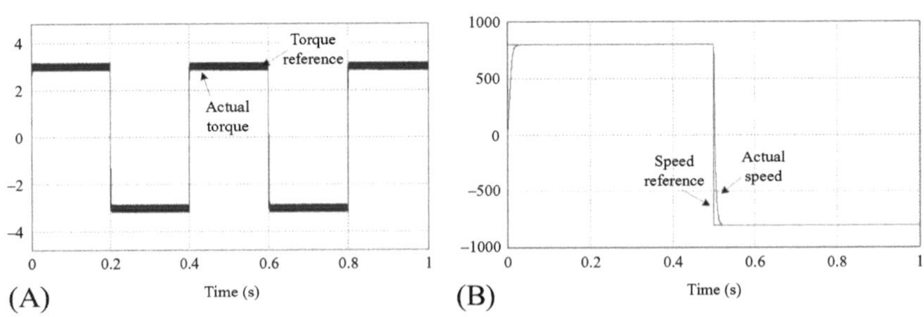

FIG. 11.25 Speed and torque response using vector control: (A) Constant speed load $\omega_{rpm} = 500$ rpm and (B) constant torque load $T_e = 1$ Nm.

11.3 CONCLUSION

In this chapter, the basic principle and control of AC motor drives such as induction motors and permanent magnet synchronous motors are described. The control methods for the induction motor drives are classified to the voltage/frequency constant control and the vector control, which is divided into direct and indirect vector control. Regarding the permanent magnet synchronous motor, there are surface-mounted and the interior permanent magnet synchronous motor depending on the position of the permanent magnet. Finally, simulation is performed in an AC motor drive system to verify the basic control of the AC motor drives.

REFERENCES

[1] I. Takahashi, T. Noguchi, A new quick-response and high-efficiency control strategy of an induction motor, IEEE Trans. Ind. Appl. IA–22 (1986) 820–827.

[2] I. Takahashi, Y. Ohmori, High-performance direct torque control of an induction motor, IEEE Trans. Ind. Appl. 25 (2) (1989) 257–264.

[3] A. Chiba, J. Asama, Influence of rotor skew in induction type bearingless motor, IEEE Trans. Magn. 48 (11) (2012) 4646–4649.

[4] P.C. Krause, O. Wasynczuk, S.D. Sudhoff, Analysis of electric machinery and drive systems, 2nd ed., Wiley, Hoboken, NJ, USA, 2002.

[5] R. Mathew, D. Houghton, W. Oghanna, Vector control techniques for induction motors, Power Electron. Drive Syst. 2 (1995) 813–818.

[6] M.P. Kazmierkowski, W. Sulkowski, A novel vector control scheme for transistor PWM inverter-fed induction motor drive, IEEE Trans. Ind. Electron. 38 (1) (1991) 41–47.

[7] S. Motlagh, S.S. Fazel, Indirect vector control of linear induction motor considering end effect, Proc. PEDSTC (Power Electronics and Drive Systems Technology), 2012, pp. 193–198.

[8] T. Rajesh, Y.S.K. Babu, Y.R. Tagore, Performance evaluation of indirect vector controlled induction motor sensorless drive, Proc. CICN (Computational Intelligence and Communication Networks), 2012, pp. 656–660.

[9] G.S. Lee, D.H. Lee, T.W. Yoon, K.B. Lee, J.H. Song, I. Choy, Speed and flux estimation for an induction motor using a parameter estimation technique, IJCAS (Int. J. Control. Autom. Syst.) 3 (1) (2005) 79–86.

[10] H. Kubota, K. Matsus, T. Nakano, DSP-based speed adaptive flux observer of induction motor, IEEE Trans. Ind. Appl. 30 (5) (1994) 1219–1224.

[11] S.K. Kim, J.S. Lee, K.B. Lee, Offset-free robust adaptive back-stepping speed control for uncertain permanent magnet synchronous motor, IEEE Trans. Power Electron. 31 (10) (2016) 7065–7076.

[12] T.M. Rowan, R.J. Kerkman, A new synchronous current regulator and an analysis of current-regulated PWM inverter, IEEE Trans. Ind. Appl. IA-22 (4) (1986) 678–690.

[13] G.F. Franklin, J.D. Powell, A. Emami-Naeini, Feedback Control of Dynamic Systems, 6th ed., Pearson, Upper Saddle River, NJ, USA, 2008.

Chapter 12

Sensorless Control of Motor Drives

Gaolin Wang, Yong Yu, Guoqiang Zhang, Bo Wang, Ming Yang and Dianguo Xu
School of Electrical Engineering and Automation, Harbin Institute of Technology, Harbin, China

12.1 INTRODUCTION

Speed-sensorless vector control for induction motor (IM) and interior permanent magnet synchronous motor (IPMSM) is popular in industrial applications. The advantages of sensorless control include low cost, reduced hardware complexity and size, improved robustness of the control system, and less maintenance requirements [1]. Fig. 12.1 shows the block diagram of the typical AC motor drive.

First, many approaches of speed-sensorless IM drives have been proposed in the past years. There are mainly two basic approaches in a sensorless control system. One approach is the signal-injection-based method [2]. The speed estimation is based on rotor slot harmonics, main inductance saturation, artificial saliency, or rotor slot leakage. The performance of this method is well at low-speed range. However, in the signal-injection-based observer, spatial salience in the motor is assumed. It is difficult to extract the correct information, and the saliency of the motor varies in different motors. Consequently, the signal-injection-based method is not often used in IM drives.

The other approach is based on the fundamental model of an IM. Many methods have been proposed [3]: model reference adaptive system, adaptive full-order observer, adaptive reduced-order observer, sliding-mode control, extended Kalman filter. The performance of the fundamental model-based method is poor when the rotor speed is close to zero or ultrahigh speed. In this section, a method based on the adaptive full-order observer is applied to solve the problem of operation close to zero and ultrahigh-speed region.

Second, the sensorless control methods of IPMSM drives can also be classified into two classes. The first class is based on the electromotive force (EMF) machine model for the middle- and the high-speed operation [4]. In particular, variable structure systems with sliding modes have been applied in sensorless control due to its simplicity in applications and good performance. Since the

Control of Power Electronic Converters and Systems. https://doi.org/10.1016/B978-0-12-805245-7.00012-3

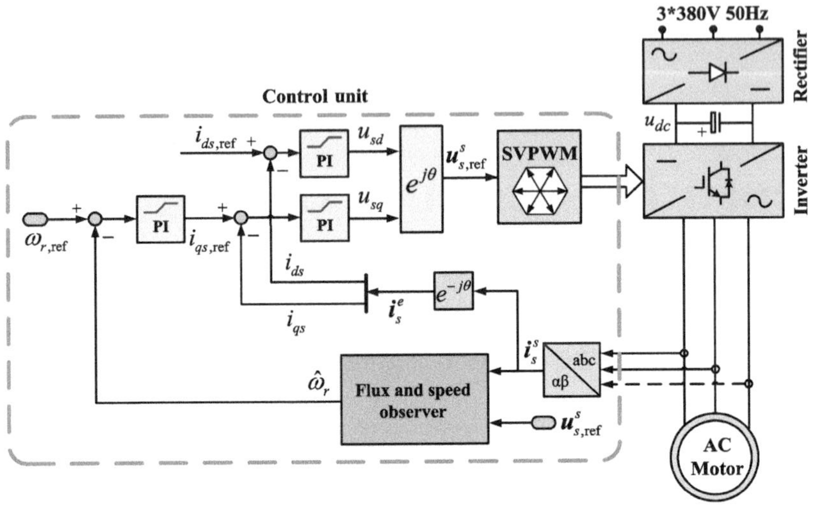

FIG. 12.1 Block diagram of the typical AC motor drive.

EMF of the IPMSM is proportional to the rotor speed, the sensorless operation in low-speed range is limited. The influencing factors include the limited accuracy of measurements, motor parameter variation, and presence of inverter nonlinearities.

The other class is the high-frequency (HF) signal injection estimation for low-speed operation [5]. The signal injection methods are based on detecting the anisotropy caused by the structure saliency or the magnetic saturation. An HF test signal is superimposed to the fundamental component, and the rotor position can be obtained from the excited HF currents. Two signal injection methods are usually adopted: one is the rotating HF voltage vector injection method in the stationary frame, and the other is the pulsating HF voltage vector injection method in the estimated rotor reference frame. In this section, a method based on the HF signal injection is applied to solve the problem of operation at low speed or even zero speed.

12.2 SENSORLESS CONTROL OF IM IN ULTRALOW-SPEED REGION

12.2.1 Principle of Adaptive Full-Order Observer

In the rotary reference frame, the dynamic model of IM is given by Eq. (12.1)

$$\begin{cases} p\vec{\lambda}_s = A_{11}\vec{\lambda}_s + A_{12}\vec{\lambda}_r + B_v\vec{v}_s \\ p\vec{\lambda}_r = A_{21}\vec{\lambda}_s + A_{22}\vec{\lambda}_r \\ \vec{i}_s = H_1\vec{\lambda}_s + H_2\vec{\lambda}_r \end{cases} \tag{12.1}$$

where $\vec{\lambda}_s = [\lambda_{sd}\ \lambda_{sq}]^T$ and $\vec{\lambda}_r = [\lambda_{rd}\ \lambda_{rq}]^T$ are the stator and rotor flux vector; $\vec{v}_s = [v_{sd}\ v_{sq}]^T$ and $\vec{i}_s = [i_{sd}\ i_{sq}]^T$ are the stator voltage and current vector; $\mathbf{A_{11}} = a_{11}\mathbf{I} + \omega_e \mathbf{J}$, $\mathbf{A_{12}} = a_{12}\mathbf{I}$, $\mathbf{A_{21}} = a_{21}\mathbf{I}$, $\mathbf{A_{22}} = a_{22}\mathbf{I} + (\omega_e - \omega_r)\mathbf{J}$, $a_{11} = \frac{-R_s}{\delta L_s}$, $a_{12} = \frac{R_s L_m}{\delta L_s L_r}$, $a_{21} = \frac{R_r L_m}{\delta L_s L_r}$, $a_{22} = \frac{-R_r}{\delta L_r}$, $b_1 = 1$, $h_1 = \frac{1}{\delta L_s}$, $h_2 = \frac{-L_m}{\delta L_s L_r}$, $\delta = 1 - \frac{L_m^2}{L_s L_r}$ R_s is the stator resistance; L_s, L_r, and L_m are the machine self and mutual inductance; ω_e is the synchronous angular speed; and $\sigma = 1 - L_m^2/(L_s L_r)$ is the leakage coefficient.

From the model of IM, adaptive full-order observer is given by Eq. (12.2):

$$\begin{cases} p\hat{\vec{\lambda}}_s = \hat{\mathbf{A}}_{11}\hat{\vec{\lambda}}_s + \mathbf{A}_{12}\hat{\vec{\lambda}}_r + \mathbf{B}_v \vec{v}_s + \mathbf{G}_1 \vec{e}_{is} \\ p\hat{\vec{\lambda}}_r = \mathbf{A}_{21}\hat{\vec{\lambda}}_s + \hat{\mathbf{A}}_{22}\hat{\vec{\lambda}}_r + \mathbf{G}_2 \vec{e}_{is} \\ \hat{\vec{i}}_s = \mathbf{H}_1 \hat{\vec{\lambda}}_s + \mathbf{H}_2 \hat{\vec{\lambda}}_r \end{cases} \tag{12.2}$$

where $\vec{e}_{is} = \left[i_{sd} - \hat{i}_{sd}, i_{sq} - \hat{i}_{sq}\right]^T$, $\hat{\mathbf{A}}_{11} = a_{11}\mathbf{I} + \hat{\omega}_e \mathbf{J}, \hat{\mathbf{A}}_{22} = a_{22}\mathbf{I} + (\hat{\omega}_e - \hat{\omega}_r)\mathbf{J}$.

Based on the error equation of stator and rotor flux in stator frame, the speed estimation algorithm can be obtained through the Lyapunov's theorem:

$$\hat{\omega}_r = \left(k_p + \frac{k_i}{s}\right)\left[\left(e_{is\alpha}\hat{\lambda}_{r\beta} - e_{is\beta}\hat{\lambda}_{r\alpha}\right) - \left(e_{\lambda r\alpha}\hat{\lambda}_{r\beta} - e_{\lambda r\beta}\hat{\lambda}_{r\alpha}\right)\right] \tag{12.3}$$

In classical speed estimation, it is also assumed that: $\vec{\lambda}_r = \hat{\vec{\lambda}}_r$. And the estimated speed is:

$$\hat{\omega}_r = \left(k_p + \frac{k_i}{s}\right)\left(e_{is\alpha}\hat{\lambda}_{r\beta} - e_{is\beta}\hat{\lambda}_{r\alpha}\right) \tag{12.4}$$

In rotary reference frame, speed estimation is:

$$\hat{\omega}_r = \left(k_p + \frac{k_i}{s}\right)\left(\hat{i}_{sq} - i_{sq}\right)\hat{\lambda}_{rd} \tag{12.5}$$

The closed-loop block diagram of speed estimation is as shown in Fig. 12.2:

The open-loop transfer function (OLTF) between actual and estimated rotor speed can be derived from Fig. 12.2 as:

$$G_{op}(s) = -\hat{\lambda}_{rd}\left(K_P + \frac{K_I}{s}\right)G_q(s) \tag{12.6}$$

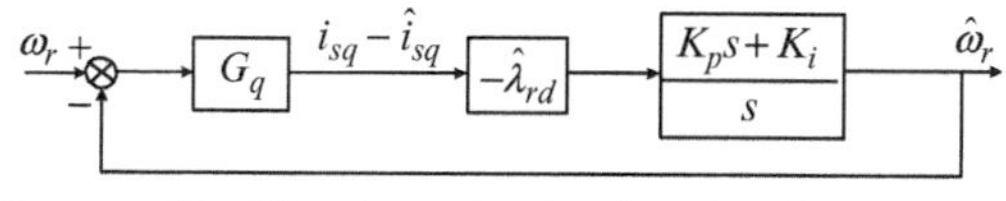

FIG. 12.2 Block diagram of traditional speed estimation algorithm.

where $G_q(s)$ is indicated as Eq. (12.12).

Based on classical speed estimation, the design method for observer feedback gains has been researched for many years and lots of designed feedback gains have been obtained. Each feedback gain has its own advantage and disadvantage. Summing up all the advantages and disadvantages of classical design methods, a robust design method is proposed and the system robustness and stability are improved.

12.2.2 Robust Speed Estimation and Robust Feedback Gains Design Method

In classical speed estimation algorithm, the estimated rotor flux error is ignored because the actual rotor flux cannot be measured in industry application. However, the estimated rotor flux error has strong influence on speed estimation stability and robustness to motor parameter deviation. It is necessary to introduce the estimated rotor flux error to speed estimation algorithm [6,7].

(1) Robust Speed Estimation With Classical Feedback Gains

Because real rotor flux cannot be measured in practical applications, the form of estimated rotor flux error in Eq. (12.3) needs to be improved. In Eq. (12.3), $e_{\lambda r\alpha} = \lambda_{r\alpha} - \hat{\lambda}_{r\alpha}$, $e_{\lambda r\beta} = \lambda_{r\beta} - \hat{\lambda}_{r\beta}$.

Rotor flux error in stationary reference frame in Eq. (12.3) is:

$$E_{flx\_s} = e_{\lambda r\alpha}\hat{\lambda}_{r\beta} - e_{\lambda r\beta}\hat{\lambda}_{r\alpha} = \lambda_{r\alpha}\hat{\lambda}_{r\beta} - \lambda_{r\beta}\hat{\lambda}_{r\alpha} \tag{12.7}$$

Extracting the same factor $\sqrt{\lambda_{r\alpha}^2 + \lambda_{r\beta}^2} \cdot \sqrt{\hat{\lambda}_{r\alpha}^2 + \hat{\lambda}_{r\beta}^2}$, Eq. (12.7) can be derived as:

$$\begin{aligned} E_{flx\_s} &= \sqrt{\lambda_{r\alpha}^2 + \lambda_{r\beta}^2} \cdot \sqrt{\hat{\lambda}_{r\alpha}^2 + \hat{\lambda}_{r\beta}^2}\left(\frac{\lambda_{r\alpha}\hat{\lambda}_{r\beta} - \lambda_{r\beta}\hat{\lambda}_{r\alpha}}{\sqrt{\lambda_{r\alpha}^2 + \lambda_{r\beta}^2} \cdot \sqrt{\hat{\lambda}_{r\alpha}^2 + \hat{\lambda}_{r\beta}^2}}\right) \\ &= \left(\left|\hat{\vec{\lambda}}_r\right|^2 + \Delta\lambda\left|\hat{\vec{\lambda}}_r\right|\right)\left[\sin\hat{\theta}\cos\theta - \cos\hat{\theta}\sin\theta\right] \end{aligned} \tag{12.8}$$

Defining θ as actual rotor flux angle, the error between actual and estimated rotor flux angle is represented by $\Delta\theta$, error between actual and estimated rotor flux amplitude is represented by $\Delta\lambda$, and Eq. (12.8) can be derived as:

$$E_{flx\_s} = -\sin\Delta\theta\left(\left|\hat{\vec{\lambda}}_r\right|^2 + \Delta\lambda\left|\hat{\vec{\lambda}}_r\right|\right) \tag{12.9}$$

According to Fig. 12.3, the rotational speed of the rotor flux vector is as same as rotation speed of stator current vector and is also equal to rotation speed of rotor reference frame. Consequently, sin $\Delta\theta$ is given as:

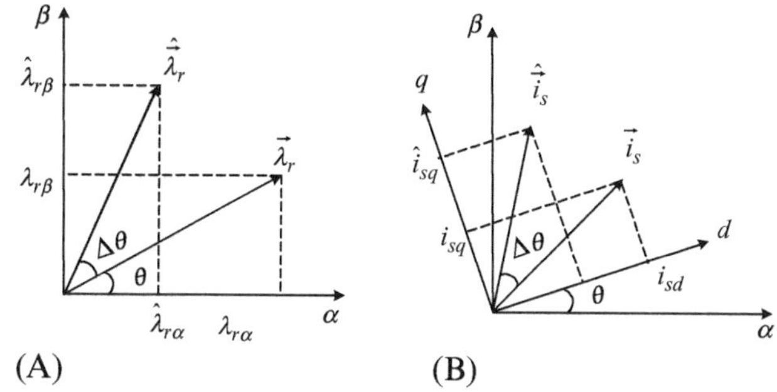

FIG. 12.3 Vector graphs of rotor flux and stator current.

$$\sin\Delta\theta = \frac{\left|i_{sq}\hat{i}_{sd} - i_{sd}\hat{i}_{sq}\right|}{\left|\vec{i}_s\right| \cdot \left|\hat{\vec{i}}_s\right|} \tag{12.10}$$

In stationary frame, $\sin\Delta\theta$ is

$$\sin\Delta\theta = \frac{i_{s\alpha}\hat{i}_{s\beta} - i_{s\beta}\hat{i}_{s\alpha}}{\sqrt{2\left|\vec{i}_s\right| \cdot \left|\hat{\vec{i}}_s\right|\left(\left|\vec{i}_s\right| \cdot \left|\hat{\vec{i}}_s\right| + i_{s\alpha}\hat{i}_{s\alpha} + i_{s\beta}\hat{i}_{s\beta}\right)}} \tag{12.11}$$

Ignoring $\Delta\lambda$ and introducing coefficient M in Eq. (12.9), the robust speed estimation algorithm is given as:

$$\hat{\omega}_r = \left(k_p + k_i\int dt\right)\left[\left(e_{is\alpha}\hat{\lambda}_{r\beta} - e_{is\beta}\hat{\lambda}_{r\alpha}\right) - M\left(i_{s\alpha}\hat{i}_{s\beta} - i_{s\beta}\hat{i}_{s\alpha}\right)\right] \tag{12.12}$$

To simplify the analysis, the speed estimation algorithm in the rotor-flux reference frame can be derived as:

$$\hat{\omega}_r = -\left(k_p + k_i\int dt\right)\left[\hat{\lambda}_{rd}\left(i_{sq} - \hat{i}_{sq}\right) - M\left(i_{sd} - \hat{i}_{sd}\right)\right] \tag{12.13}$$

Fig. 12.4 depicts the block diagram of proposed speed estimation algorithm. Compared with classical speed estimation algorithm, the transfer function $G_d(s)$ and coefficient M are added.

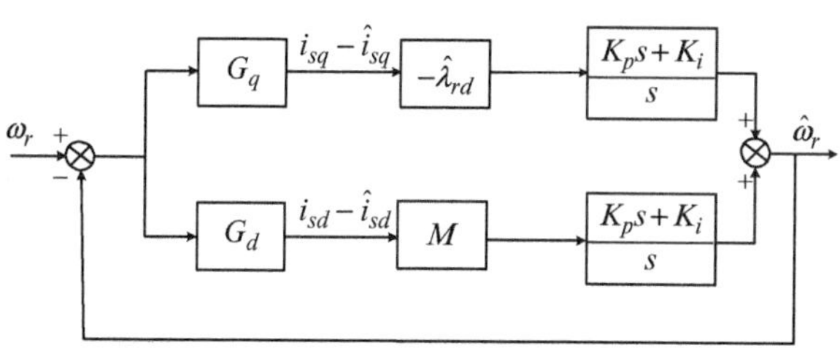

FIG. 12.4 Block diagram of proposed speed estimation algorithm.

As indicated in Fig. 12.4, the OLTF G_{op_new} of the estimated speed is

$$G_{op\_new}(s) = \left(K_P + \frac{K_I}{s}\right)\left[-\hat{\lambda}_{rd}G_q(s) + MG_d(s)\right] \tag{12.14}$$

It is well known that when the feedback gains are zero, there are unstable regions of the speed estimation when the motor runs at low-speed regenerating range. However, if M is designed properly, the speed estimation can be stable at all speed ranges. Considering the stability of speed estimation, the real part of zeros and poles of OLTF G_{op_new} have to be negative. In order to find out the value of M that can guarantee the stability of speed estimation.

(2) Robust Speed Estimation with Corresponding Robust Feedback Gains

To design the feedback gains, the stability of speed estimation is considered first. In the same way, to guarantee the stability of speed estimation, all the zeros of Eq. (12.14) have to be in the left plane.

For simplicity, $M = \hat{\lambda}_{rd}$ and Eq. (12.14) is improved as:

$$G_{op\_new}(s) = \frac{(K_P s + K_I)}{Bs}\left[s^3 + p_2 s^2 + p_1 s + p_0\right] \tag{12.15}$$

where $p_2 = q_2 - d_2 = A_3\omega_e + A_2$, $p_1 = q_1 - d_1 = \omega_e^2 + A_1$, $p_0 = q_0 - d_0 = A_3\omega_e^3 + A_2\omega_e^2 + A_0\omega_e$.

With Routh Table [6], the necessary and sufficient conditions for the stability of speed estimation can be obtained as:

$$\begin{cases} A_3\omega_e + A_2 > 0 \\ A_1 - \dfrac{A_0\omega_e}{A_3\omega_e + A_2} > 0 \\ \omega_e^2(A_3\omega_e + A_2) + A_0\omega_e > 0 \end{cases} \tag{12.16}$$

To satisfy Eq. (12.16), condition (12.17) is obtained as:

$$\begin{cases} A_0 = 0 \\ A_1 > 0 \\ A_3\omega_e + A_2 > 0 \end{cases} \tag{12.17}$$

(a) $A_0 = 0$. Based on this equation, the relationship between g_1 and g_2 can be obtained as:

$$g_2 = \frac{\delta L_s L_r^2\omega_r + L_m R_r}{L_r L_m \omega_r - \delta L_s L_r R_r}(R_s + g_1) \tag{12.18}$$

(b) $A_1 > 0$. Based on this condition, a range of g_2 for the speed estimation stability can be obtained as:

$$\frac{g_2}{L_r\omega_r+\dfrac{L_mR_r}{\delta L_sL_r}}<0 \tag{12.19}$$

When $-100\pi < \omega_r < 100\pi$, the denominator of Eq. (12.19) is always positive. So Eq. (12.19) can be satisfied as long as g_2 is negative. However, this feedback gain can only be applied in the fundamental speed range (within ±50 Hz).

(c) $A_3\omega_e + A_2 > 0$. Based on this condition the relationship between slip frequency and feedback gains can be obtained:

$$\omega_e-\omega_r>\left(\frac{1}{\delta L_s}-\frac{\dfrac{L_r}{\delta L_s}\omega_r-\dfrac{L_rR_r}{L_m}}{L_r\omega_r+\dfrac{L_mR_r}{\delta L_sL_r}}\right)g_2-\frac{L_m}{\delta L_sL_r}g_4-\frac{R_rL_s-L_mg_3}{L_m} \tag{12.20}$$

Similarly when $-100\pi < \omega_r < 100\pi$ (within ±50 Hz)

$$\frac{1}{\delta L_s}-\frac{\dfrac{L_r}{\delta L_s}\omega_r-\dfrac{L_rR_r}{L_m}}{L_r\omega_r+\dfrac{L_mR_r}{\delta L_sL_r}}>0 \tag{12.21}$$

Because g_2 is negative, condition (12.20) can be satisfied as long as $g_3 < 0$, $g_4 > 0$ and all right terms of Eq. (12.20) are less than the minimum slip frequency, which is negative because the motor runs in regenerating mode.

Finally, the necessary and sufficient conditions for the stability of speed estimation can be obtained as:

$$\begin{cases} g_1=-L_mL_r\omega_r+\delta L_sL_rR_r-R_s \\ g_2=-\delta L_sL_r^2\omega_r-L_mR_r \\ g_3=k \\ g_4=-k \end{cases} \tag{12.22}$$

where k is the minimum slip frequency, which is negative. In this chapter, $k = -15$.

12.2.3 Experimental Results

The proposed methods are verified using an experimental platform with STM32F103 ARM. The experimental platform, adopting two 7.5 kW IMs. One is the tested motor and is controlled by sensorless vector control. The other one is the load motor offering torque, and it is controlled with a speed encoder.

Fig. 12.5 indicates the experiments of speed estimation stability. Because the unstable region is always in the low-speed regenerating range, the motor

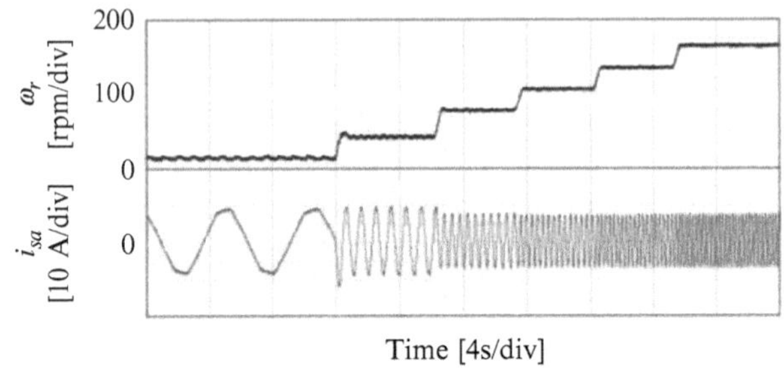

FIG. 12.5 Stability of speed estimation at low-speed regenerating range, 100% rated load.

is controlled in regenerating mode and the load is 100% rated load. The reference speed is in sequence of 15, 30, 60, 90, 120, and 150 rpm. In Fig. 12.4, the motor runs at low-speed regenerating range with the proposed method, and the estimation is stable during the runtime.

Fig.12.6 indicates the experiments that the reference speed is 90 rpm, but the load torque is decreased from 120% rated load to −120% rated load. When the load torque is negative the motor is running in regenerating mode. The stability of the speed estimation can be guaranteed in the low-speed regenerating range when the two proposed methods are used. Consequently with the same external or internal interrupt, the estimated speed error in regenerating mode is a little bigger than the error in the motoring mode.

Fig. 12.7 indicates the experiment of system robustness to stator resistance deviation. The motor runs in regenerating mode. The reference is 15 rpm and the load is 120% rated load. When the proposed method is used, the system can endure the 50% stator resistance error and still keep stable. Similarly, Fig. 12.8 indicates the experiment of system robustness to rotor resistance deviation. The motor runs in regenerating mode. The reference is 15 rpm and the load is 120%

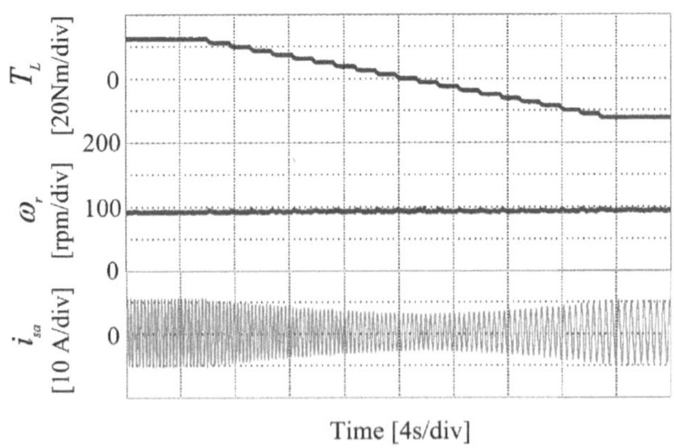

FIG. 12.6 Stability of speed estimation with variable load torque.

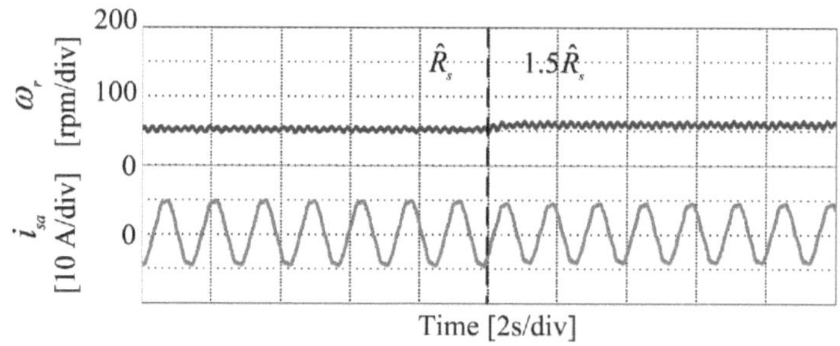

FIG. 12.7 Robustness to stator resistance deviation at regenerating range, 15 rpm reference speed, 120% rated load.

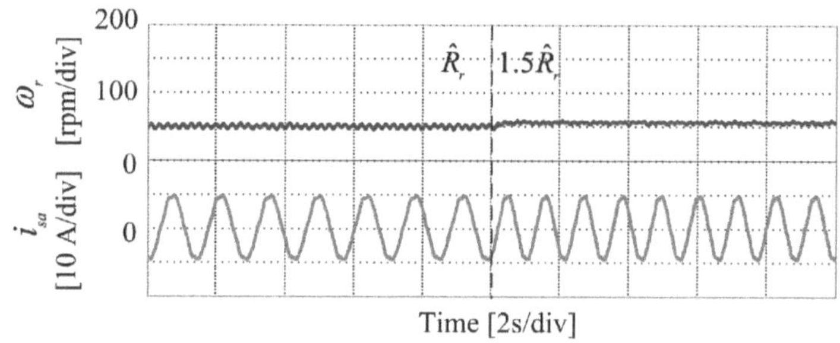

FIG. 12.8 Robustness to rotor resistance deviation at regenerating range, 15 rpm reference speed, 120% rated load.

rated load. When the rotor resistance is accurate, the actual speed is 30 rpm with the proposed method. When the rotor resistance increases by 50%, the actual speed becomes 35 rpm. The robustness of the proposed method to resistances deviation is strong.

12.3 SENSORLESS CONTROL OF IM IN HIGH-SPEED REGION

12.3.1 Principle of IM Field-Weakening Control

The stator voltage model of an IM viewed from the synchronous d-q frame is shown in Eq. (12.23)

$$\begin{cases} u_{ds} = R_s i_{ds} + \sigma L_s \dfrac{di_{ds}}{dt} - \omega_e \sigma L_s i_{qs} + \dfrac{L_m}{L_r}\dfrac{d\lambda_{dr}}{dt} - \omega_e \dfrac{L_m}{L_r}\lambda_{qr} \\ u_{qs} = R_s i_{qs} + \sigma L_s \dfrac{di_{qs}}{dt} + \omega_e \sigma L_s i_{ds} + \dfrac{L_m}{L_r}\dfrac{d\lambda_{qr}}{dt} + \omega_e \dfrac{L_m}{L_r}\lambda_{dr} \end{cases} \quad (12.23)$$

where u_{ds} and u_{qs} are the stator voltage d-q components; i_{ds} and i_{qs} are the stator current d-q components; λ_{dr} and λ_{qr} are the rotor flux d-q components; R_s

is the stator resistance; L_s, L_r, and L_m are the machine self- and mutual inductance; ω_e is the synchronous angular speed; and $\sigma = 1 - L_m^2/(L_s L_r)$ is the leakage coefficient.

Under the steady-state RFOC, $\lambda_{qr} = 0$ and $\lambda_{dr} = L_m i_{ds}$ are true, and the derivative terms of Eq. (12.23) are 0, so Eq. (12.23) can be simplified to

$$\begin{cases} u_{ds} = R_s i_{ds} - \omega_e \sigma L_s i_{qs} \\ u_{qs} = R_s i_{qs} + \omega_e L_s i_{ds} \end{cases} \tag{12.24}$$

The IM electromagnetic torque is expressed as

$$T_e = \frac{3}{2}\frac{P}{2}\frac{L_m^2}{L_r} i_{ds} i_{qs} \tag{12.25}$$

where P is the number of poles.

In the field-weakening region, the maximum torque capacity is limited by both the voltage and current constraints from the inverter and IM. First, the voltage constraint is given in Eq. (12.26)

$$u_{ds}^2 + u_{qs}^2 \le u_{s,\max}^2 \tag{12.26}$$

where the maximum inverter output voltage $u_{s,\max}$ is limited by the DC-link voltage and pulse width modulation (PWM) strategy. The space vector PWM (SVPWM) technique is selected here due to its high-voltage utilization ratio and easy realization. Without overmodulation, $u_{s,\max} = u_{dc}/\sqrt{3}$. Substituting Eq. (12.24) into Eq. (12.26) leads to the following voltage constraint

$$\left(R_s i_{ds} - \omega_e \sigma L_s i_{qs}\right)^2 + \left(R_s i_{qs} + \omega_e L_s i_{ds}\right)^2 \le u_{s,\max}^2 \tag{12.27}$$

Second, the current constraint is determined by the machine and inverter maximum allowable current $i_{s,\max}$

$$i_{ds}^2 + i_{qs}^2 \le i_{s,\max}^2 \tag{12.28}$$

The voltage and current constraints with the constant torque curves in the current trajectory ($i_{ds} - i_{qs}$) plane are shown in Fig. 12.9A. The voltage constraint boundary is an ellipse with its center at the origin. As the machine speed increases or the DC-link voltage decreases, the voltage constraint ellipse shrinks toward the origin. The current constraint boundary is a circle with its center at the origin. The constant torque curves are the inverse proportion function curves in the first and fourth quadrants. The curves in the first quadrant imply that the machine operates in the motoring mode, while the curves in the fourth quadrant imply that the machine operates in the generating mode. As the torque increases, the curves move away from the origin.

Since an IM has to operate under both the voltage and current constraints, the available operating region in Fig. 12.9A is the overlap of voltage constraint ellipse and current constraint circle. As the machine speed increases, the shrinking voltage constraint ellipse produces a reduced overlap, which means that the

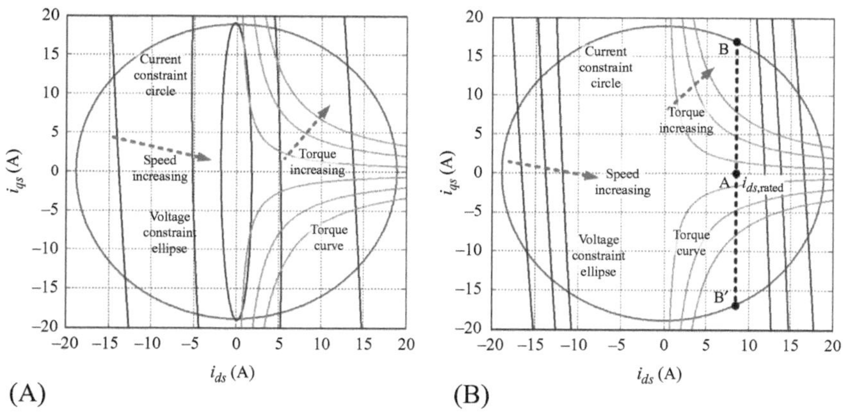

FIG. 12.9 (A) Voltage and current constraints. (B) Constraints in constant torque region.

operating region is further limited. For a maximum torque capacity, the operating point should be on the overlap edge.

(1) Constant Torque Region

In this region, the machine speed is lower than the base speed. As shown in Fig. 12.9B, the available operating region is only limited by the current constraint circle. The current locus for maximum torque capacity is point B in the motoring mode (the generating operation is point B′). So the field current remains at its rated value $i_{ds,\text{rated}}$ in this region. According to Eq. (12.28), the torque current is limited by $|i_{qs,\text{limit}}|$

$$\left|i_{qs,\text{limit}}\right| = \sqrt{i_{s,\max}^2 - i_{ds,\text{rated}}^2} \tag{12.29}$$

(2) Field-Weakening Region 1

As Fig. 12.10A shows, the voltage constraint ellipse shrinks toward the origin with the increase of machine speed. When the speed exceeds the base speed, the available operating region is limited by the voltage and current constraints simultaneously. The maximum torque operating point is the intersection of voltage constraint ellipse and current constraint circle, so the field current needs to be reduced. The maximum torque in the motoring mode is represented by curve B-C, while in the generating mode it is represented by curve B′-C′. Based on Eq. (12.28), the torque current limitation $|i_{qs,\text{limit1}}|$ is calculated as

$$\left|i_{qs,\text{limit1}}\right| = \sqrt{i_{s,\max}^2 - i_{ds}^2} \tag{12.30}$$

(3) Field-Weakening Region 2

As the machine speed further increases, the voltage constraint ellipse shrinks into the current constraint circle, as shown in Fig. 12.10B. Thus the maximum

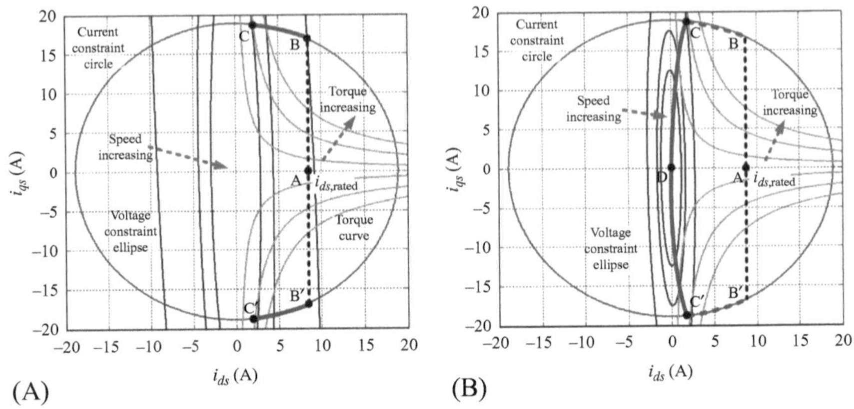

FIG. 12.10 (A) Constraints in field-weakening region 1. (B) Constraints in field-weakening region 2.

torque operating point is on the edge of voltage constraint ellipse, where the field current needs to be further reduced. The maximum torque locus in the motoring mode is curve C-D, while in the generating mode it is curve C′-D. The maximum torque current can be obtained from Eq. (12.27). However, the calculation result is too complicated for online realization, and thus a simple and useful method is adopted. In the high-speed region, the stator resistance voltage drop can be ignored [9], so Eq. (12.27) is transformed into

$$i_{qs}^2 + (i_{ds}/\sigma)^2 \le u_{s,\max}^2/(\sigma L_s \omega_e)^2 \tag{12.31}$$

If $i_{qs} < i_{ds}/\sigma$ is satisfied, Eq. (12.31) is true. To lower the impact of i_{ds} perturbation, $\lambda_{dr} = L_m i_{ds}$ is applied. So the torque current limitation $|i_{qs,\text{limit2}}|$ is eventually expressed as

$$|i_{qs,\text{limit2}}| = \lambda_{dr}/(\sigma L_m) \tag{12.32}$$

12.3.2 Robust Speed-Sensorless Field-Weakening Control

Fig. 12.11 shows the block diagram of the proposed speed-sensorless IM field-weakening strategy based on RFOC. The contributions cover two different parts. First, a robust field-weakening controller is presented to achieve maximum torque capacity in the field-weakening region. Second, a discrete speed adaptive full-order observer is used to estimate the machine speed over the whole speed range [8].

(1) Robust Field-Weakening Control

According to the analysis of maximum torque selection in Figs. 12.9 and 12.10, the field current i_{ds} should be adjusted to ensure an output voltage margin for the

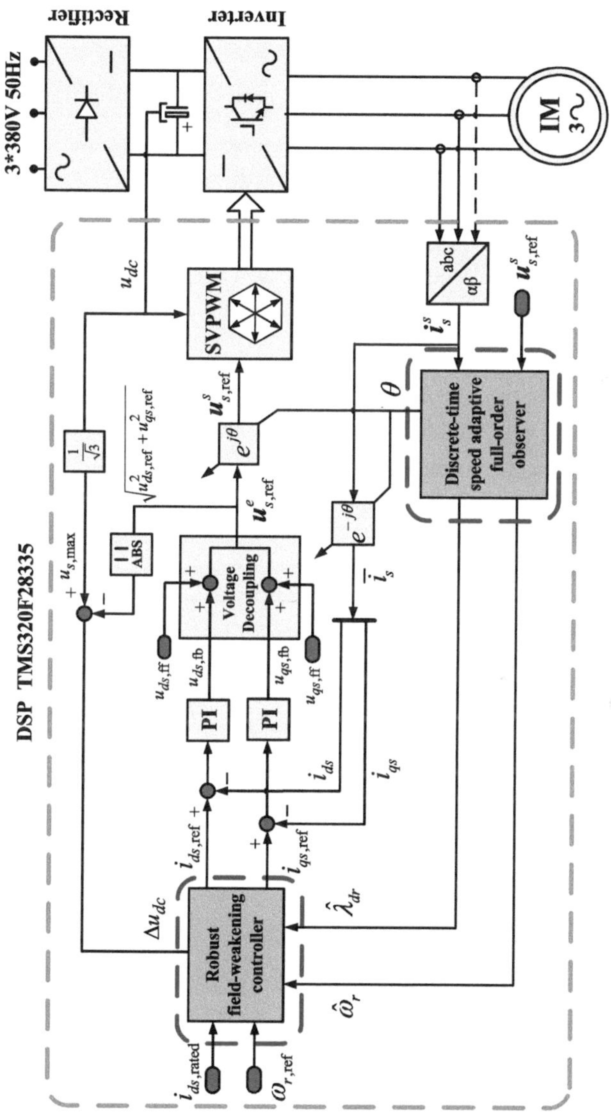

FIG. 12.11 Proposed speed-sensorless IM field-weakening strategy based on RFOC.

machine speed increase, and the torque current i_{qs} needs to be clamped. Fig. 12.12A shows the block diagram of the proposed robust field-weakening strategy.

First, the field current reference $i_{ds,ref}$ is regulated by the basic strategy and the voltage loop simultaneously. And thus

$$i_{ds,\text{ref}} = i_{BS} + i_{VL} \tag{12.33}$$

where i_{BS} is the basic strategy output; i_{VL} is the voltage loop output; and

$$\begin{cases} i_{BS} = i_{ds,\text{rated}} & \omega_{r,\text{ref}} \le \omega_{base} \\ i_{BS} = i_{ds,\text{rated}}\omega_{base}/\omega_{r,\text{ref}} & \omega_{r,\text{ref}} > \omega_{base} \end{cases}$$

The basic strategy is essentially a rough open-loop regulation. The base speed ω_{base} can be calculated by solving Eqs. (12.27), (12.28). However, this calculation not only increases the complexity of algorithm but also reduces the system robustness to machine parameter variations. For simplicity, the machine-rated speed serves as the base speed here, but this will lead to the unnecessary high flux level, and the maximum torque capability cannot be achieved. To obtain the maximum torque capability over the whole speed range, a voltage loop whose maximum output is limited to 0 is added to the basic strategy.

In the constant torque region, since the inverter output reference is less than $u_{s,\max}$, the voltage loop output is limited to 0, which means that $i_{ds,ref}$ is solely provided by the basic strategy. As for the field-weakening region, if the inverter output reference exceeds $u_{s,\max}$, the voltage loop will be activated to reduce the unnecessary high flux level, and Eq. (12.34) becomes

$$i_{ds,ref} = i_{BS} + i_{VL} = i_{BS} + k_{pv}\Delta u_{dc} + k_{iv}\int \Delta u_{dc}dt \tag{12.34}$$

where $\Delta u_{dc} = u_{s,\boldsymbol{max}} - \sqrt{u_{ds,ref}^2 + u_{qs,ref}^2}$; and k_{pv} and k_{iv} are the coefficients of voltage loop. The selection of k_{pv} and k_{iv} should take both stability and dynamic properties into consideration. Here a small k_{iv} is initially set, and then k_{pv} is selected to improve the dynamic properties. Therefore, the voltage loop eventually keeps the output voltage around $u_{s,\max}$ with the proper flux level for the maximum torque capability.

Second, i_{qs} is clamped according to the operating mode. The flow chart of maximum torque current $|i_{qs,\max}|$ selection is shown in Fig. 12.12B. When the machine operates in the constant and field-weakening region 1, $|i_{qs,\max}|$ is given by Eqs. (12.29), (12.30), respectively, according to the current constraint. As the operating frequency further increases, the machine ultimately enters into field-weakening region 2, where $|i_{qs,\max}|$ is principally limited by the voltage constraint. As shown in Fig.12.10B, since the voltage constraint ellipse is inside the current constraint circle, the voltage limitation $|i_{qs,\text{limit2}}|$ is less than the current limitation $|i_{qs,\text{ limit1}}|$. Thus when $|i_{qs,\text{ limit2}}| < |i_{qs,\text{ limit1}}|$, the machine operates in field-weakening region 2, and $|i_{qs,\max}|$ is given by Eq. (12.32).

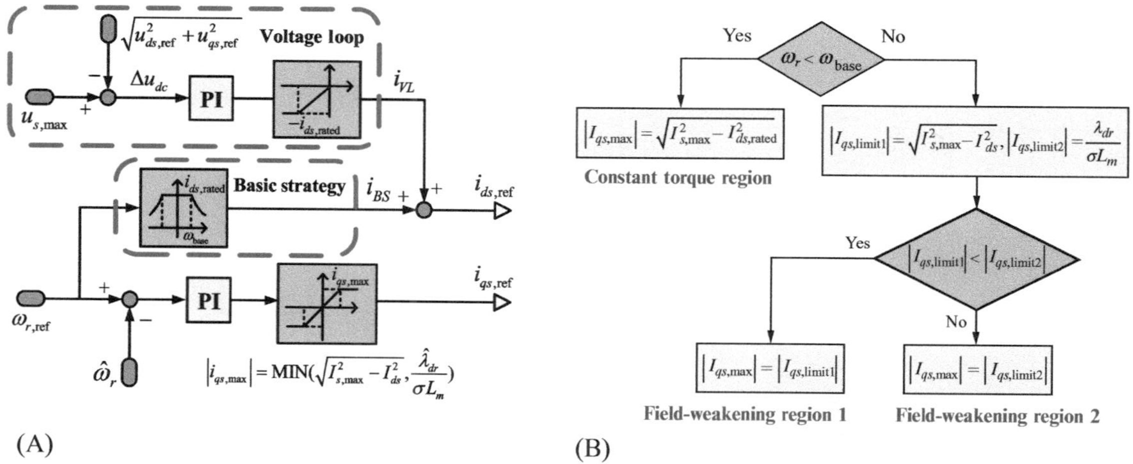

FIG. 12.12 (A) Proposed robust field-weakening strategy. (B) Torque current clamp selection.

Based on all the above summaries, the advantages of proposed approach can be concluded as follows: (1) it achieves the maximum torque in the steady state over the field-weakening region; (2) it has no need of complicated flux level calculation, look-up tables or additional two or more regulators; (3) it is robust to parameter variations due to its independent of the machine parameters. However, the proposed approach is not a perfect solution because the maximum torque capability can only be guaranteed in the steady state. When the machine is accelerating, if the rotor actual speed loses track of its reference, the maximum torque capability cannot attain due to the sharply reducing of $i_{ds,\text{ref}}$.

(2) Discrete Speed Adaptive Full-Order Observer

A speed adaptive full-order observer in the continuous domain is designed as

$$d\hat{\mathbf{x}}/dt = \hat{\mathbf{A}}\hat{\mathbf{x}} + \mathbf{B}\mathbf{u}_s + \mathbf{H}\left(\mathbf{i}_s - \hat{\mathbf{i}}_s\right) \tag{12.35}$$

where ^ is the estimated value; $\mathbf{x} = [\boldsymbol{\lambda}_s \ \boldsymbol{\lambda}_r]^T$; $\mathbf{A} = \begin{bmatrix} \mathbf{A}_{11} & \mathbf{A}_{12} \\ \mathbf{A}_{21} & \mathbf{A}_{22} \end{bmatrix}$; $\mathbf{B} = [\mathbf{I} \ \ 0]^T$; and $\boldsymbol{H}$ is the observer gain.

The speed estimation algorithm is

$$\hat{\omega}_r = \left(k_{p\omega} + k_{i\omega}/s\right)\left(e_{i_{\alpha s}}\hat{\lambda}_{\beta r} - e_{i_{\beta s}}\hat{\lambda}_{\alpha r}\right) \tag{12.36}$$

where $k_{p\omega}$ and $k_{i\omega}$ are the speed adaptive algorithm coefficients. For this zero gain observer, since the current error servers as the feedback through the speed estimation, it can be considered as a closed-loop observer. The observer can also be designed with the proper selection of $\boldsymbol{H}$ to enhance the proposed observer robustness to parameter variations and measurement noises. The block diagram of the proposed speed adaptive full-order observer is shown in Fig. 12.13.

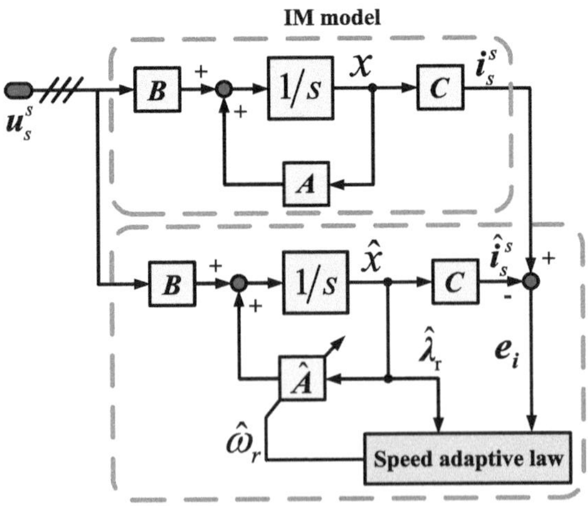

FIG. 12.13 Block diagram of the proposed speed adaptive full-order observer.

In a microprocessor-based control system, the continuous observer has to be discretized for digital implementation. To solve the conflict between stability, accuracy, and computational burden, a modified Euler (ME) approximation is used to discretize the observer.

The discrete full-order observer with ME approximation is given by Eqs. (12.37)–(12.39)

$$\hat{\mathbf{x}}^{k+1} = \hat{\mathbf{x}}^k + (T_s/2)\left(\mathbf{K}^k + \mathbf{K}^{k+1}\right) \tag{12.37}$$

$$\mathbf{K}^k = \hat{\mathbf{A}}\hat{\mathbf{x}}^k + \mathbf{B}\mathbf{u}_s^k \tag{12.38}$$

$$\mathbf{K}^{k+1} = \hat{\mathbf{A}}\left(\hat{\mathbf{x}}^k + T_s\mathbf{K}^k\right) + \mathbf{B}\mathbf{u}_s{}^{k+1} \tag{12.39}$$

It can be seen that ME approximation employs a similar idea with bilinear approximation. The difference is that ME approximation uses the "prediction-correction" scheme to avoid the complicated matrix inversion calculation. The discrete observer with ME approximation is readily implemented. To analyze the stability of proposed method, the characteristic equation of (12.37) is derived using Eqs. (12.38), (12.39)

$$\det\left(z\mathbf{I} - \left(\mathbf{I} + T_s\hat{\mathbf{A}} + T_s^2\hat{\mathbf{A}}^2/2\right)\right) = 0. \tag{12.40}$$

The poles migration of Eq. (12.40) as IM speed increases (0–500 Hz) in different discrete steps is shown in Fig. 12.14. At $T_s = 0.125$ ms, all the poles are inside the unit circle over the whole speed range, so the discrete observer is

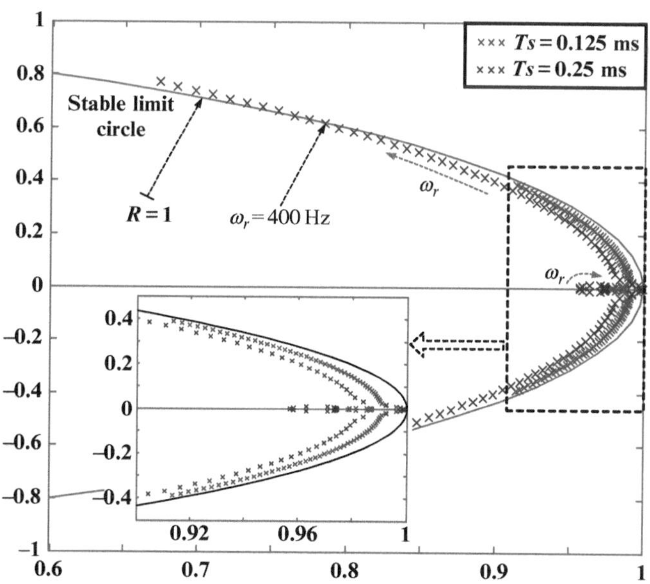

FIG. 12.14 Poles migration using discrete observer with ME approximation.

stable. However, the proposed ME approximation cannot always guarantee the stability of the discrete observer. At $T_s = 0.25$ ms, the discrete observer tends to be unstable as IM speed increases. When the operating frequency exceeds 400 Hz, the discrete observer loses its stability. In this critical stable point, f_{sample}/f_{out} is about 10. Still, the stable operating range of proposed observer is completely enough for the practical applications in this paper, so the ME approximation-based observer is an acceptable compromise.

12.3.3 Experimental Results

The discussed algorithm has been assessed experimentally in an arrangement, which is composed by the followings: Two IMs with the same model and rating are applied, and the motor parameters are given in Table 12.1. One is used for testing the proposed algorithm, while the other is controlled by the loading algorithm. The load value is set by a panel input. The rectifier rectifies the input alternating current into the direct current for the DC-link, while the inverter provides the IM drive alternating current. A TMS320F28335 Texas Instruments DSP serves as the control core. The DSP is a 32-bit fixed-point microcontroller with 150 MHz of system frequency, 256 K of on-chip flash, and 34 K of SARAM. High-speed A/D and D/A converters are used for data conversion. Two-phase currents are measured for control, and the SVPWM is utilized to generate the inverter switching signals. A linked computer realizes the proposed

TABLE 12.1 Parameters of the Induction Machine

Symbol	Quantity	Value
P_N	Rated power	3.7 kW
V_N	Rated voltage	380 V
I_N	Rated current	8.9 A
f_N	Rated frequency	50 Hz
T_N	Rated torque	23.6 N·m
n_p	Pole pair	2
R_s	Stator resistance	1.142 Ω
R_r	Rotor resistance	0.825 Ω
L_m	Mutual inductance	118.9 mH
L_s, L_r	Stator, rotor inductance	124.4 mH
J	Inertia	0.0123 kg·m^2
$I_{s,\max}$	Maximum *RMS* current limit	13.35 A

algorithm and downloads the program to the DSP through an online emulator. An oscilloscope monitors the control variables. The RFOC is applied as the basic control algorithm. When the machine operates below the base speed, the field current reference is set to 8.48 A via the synchronous rotation coordination transformation. The PWM switching frequency is 8 kHz, and the sampling time of control is 0.125 ms.

The proposed robust field-weakening algorithm with ME approximation-based observer is performed. Fig. 12.15 shows behavior of the algorithms during a speed step change from 0 to 7500 rpm. The machine is initially at standstill, and the phase current is zero. After enabling the speed reference command, the machine operates in a premagnetization state for 0.2 s, after which it starts up in the constant torque region, and the phase current reaches its maximum value. As the speed increases, the machine enters field-weakening region 1, where the proposed algorithm modifies i_{ds} and i_{qs} according to Fig. 12.12A. Here, i_{ds} is excessively decreased to obtain more sufficient voltage margin for acceleration, which means that the proposed strategy can only

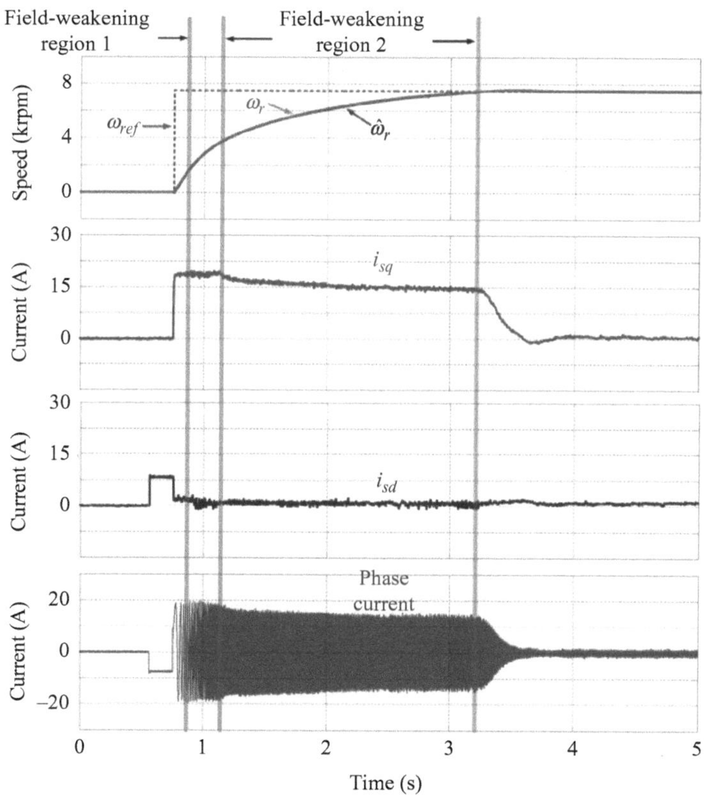

FIG. 12.15 Behavior of the proposed scheme during a speed step change from 0 to 7500 rpm.

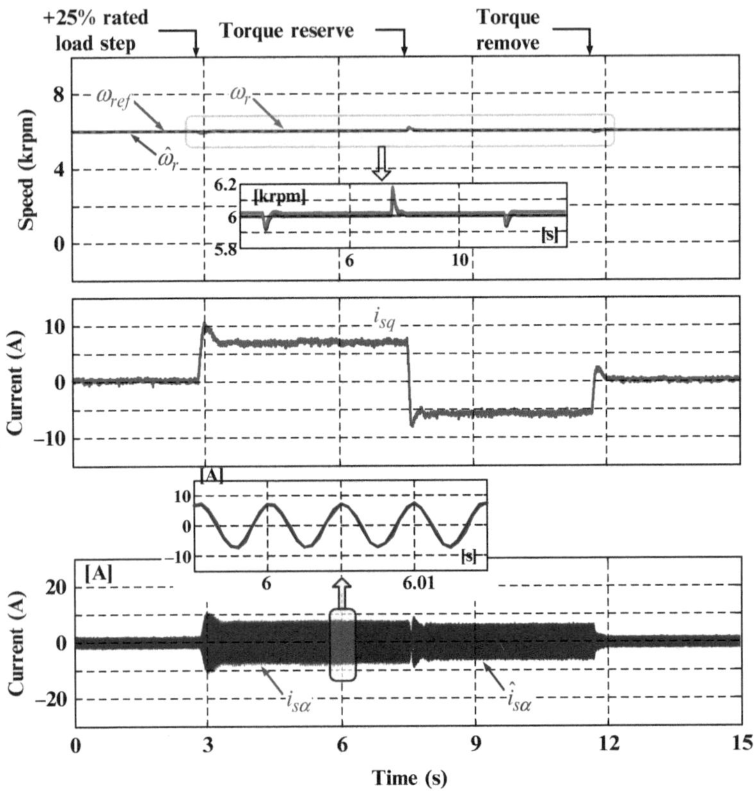

FIG. 12.16 Behavior of the proposed scheme during the load step change.

achieve the maximum torque capability in the steady state. Further, the machine enters field-weakening region 2, where i_{qs} is reduced by the clamp of maximum allowable torque current. Finally, the machine speed is stabilized at the target speed, and the maximum torque capability can be achieved.

Fig. 12.16 shows the behavior of the proposed robust field-weakening control with ME approximation-based observer during load step changes at 6000 rpm. The magnitude of load torque is 20% rated load, and the change time constant is about 0.1 s. The machine operates in motoring mode with +20% rated load, and in generating mode with −20% rated load. During the entire operation, the estimated speed $\hat{\omega}_r$ and stator current α component $\hat{i}_{s\alpha}$ follow the actual value accurately, which indicates that the proposed method has strong robustness.

12.4 HIGH-FREQUENCY SIGNAL INJECTION-BASED SENSORLESS CONTROL OF PMSM DRIVES

The vector control scheme of the sensorless IPMSM drive using the high-frequency signal injection-based method is shown in Fig. 12.17.

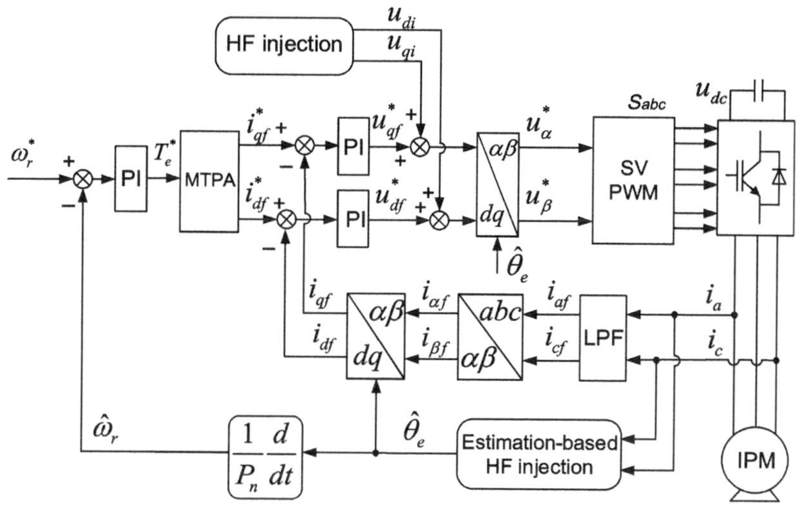

FIG. 12.17 High-frequency signal injection-based sensorless IPMSM drive.

12.4.1 Principle of Position Estimation

A HF voltage is injected into the stator windings. Assuming that the frequency is far higher than the fundamental frequency, the HF voltage model can be expressed as

$$\begin{bmatrix} u_{\alpha i} \\ u_{\beta i} \end{bmatrix} = \begin{bmatrix} L_0 + L_1 \cos(2\theta_e) & L_1 \sin(2\theta_e) \\ L_1 \sin(2\theta_e) & L_0 - L_1 \cos(2\theta_e) \end{bmatrix} \frac{d}{dt} \begin{bmatrix} i_{\alpha i} \\ i_{\beta i} \end{bmatrix} \tag{12.41}$$

where the subscript "*i*" is the corresponding HF component.

To analyze the characteristic of the high-frequency voltage model conveniently, Eq. (12.41) can be rewritten as follows

$$\frac{d}{dt} \begin{bmatrix} i_{\alpha i} \\ i_{\beta i} \end{bmatrix} = k_1 \begin{bmatrix} L_0 - L_1 \cos(2\theta_e) & -L_1 \sin(2\theta_e) \\ -L_1 \sin(2\theta_e) & L_0 + L_1 \cos(2\theta_e) \end{bmatrix} \begin{bmatrix} u_{\alpha i} \\ u_{\beta i} \end{bmatrix} \tag{12.42}$$

where $k_1 = 1/(L_0^2 - L_1^2)$.

Continually, transform Eq. (12.42) to the *d-q* rotating reference frame, and the expression can be given as

$$\frac{d}{dt} \begin{bmatrix} i_{di} \\ i_{qi} \end{bmatrix} = k_1 \begin{bmatrix} L_0 - L_1 & 0 \\ 0 & L_0 + L_1 \end{bmatrix} \begin{bmatrix} u_{di} \\ u_{qi} \end{bmatrix} \tag{12.43}$$

Defining the estimated rotating frame as $d^e - q^e$, then Eq. (12.43) can be transformed into the estimated frame as

$$\frac{d}{dt} \begin{bmatrix} i_{dei} \\ i_{qei} \end{bmatrix} = k_1 \begin{bmatrix} L_0 - L_1 \cos(2\Delta\theta_e) & L_1 \sin(2\Delta\theta_e) \\ L_1 \sin(2\Delta\theta_e) & L_0 + L_1 \cos(2\Delta\theta_e) \end{bmatrix} \begin{bmatrix} u_{dei} \\ u_{qei} \end{bmatrix} \tag{12.44}$$

where the subscript "*e*" means the corresponding component in the estimated rotating frame, and $\Delta\theta_e$ means the position estimation error.

For the sensorless IPMSM drive, the high-frequency voltage signal is injected at d^e-axis, and it can be expressed as

$$\begin{bmatrix} u_{dei} \\ u_{qei} \end{bmatrix} = \begin{bmatrix} u_i \\ 0 \end{bmatrix} = \begin{bmatrix} U_i \cdot \sin\omega_i t \\ 0 \end{bmatrix} \tag{12.45}$$

In order to extract the rotor position from the high-frequency current, a measured frame $d^m - q^m$ is defined. The $d^m - q^m$ frame lags the $d^e - q^e$ frame with θ_0. And Eq. (12.43) can be transformed to the measure frame as follows:

$$\frac{d}{dt}\begin{bmatrix} i_{dmi} \\ i_{qmi} \end{bmatrix} = k_1 \begin{bmatrix} L_0 - L_1\cos 2(\Delta\theta_e - \theta_0) & L_1 \sin 2(\Delta\theta_e - \theta_0) \\ L_1 \sin 2(\Delta\theta_e - \theta_0) & L_0 + L_1 \cos 2(\Delta\theta_e - \theta_0) \end{bmatrix} \begin{bmatrix} u_{dmi} \\ u_{qmi} \end{bmatrix} \tag{12.46}$$

In a special case, if the measured frame lags the estimation frame with 0.25π, then Eq. (12.46) can be simplified to

$$\frac{d}{dt}\left(i_{qmi} - i_{dmi}\right) = \frac{U_i \sin\omega_i t}{\sqrt{2}\left(L_0{}^2 - L_1{}^2\right)} \cdot 2L_1 \sin 2\Delta\theta_e \tag{12.47}$$

If the rotor position estimation error $\Delta\theta_e$ is sufficiently small, then Eq. (12.47) can be approximated as

$$\frac{d}{dt}\left(i_{qmi} - i_{dmi}\right) \approx \frac{2\sqrt{2}L_1 U_i \sin\omega_i t}{L_0{}^2 - L_1{}^2}\Delta\theta_e \tag{12.48}$$

As a result, the equivalent estimation error signal ε for the rotor position observer can be obtained from the magnitude of the measured high-frequency current components according to Eq. (12.48), which is given by

$$\varepsilon = \frac{\omega_i\left(L_1{}^2 - L_0{}^2\right)}{2\sqrt{2}L_1 U_i}\left(I_{qmi} - I_{dmi}\right) \triangleq k_2\left(I_{qmi} - I_{dmi}\right) \tag{12.49}$$

Then, the normalized estimation error signal $\varepsilon_i = \varepsilon / I_{mi}$ is fed to the observer for the position and speed estimation.

12.4.2 Signal Processes for Obtaining the Equivalent Position Estimation Error

Based on the HF voltage signal injection estimation principle, the HF current should be extracted first. Then, the position estimation error signal can be acquired from the magnitude of HF current components in the measured axes. The corresponding signal processes is shown in Fig. 12.18.

In Fig. 12.18, both the fundamental component and the PWM noise are eliminated by a synchronous rotating band pass filter (BPF), and the rotating

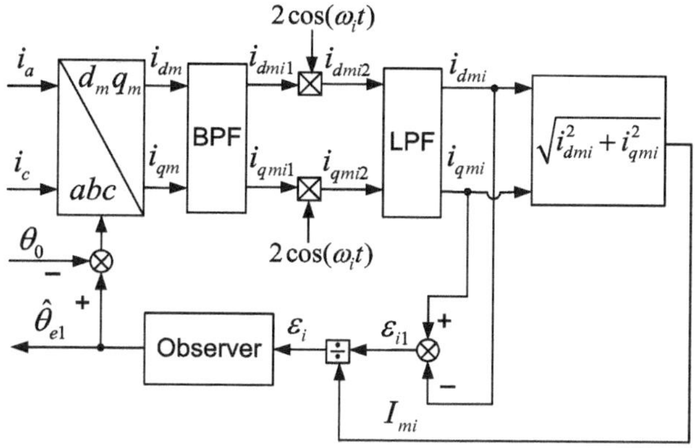

FIG. 12.18 Signal processes for rotor position estimation using high-frequency signal injection.

vector angle is set as $\hat{\theta}_{e1} - \theta_0$. The BPF may lead to the phase delay and the magnitude attenuation of the HF current signal. So the design for the digital BPF is crucial. In applications, the HF signal injection is only adopted for low-speed operation. When the machine operates in the middle- and high-speed ranges, the position observer based on the EMF model is adopted. Since the frequency of the injection signal is 1 kHz in this scheme, so the pass band of the second order Butterworth BPF is set as [925 Hz, 1075 Hz]. Therefore, the differential equation of the BPF is given by

$$H_1(z) = 0.045\frac{1 - z^{-2}}{1 - 1.5469z^{-1} + 0.9099z^{-2}} \tag{12.50}$$

In this way, the HF current components in the measured rotating frame (i_{dmi} and i_{qmi}) can be obtained. Then both the two components are multiplied by 2 cos $\omega_i t$. Continually, a second-order Butterworth low-pass filter (LPF) is used to eliminate the HF components. The cutoff frequency of the LPF is set to 75 Hz. Then the expression of the digital LPF can be given as

$$H_2(z) = 0.000537\frac{1 + 2z^{-1} + z^{-2}}{1 - 1.93338z^{-1} + 0.93553z^{-2}} \tag{12.51}$$

To mitigate the influence of the nonlinearity, the equivalent position error signal ε is normalized by its magnitude I_{mi}. And the normalized position error signal ε_i is used to acquire the rotor position.

12.4.3 Design of Position Observer

According to the normalized position error signal ε_i, a robust rotor position observer based on the IPMSM mechanical motion model is constructed [10]. Position error feedback scheme can be done in many different ways, which play an important role in improving the robustness to the load disturbance.

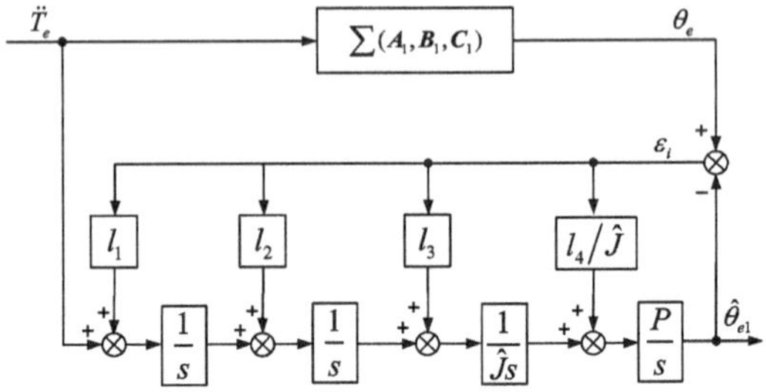

FIG. 12.19 The equivalent structure of IPMSM position observer based on the high-frequency signal injection.

Fig. 12.19 shows the equivalent structure of the proposed robust rotor position observer for sensorless IPMSM. In the position observer, a linear feedback control method based on the position error signal ε_i is adopted. The differential item of electromagnetic torque $\ddot{T}_e$ can be treated as an equivalent input of the observer. In this way, the position observer robustness can be improved when the load torque changes.

The equivalent observer model of IPMSM in Fig. 12.19 can be expressed as follows

$$\dot{\mathbf{X}} = \mathbf{A}_1\mathbf{X} + \mathbf{B}_1\mathbf{u} \tag{12.52}$$

$$\mathbf{y} = \mathbf{C}_1\mathbf{X} \tag{12.53}$$

where $\mathbf{X} = \begin{bmatrix} \dot{T}_e & T_e & \omega_r & \theta_r \end{bmatrix}^{\mathrm{T}}$, $\mathbf{u} = \ddot{T}_e$, $\mathbf{y} = \theta_r$,

$$\mathbf{A}_1 = \begin{bmatrix} 0 & 0 & 0 & 0 \\ 1 & 0 & 0 & 0 \\ 0 & \frac{1}{J} & 0 & 0 \\ 0 & 0 & 1 & 0 \end{bmatrix}, \quad \mathbf{B}_1 = \begin{bmatrix} 1 \\ 0 \\ 0 \\ 0 \end{bmatrix}, \quad \mathbf{C}_1 = \begin{bmatrix} 0 \\ 0 \\ 0 \\ 1 \end{bmatrix}^{\mathrm{T}}$$

θ_r is the rotor mechanical position, and $\theta_r = \theta_e / P_n$, T_e can be calculated from

$$T_e = P_n\left[\psi_f i_q + \left(L_d - L_q\right) i_d i_q\right] \tag{12.54}$$

According to the mechanical motion model of IPMSM, the proposed position observer can be expressed as

$$\dot{\hat{\mathbf{X}}} = \begin{bmatrix} 0 & 0 & 0 & 0 \\ 1 & 0 & 0 & 0 \\ 0 & \frac{1}{\hat{J}} & 0 & 0 \\ 0 & 0 & 1 & 0 \end{bmatrix} \hat{\mathbf{X}} + \begin{bmatrix} 1 \\ 0 \\ 0 \\ 0 \end{bmatrix} \mathbf{u} + \begin{bmatrix} l_1 \\ l_2 \\ \frac{l_3}{\hat{J}} \\ \frac{l_4}{\hat{J}} \end{bmatrix} (\mathbf{y} - \hat{\mathbf{y}}) \tag{12.55}$$

Here, the feedback matrix of the position observer is defined as $\mathbf{L}=\left[l_1\ \ l_2\ \ l_3/\hat{J}\ \ l_4/\hat{J}]^{\mathrm{T}}\right]^{\mathrm{T}}$.

According to the characteristics of the observer, both step and slope disturbances will not result in a stationary estimation error. So the robustness of the rotor position observer can be improved to avoid the estimated position converging to the opposite direction of the actual position. The feedback matrix coefficient of robust position observer can be designed by using the pole assignment method. All the eigenvalues of the dynamic matrix ***A-LC*** should be negative.

12.4.4 Experimental Results

The robust hybrid position estimation scheme has been verified at a 2.2 kW IPMSM sensorless vector controlled drive based on a DSP as shown in Fig. 12.20. The rated parameters of the IPMSM are listed as follows: 380 V, 5.0 A, 14 Nm, 75 Hz, 1500 r/min, $P_n = 3$, $R_s = 2.75\ \Omega$, $L_d = 45$ mH, $L_q = 60$ mH, $\psi_f = 0.48$ Wb. Texas Instruments TMS320F2808 DSP is adopted to execute the whole sensorless control algorithm. An IM is mechanically

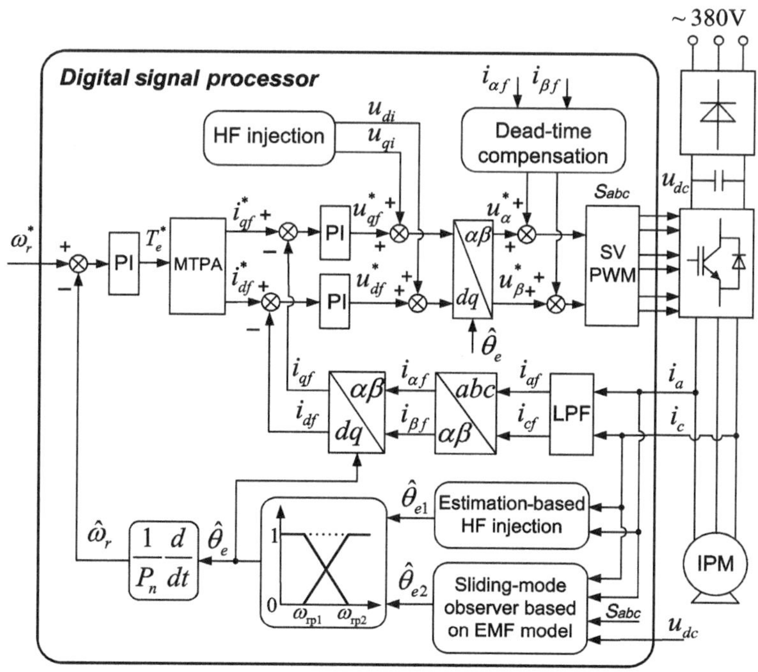

FIG. 12.20 Field-oriented control scheme of sensorless IPMSM based on DSP.

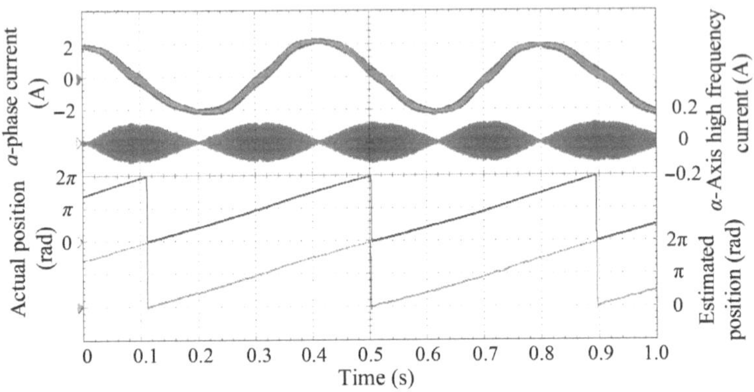

FIG. 12.21 Signal process results of the high-frequency signal injection at 50 r/min with 25% rated load.

coupled with the IPMSM to produce the load torque. The PWM switching frequency of the inverter is 10 kHz, and the dead time is set to 3.2 μs. The magnitude and the frequency of the injected high-frequency voltage are 57 V and 1 kHz, respectively. An encoder (ECN1113) is used to obtain the actual position that is solely used for comparison and not for control purposes. The current loop is corrected to a typical I-type system, and the speed loop is corrected to a typical II-type system. The parameters of the current PI regulator are $K_{p2} = 4$ and $K_{i2} = 25$, respectively. The parameters of the speed PI regulator are $K_{p1} = 35$ and $K_{i1} = 10$, respectively.

Fig. 12.21 shows the experimental results of the digital signal process for the HF signal injection at 50 r/min with 25% rated load. The waveforms are a-phase current, the obtained α-axis HF current signal, the actual and the estimated position, respectively. According to the sample value of stator current, HF current signal is obtained through the digital filters, and the estimated position can be obtained. From the waveforms of the HF current and the rotor position, it can be seen that the rotor position information is modulated at the magnitude of the current signal.

Fig. 12.22 shows the estimated speed and the estimation errors of the sensorless IPMSM control drive at zero speed with step rated load disturbance. It can be seen that the speed estimation error is within ±10 r/min during the transients, and the estimated speed tracks the actual speed well. The IPMSM sensorless drive can operate at zero speed with full load disturbance.

Fig. 12.23 shows the transients of the speed reversal operation at ±20 r/min with rated load. The estimated position follows the actual position well during both the motoring and the regenerating operations. The estimation error is within $\pm 0.05\,\pi$. The speed estimation error is within ±8 r/min during the transients and the steady state. The sensorless IPMSM vector control drive can operate at low speed regenerating mode with good stability.

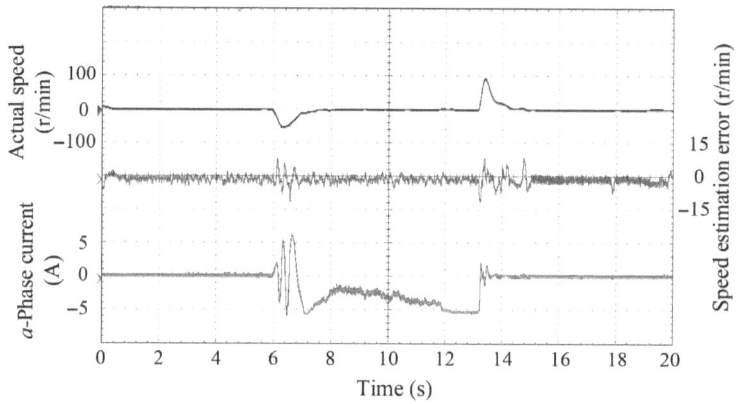

FIG. 12.22 Zero-speed sensorless operation with step rated load disturbance.

12.5 EMF-MODEL-BASED SENSORLESS CONTROL OF PMSM DRIVES

12.5.1 Active Flux-Based Back-EMF Observer

The vector control scheme of the sensorless IPMSM drive based on a position observer adopting an extended EMF model is shown in Fig. 12.24.

The voltage model of IPMSM in the α-β stationary reference frame can be expressed as

$$\begin{bmatrix} v_\alpha \\ v_\beta \end{bmatrix} = \begin{pmatrix} R+pL_d & \omega_e L_d - L_q) \\ -\omega_e (L_d - L_q) & R+pL_d \end{pmatrix} \begin{bmatrix} i_\alpha \\ i_\beta \end{bmatrix} + \begin{bmatrix} e_\alpha \\ e_\beta \end{bmatrix} \tag{12.56}$$

where v_α and v_β are the α-β axis stator voltages, i_α and i_β are the α-β axis stator currents, L_d and L_q are the d-q axis inductances, R is the stator resistance, p is the differential operator, and ω_e is the rotor electrical speed, respectively, $[e_\alpha\, e_\beta]^{\mathrm{T}} = E_{ex}[-\sin\theta_e \cos\theta_e]^{\mathrm{T}}$, where θ_e is the rotor position, E_{ex} is the extended EMF amplitude which is defined by [4],

$$E_{ex} = \omega_e\left[\left(L_d - L_q\right)i_d + \lambda_{PM}\right] - \left(L_d - L_q\right)\left(di_q/dt\right) \tag{12.57}$$

where λ_{PM} is the permanent magnet flux.

Eq. (12.56) can be written in vector form as follows:

$$\mathbf{v}_{\alpha\beta} = (R + pL_d)\mathbf{i}_{\alpha\beta} - j\omega_e\left(L_d - L_q\right)\mathbf{i}_{\alpha\beta} + \mathbf{e}_{\alpha\beta} \tag{12.58}$$

where $\boldsymbol{i}_{\alpha\beta} = i_\alpha + ji_\beta$, $\mathbf{v}_{\alpha\beta} = v_\alpha + jv_\beta$, $\boldsymbol{e}_{\alpha\beta} = e_\alpha + je_\beta$, and j denotes the imaginary unit.

Taking the Laplace transform of Eq. (12.58), the following can be obtained

$$\boldsymbol{e}_{\alpha\beta} = \boldsymbol{v}_{\alpha\beta} + j\omega_e\left(L_d - L_q\right)\boldsymbol{i}_{\alpha\beta} - (R + L_d s)\boldsymbol{i}_{\alpha\beta} \tag{12.59}$$

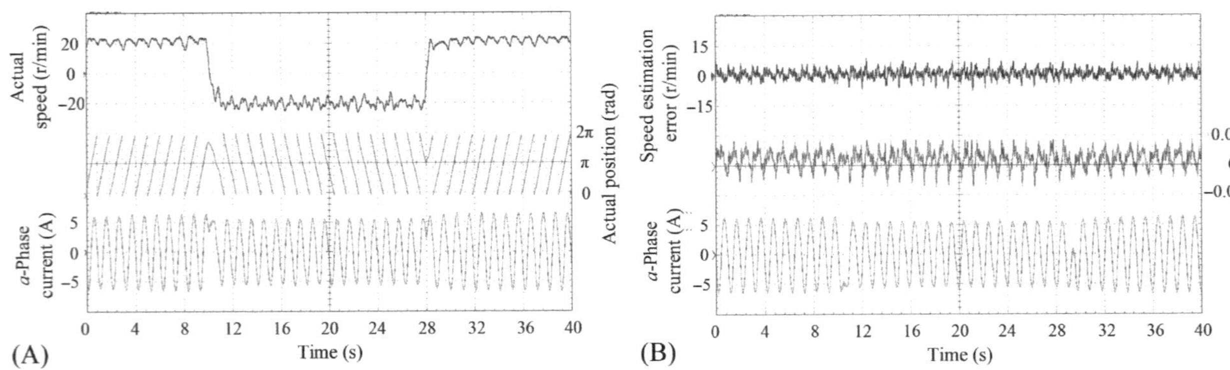

FIG. 12.23 Motoring and regenerating sensorless operation at ±20 r/min with rated load: (A) actual speed and position and (B) speed and position estimation error.

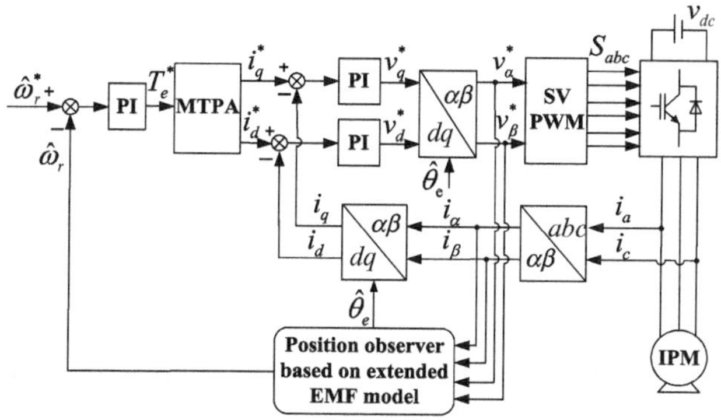

FIG. 12.24 EMF model-based sensorless IPMSM drive.

By adopting the state filter method [13], the EMF can be estimated as follows:

$$\hat{\boldsymbol{e}}_{\alpha\beta} = \frac{(k_p s + k_i)(L_d s + R)}{L_d s^2 + (R + k_p)s + k_i}\left(\frac{\boldsymbol{v}^*_{\alpha\beta} + j\hat{\omega}_e(L_d - L_q)\boldsymbol{i}_{\alpha\beta}}{L_d s + R} - \boldsymbol{i}_{\alpha\beta}\right) \tag{12.60}$$

where "^" denotes the estimated value, $\boldsymbol{v}^*_{\alpha\beta}$ denotes the reference voltage vector, k_p and k_i are the proportional and integral gains of the EMF estimator, respectively.

The gains of the EMF estimator are selected by using the pole-zero cancellation method as follows:

$$k_p = L_d\omega_c, k_i = R\omega_c \tag{12.61}$$

where ω_c is the bandwidth of the EMF estimator. By substituting Eq. (12.60) into Eq. (12.61), the following can be obtained

$$\frac{\hat{\boldsymbol{e}}_{\alpha\beta}}{\boldsymbol{e}_{\alpha\beta}} = \frac{\omega_c}{s + \omega_c} \tag{12.62}$$

So it is clear that the characteristics of the EMF estimator are the same given a first-order low-pass filter.

12.5.2 Analysis of Position Estimation Error

12.5.2.1 Analysis of the Harmonic Components in Extended EMF Estimates

A voltage difference between the reference and the actual values is introduced due to the dead-time effect, which results in stator current distortion. In addition, the flux spatial harmonics of IPMSM can make the extended EMF distorted. Both of the two cases result in the $(6k \pm 1)$th harmonic components in the stator currents, which can be expressed as

$$\begin{aligned} i_x &= I_1 \sin(\omega_e t + \theta_1 + i \cdot 2\pi/3) + I_{6k-1} \sin[-(6k-1)\omega_e t + \theta_{6k-1} + i \cdot 2\pi/3] \\ &\quad + I_{6k+1} \sin[(6k+1)\omega_e t + \theta_{6k+1} + i \cdot 2\pi/3] \\ &= I_1 \sin(\omega_e t + \theta_1 + i \cdot 2\pi/3) + \sum_{k=1}^{n} I_{6k\pm1} \sin[\pm(6k \pm 1)\omega_e t + \theta_{6k\pm1} + i \cdot 2\pi/3] \end{aligned} \tag{12.63}$$

where the subscript x represents a, b, or c phase, of which corresponding i is 0, 1, or 2. I_1, I_{6k-1}, and I_{6k+1} are the amplitudes of the fundamental, $(6k-1)$th and $(6k+1)$th stator current harmonics, respectively. θ_1, θ_{6k-1}, and θ_{6k+1} are the initial phases of the fundamental, $(6k-1)$th and $(6k+1)$th stator current harmonics, respectively.

Using the Clarke and Park transformation, the stator currents in the d-q rotating reference frame can be expressed as

$$i_d = I_1 \sin\theta_1 + \sum_{k=0}^{n} I_{6k\pm1} \sin(\pm 6k\omega_e t + \theta_{6k\pm1}) \tag{12.64}$$

$$i_q = -I_1 \cos\theta_1 - \sum_{k=0}^{n} I_{6k\pm1} \cos(\pm 6k\omega_e t + \theta_{6k\pm1}) \tag{12.65}$$

According to Eqs. (12.64), (12.65), the $(6k-1)$th and $(6k+1)$th harmonics in the α-β axes correspond to the $(6k)$th harmonics in the d-q axes. Substituting Eqs. (12.64), (12.65) into Eq. (12.57), and only considering the $(6k-1)$th and $(6k+1)$th harmonics, the extended EMF in the α-β axes can be obtained as

$$\begin{aligned} e_\alpha &= -E_{ex} \sin\omega_e t \\ &= (L_d - L_q)\omega_e - \left(I_1 \sin\theta_1 + \frac{\lambda_{PM}}{L_d - L_q}\right) \sin\omega_e t \\ &\mp \sum_{k=0}^{n} \frac{6k \mp 1}{2} I_{6k\pm1} [\cos(\pm(6k \pm 1)\omega_e t + \theta_{6k\pm1}) \\ &- \cos(\pm(6k \mp 1)\omega_e t + \theta_{6k\pm1})] \end{aligned} \tag{12.66}$$

$$\begin{aligned} e_\beta &= E_{ex} \cos\omega_e t \\ &= (L_d - L_q)\omega_e \left\{ \left(I_1 \sin\theta_1 + \frac{\lambda_{PM}}{L_d - L_q}\right) \cos\omega_e t \right. \\ &\mp \sum_{k=0}^{n} \frac{6k \mp 1}{2} I_{6k\pm1} [\sin(\pm(6k \pm 1)\omega_e t + \theta_{6k\pm1}) \\ &\left. + \sin(\pm(6k \mp 1)\omega_e t + \theta_{6k\pm1})] \right\} \end{aligned} \tag{12.67}$$

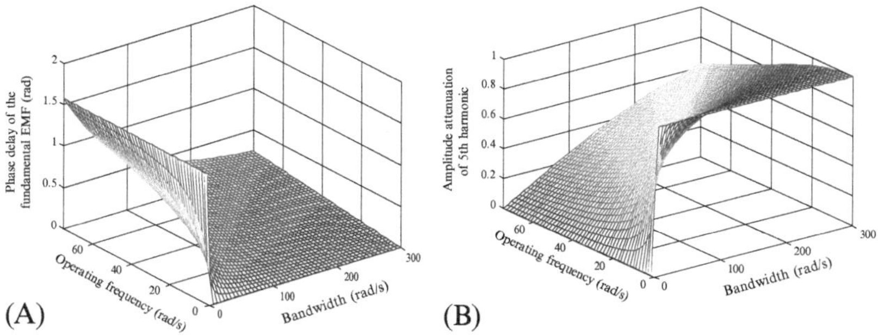

FIG. 12.25 Phase delay and harmonic suppression related to the bandwidth ω_c and operating frequency ω_e. (A) Phase delay of the fundamental EMF. (B) Amplitude suppression of the 5th harmonic component.

In practical applications, the 5th and 7th harmonics are the dominant components.

It is known that the back-EMF estimator produces phase delay between the estimated and actual values. On the other hand, it can suppress high-order harmonics to some extent. The phase delay and the harmonic suppression are related to the bandwidth ω_c and the operating frequency ω_e (as shown in Fig. 12.25). It can be seen that for a certain ω_e, the larger ω_c corresponds to a smaller phase delay of the fundamental EMF estimate but corresponds to a weaker capability of suppressing harmonics. In order to obtain a smaller phase delay, the bandwidth ω_c should be a relative large value, resulting in harmonic components in the back-EMF estimate [11].

12.5.2.2 Position Estimation Error Due to the Harmonics in the Estimated EMF

From the above analysis, it can be noted that the EMF estimates contain $(6k \pm 1)$th harmonics. Under this condition, the voltage model of IPMSM can be written as follows:

$$\begin{bmatrix} v_\alpha \\ v_\beta \end{bmatrix} = \begin{pmatrix} R+pL_d & \omega_e L_d - L_q) \\ -\omega_e (L_d - L_q) & R+pL_d \end{pmatrix} \begin{bmatrix} i_\alpha \\ i_\beta \end{bmatrix} + \boldsymbol{e}_f' + \boldsymbol{e}_h' \tag{12.68}$$

where $\boldsymbol{e}_h'$ and $\boldsymbol{e}_f'$ denote the $(6k \pm 1)$th harmonics and the fundamental component in the EMF estimates, which can be expressed as

$$\boldsymbol{e}'_f = \begin{bmatrix} e'_{f\alpha} \\ e'_{f\beta} \end{bmatrix} = \begin{bmatrix} -e'_1 \sin(\omega_e t + \theta_{e1}) \\ e'_1 \cos(\omega_e t + \theta_{e1}) \end{bmatrix} \tag{12.69}$$

$$\boldsymbol{e}_h' = \begin{bmatrix} e_{h\alpha}' \\ e_{h\beta}' \end{bmatrix} = \begin{bmatrix} -e_{6k-1}' \sin\left(-(6k-1)\omega_e t + \theta_{e(6k-1)}\right) \\ e_{6k-1}' \cos\left(-(6k-1)\omega_e t + \theta_{e(6k-1)}\right) \end{bmatrix} + \begin{bmatrix} -e_{6k+1}' \sin\left((6k+1)\omega_e t + \theta_{e(6k+1)}\right) \\ e_{6k+1}' \cos\left((6k+1)\omega_e t + \theta_{e(6k+1)}\right) \end{bmatrix} \tag{12.70}$$

where $e_{6k-1}{}'$ and $e_{6k+1}{}'$ represent the amplitudes of the $(6k-1)$th and $(6k+1)$th harmonic components and $\theta_{e(6k-1)}$ and $\theta_{e(6k+1)}$ represent the corresponding initial phases, respectively.

When the EMFs $e_\alpha{}'$ ($e_\alpha{}' = e_{f\alpha}{}' + e_{h\alpha}{}'$) and $e_\beta{}'$ ($e_\beta{}' = e_{f\beta}{}' + e_{h\beta}{}'$) are obtained, the quadrature PLL is adopted to extract the information of the rotor position and speed. The phase difference is obtained by the following method:

$$\begin{aligned} \varepsilon_\theta &= -e_\alpha' \cos\hat{\theta}_e - e_\beta' \sin\hat{\theta}_e \\ &= -\left(e_{f\alpha}' + e_{h\alpha}'\right)\cos\hat{\theta}_e - \left(e_{f\beta}' + e_{h\beta}'\right)\sin\hat{\theta}_e \\ &= e_1' \sin\left[(\omega_e - \hat{\omega}_e)t + \theta_{e1} - \hat{\theta}_{ei}\right] \\ &\quad + \sum_{k=1}^{n} e_{6k-1}' \sin\left[(-(6k-1)\omega_e - \hat{\omega}_e)t + \theta_{e(6k-1)} - \hat{\theta}_{ei}\right] \\ &\quad + \sum_{k=1}^{n} e_{6k+1}' \sin\left[((6k+1)\omega_e - \hat{\omega}_e)t + \theta_{e(6k+1)} - \hat{\theta}_{ei}\right] \end{aligned} \tag{12.71}$$

From (12.71), it can be seen that the phase difference contains one difference-frequency term and two high-frequency terms. When the position estimator converges, (12.71) can be simplified

$$\varepsilon_\theta = e_1' \sin\left(\theta_{e1} - \hat{\theta}_{ei}\right) \pm \sum_{k=1}^{n} e_{6k}' \sin\left(6k\omega_e t + \theta_{e(6k)}\right) \tag{12.72}$$

in which $e_{6k}{}'$ represents the amplitude of the equivalent $6k$th EMF harmonics and $\theta_{e(6k)}$ represents the initial phase. The $6k$th EMF harmonics will display as additional errors which exist in the phase difference. As a result, the rotor position obtained from the quadrature PLL, whose input signal is the phase difference contains $6k$th harmonics.

12.5.3 Adaptive Vector Filter Method for Position Error Harmonic Fluctuation Elimination

To eliminate the position error harmonic fluctuations, and hence improve the position estimation accuracy for sensorless IPMSM drives, a multiple-adaptive vector filter (AVF)-based cross-feedback network (CFN) strategy is proposed [12]. In the proposed approach, the multiple-AVFs can work together inside a CFN, achieving back-EMF harmonic decoupling and compensating.

12.5.3.1 Adaptive Vector Filter

AVF has been a good candidate for harmonic detection and provides promising characteristics of positive- and negative-frequency selectivity [12]. The block diagrams of AVF in both scalar and vector forms are shown in Fig. 12.26.

From Fig. 12.26, the transfer function in the scalar form of AVF is depicted by

$$\begin{bmatrix} y_\alpha \\ y_\beta \end{bmatrix} = \frac{k_f\omega_0}{\left(s+k_f\omega_0\right)^2+\omega_0^2}\begin{bmatrix} s+k_f\omega_0 & -\omega_0 \\ -\omega_0 & s+k_f\omega_0 \end{bmatrix}\begin{bmatrix} y_\alpha \\ y_\beta \end{bmatrix} \tag{12.73}$$

where $\mathbf{x}_{\alpha\beta}=\left[x_\alpha \;\; x_\beta\right]^{\mathrm{T}}$ and $\mathbf{y}_{\alpha\beta}=\left[y_\alpha \;\; y_\beta\right]^{\mathrm{T}}$ denote the input vector and the desired output vector, ω_0 is the resonance frequency, and k_f is the gain of AVF.

Then, through the complex vector concept, introducing

$$\begin{aligned} \mathbf{X}_{\alpha\beta} &= x_\alpha + jx_\beta \\ \mathbf{Y}_{\alpha\beta} &= y_\alpha + jy_\beta \end{aligned} \tag{12.74}$$

the transfer function in vector form can be obtained as

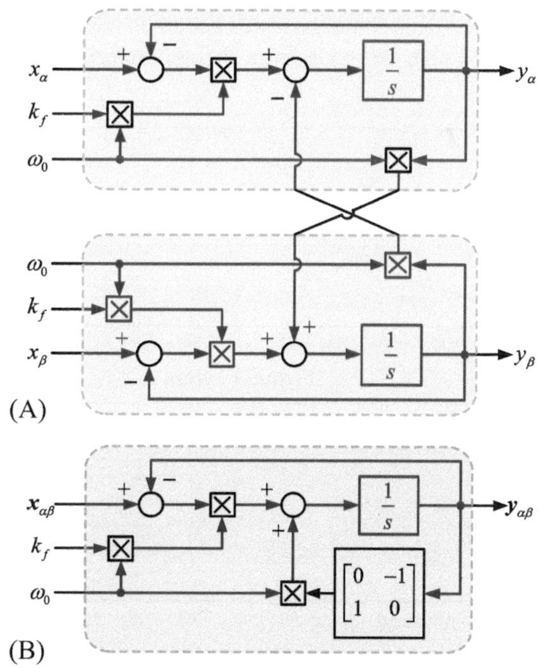

FIG. 12.26 Block diagram of AVF. (A) Scalar form. (B) Vector form.

$$\mathbf{H}_{\mathrm{AVF}}(s) = \frac{\mathbf{Y}_{\alpha\beta}}{\mathbf{X}_{\alpha\beta}} = \frac{k_f\omega_0(s + k_f\omega_0 + j\omega_0)}{(s + k_f\omega_0)^2 + \omega_0^2} \quad (12.75)$$

with the characteristic equation roots

$$s_{1,\,2} = -k_f\omega_0 \pm j\omega_0 \quad (12.76)$$

As it can be seen, the two complex roots of the characteristic equation could always be located in the left-half s-plane if $k_f\omega_0 > 0$, which guarantees the stability of AVF system.

The use of the complex vector concept simplifies the model of AVF from a multiple-input and multiple-output system to an equivalent single-input and single-output complex vector system, which reduces the order and the number of system inputs and outputs by one-half. The performance of this complex vector model can be evaluated using generalized forms of frequency-response function (FRF) [14]. The complex vector FRF could be exhibited with both positive and negative frequencies since it is possible for complex vectors to rotate both forward (positive frequencies) and backward (negative frequencies).

Fig. 12.27 shows the complex vector FRF of AVF under different k_f with ω_0 setting to 30 Hz. As it can be seen, for any value of k_f, AVF has a unity gain and a zero phase shift at the frequency ω_0, and for any other frequencies, the input signal can be attenuated. The characteristics guarantee the extraction of the desired frequency content. Besides, it can be noted that the filter is more selective with smaller k_f, but the bandwidth becomes narrower, and vice versa. This conclusion could also be made from the characteristic equation of AVF.

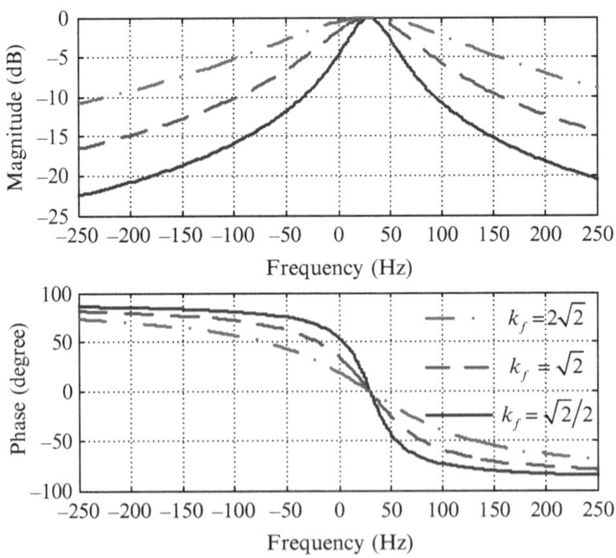

FIG. 12.27 Complex vector FRF of AVF with different k_f.

In practical applications, a trade-off between harmonic selectivity and convergence speed should be taken into consideration. Moreover, it can be noticed that AVF has the unique feature of both frequency and polarity-selective properties, which facilitates the extraction of the fundamental contents in the back-EMF with distorted harmonics for sensorless PMSM drives.

12.5.3.2 Multiple-AVF CFN

As it has been noted that AVF has unity gain and zero phase shift only for the tuned frequency. For any other frequencies, the input signal could be attenuated. It can be observed from Fig. 12.27 that attenuation for other frequency contents depends on the selected gain k_f. Accordingly, with regard to the estimated back-EMF with several frequency contents for IPMSM drives, the value of k_f should be decreased to offer the desired attenuation. As a counterpart, lower values of k_f imply longer settling time, which might lead to substantial distortion on the fundamental content and hence degrade the performance of the system. Therefore, the CFN is utilized, in which the multiple-AVFs tuned for separate frequency contents are arranged in parallel working in a collaborative way. This CFN can be understood as a set of selective and adaptive filters and mainly consists of subtracting the undesired frequency contents at the input of each AVF. Only taking the −5th and 7th harmonic contents with larger amplitudes into account, the proposed multiple-AVF CFN shown in Fig. 12.28 is used to extract the fundamental content in the estimated back-EMF.

As shown in Fig. 12.28, the CFN consists of three AVFs, where $\boldsymbol{H}_n(s)$ is a customized version of the transfer function shown in Eq. (12.75), where the center frequency is given by $n\omega_0$, and n $\{1, -5, 7\}$ is the fundamental and harmonic order for the AVF$_n$ block, i.e.,

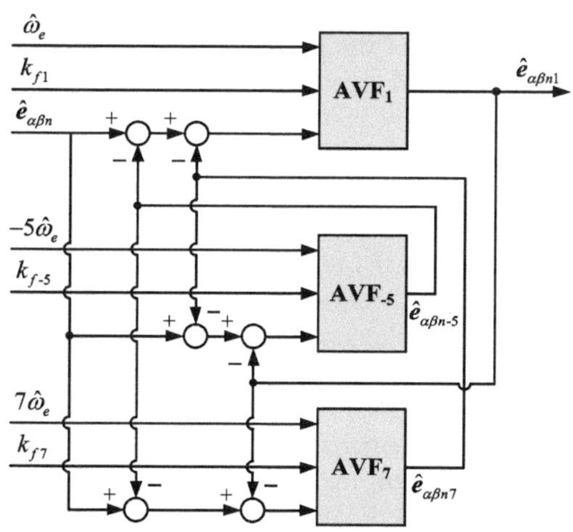

FIG. 12.28 Multiple-AVF CFN.

$$\begin{aligned}\hat{\mathbf{e}}_{\alpha\beta n1} &= \mathbf{H}_1(s)\left(\hat{\mathbf{e}}_{\alpha\beta n} - \hat{\mathbf{e}}_{\alpha\beta n-5} - \hat{\mathbf{e}}_{\alpha\beta n7}\right)\\ \hat{\mathbf{e}}_{\alpha\beta n-5} &= \mathbf{H}_{-5}(s)\left(\hat{\mathbf{e}}_{\alpha\beta n} - \hat{\mathbf{e}}_{\alpha\beta n1} - \hat{\mathbf{e}}_{\alpha\beta n7}\right)\\ \hat{\mathbf{e}}_{\alpha\beta n7} &= \mathbf{H}_7(s)\left(\hat{\mathbf{e}}_{\alpha\beta n} - \hat{\mathbf{e}}_{\alpha\beta n1} - \hat{\mathbf{e}}_{\alpha\beta n-5}\right)\end{aligned} \tag{12.77}$$

where $\hat{\mathbf{e}}_{\alpha\beta n1}$, $\hat{\mathbf{e}}_{\alpha\beta n-5}$, and $\hat{\mathbf{e}}_{\alpha\beta n7}$ denote the fundamental, −5th and 7th harmonic back-EMF estimates.

Hence, the transfer function of the proposed multiple-AVF CFN can be given as

$$\mathbf{H}_{\text{CFN}}(s) = \mathbf{H}_1 \frac{1 - \mathbf{H}_{-5} - \mathbf{H}_7 + 2\mathbf{H}_{-5}\mathbf{H}_7}{1 - \mathbf{H}_{-5}\mathbf{H}_7 - \mathbf{H}_1\mathbf{H}_{-5} - \mathbf{H}_1\mathbf{H}_7 + 2\mathbf{H}_1\mathbf{H}_{-5}\mathbf{H}_7} \tag{12.78}$$

with

$$\mathbf{H}_n(s) = \frac{k_{fn}(n\omega_0)\left(s + k_{fn}(n\omega_0) + j(n\omega_0)\right)}{\left(s + k_{fn}(n\omega_0)\right)^2 + (n\omega_0)^2} \tag{12.79}$$

where k_{fn} is the gain of AVF tuned for the selective nth back-EMF harmonic content.

From Eqs. (12.77) to (12.79), it can be concluded that the CFN structure can be able to extract fundamental and multiple harmonic contents (−5th and 7th harmonics in back-EMF estimates) using multiple AVFs connected in parallel. The extracted signals are cross-feedback canceled from the inputs of the multiple AVFs in order to obtain the fundamental and harmonic contents without harmonic distortion.

Through above analysis, it can be concluded that the input signal of each AVF is calculated by subtracting the output of all the rest of AVFs from the original input signal. In this way, after a transient process, the input signal of each AVF is cleaned up from the harmonic contents detected by the rest of AVFs, which will compensate the distortion at its output.

The complex vector FRF of the proposed multiple-AVF CFN is shown in Fig. 12.29 when the IPMSM operates at 10 Hz. As it can be seen, the multiple-AVF CFN has a unity gain and a zero phase shift at the frequency ω_0, and it has infinite gain at the −5th and 7th harmonics allowing for the removal of the undesired harmonic contents. The frequency and polarity-selective characteristics guarantee the extraction of fundamental content in back-EMF and elimination of the undesired $\pm(6k \pm 1)$th harmonic distortions for sensorless PMSM drives. Moreover, the computational burden can be reduced one-half for implementation due to the polarity-selective characteristic [12].

The proposed multiple-AVF CFN is plugged into the full-order Luenberger back-EMF observer with a normalized PLL. The normalized estimated back-EMF serves as the input vector, the estimated speed serves as the resonance frequency, and the output back-EMF vector after filtering is utilized to obtain the rotor position and speed estimates through the PLL.

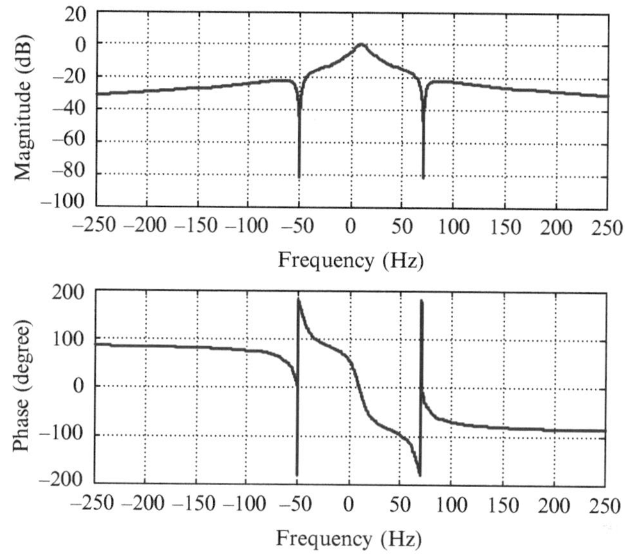

FIG. 12.29 Complex vector FRF of the proposed multiple-AVF CFN.

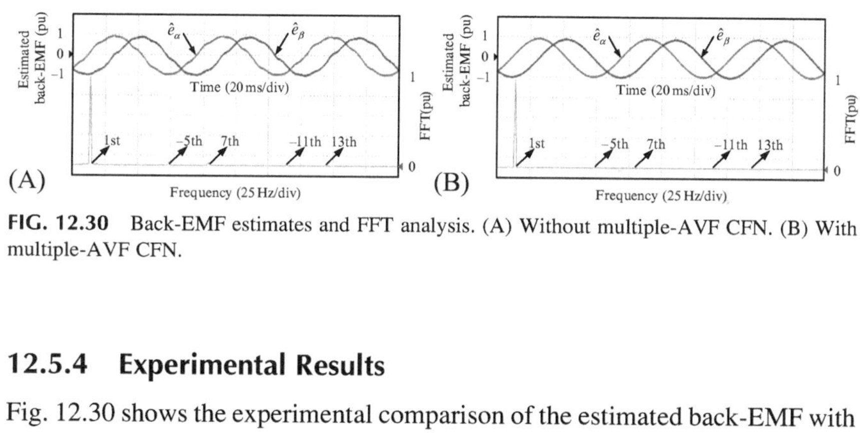

FIG. 12.30 Back-EMF estimates and FFT analysis. (A) Without multiple-AVF CFN. (B) With multiple-AVF CFN.

12.5.4 Experimental Results

Fig. 12.30 shows the experimental comparison of the estimated back-EMF with normalization and FFT spectrum analysis before and after using the proposed multiple-AVF CFN at 300 rpm under 75% rated load. From the experimental results, it can be seen obviously that the -5th, 7th, -11th, and 13th harmonic contents in the estimated back-EMF can be observed without utilizing the proposed multiple-AVF CFN, and the -5th and 7th harmonic contents take the critical role of the position estimation due to larger amplitudes. After employing the proposed multiple-AVF CFN, the -5th and 7th harmonic contents can be completely eliminated and hence the back-EMF estimate becomes more sinusoidal, but the -11th and 13th harmonic contents with smaller amplitudes still remain since they are not compensated in the test.

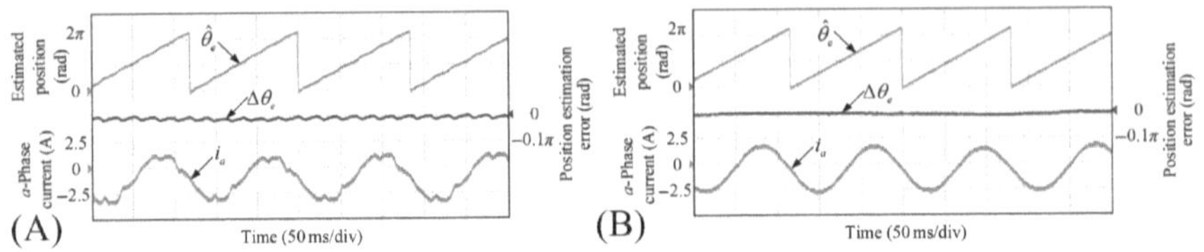

FIG. 12.31 Comparison of estimated rotor position and error at 150 rpm (10% rated speed). (A) Without multiple-AVF CFN. (B) With multiple-AVF CFN.

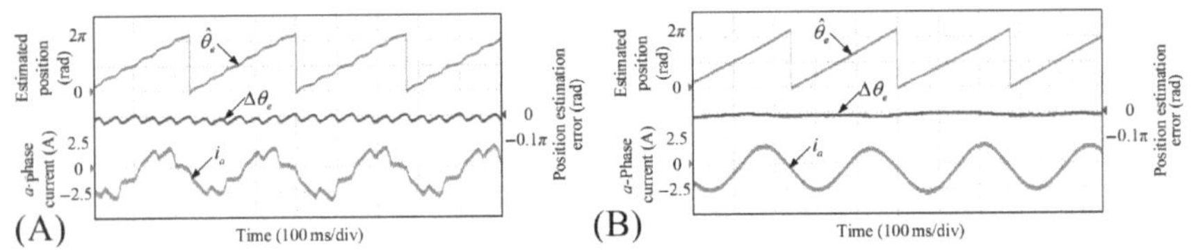

FIG. 12.32 Comparison of estimated rotor position and error at 75 rpm (5% rated speed). (A) Without multiple-AVF CFN. (B) With multiple-AVF CFN.

Since the inverter nonlinearities can behave more critically in low-speed range, the signal-to-noise ratio of back-EMF decreases as operating speed decreases. Hence, the comparison tests of estimated position and error in low-speed operation were carried out. The experimental results at 150 rpm (10% rated speed) and 75 rpm (5% rated speed) are shown in Figs. 12.31 and 12.32, respectively. As it can be seen, the estimated position is distorted with a remarkable 6th harmonic fluctuation up to 0.025π at 150 rpm and 0.045π at 75 rpm, respectively. In contrast, with the proposed multiple-AVF CFN, the estimated rotor position is in good agreement with the measured position and the 6th position error harmonic fluctuation is completely eliminated. Moreover, the phase current becomes more sinusoidal without distortions. Hereby, it can be concluded that the proposed multiple-AVF CFN is effective in improving the accuracy of the position estimation at low speed.

REFERENCES

[1] J. Holtz, Sensorless control of induction machines—with or without signal injection, IEEE Trans. Ind. Electron. 53 (1) (2006) 7–30.

[2] V.-M. Leppanen, J. Luomi, Speed-sensorless induction machine control for zero speed and frequency, IEEE Trans. Ind. Electron. 51 (5) (Oct. 2004) 1041–1047.

[3] S. Suwankawin, S. Sangwongwanich, Design strategy of an adaptive full-order observer for speed-sensorless induction-motor drives-tracking performance and stabilization, IEEE Trans. Ind. Electron. 53 (1) (2006) 96–119.

[4] Z.Q. Chen, M. Tomita, S. Doki, S. Okuma, An extended electromotive force model for sensorless control of interior permanent magnet synchronous motors, IEEE Trans. Ind. Electron. 50 (2) (2003) 288–295.

[5] Z.Q. Zhu, L.M. Gong, Investigation of effectiveness of sensorless operation in carrier-signal-injection-based sensorless-control methods, IEEE Trans. Ind. Electron. 58 (8) (2011) 3431–3439.

[6] W. Sun, Y. Yu, G. Wang, D. Xu, Design method of adaptive full order observer with or without estimated flux error in speed estimation algorithm, IEEE Trans. Power Electron. 31 (3) (2016) 2609–2626.

[7] W. Sun, J. Gao, X. Liu, Y. Yu, G. Wang, D. Xu, Inverter nonlinear error compensation using feedback gains and self-tuning estimated current error in adaptive full-order observer, IEEE Trans. Ind. Appl. 52 (1) (2016) 472–482.

[8] B. Wang, Y. Zhao, Y. Yu, G. Wang, D. Xu, Speed-sensorless induction machine control in the field-weakening region using discrete speed-adaptive full-order observer, IEEE Trans. Power Electron. 31 (8) (2016) 5759–5773.

[9] F. Briz, M.W. Degner, R.D. Lorenz, Analysis and design of current regulators using complex vectors, IEEE Trans. Ind. Appl. 36 (3) (2000) 817–825.

[10] G. Wang, R. Yang, D. Xu, DSP-based control of sensorless IPMSM drives for wide-speed-range operation, IEEE Trans. Ind. Electron. 60 (2) (2013) 720–727.

[11] G. Wang, L. Ding, Z. Li, J. Xu, G. Zhang, H. Zhan, R. Ni, D. Xu, Enhanced position observer using second-order generalized integrator for sensorless interior permanent magnet synchronous motor drives, IEEE Trans. Energy Convers. 29 (2) (2014) 486–495.

[12] G.Q. Zhang, G.L. Wang, D.G. Xu, R.G. Ni, C.H. Jia, Multiple-AVF cross-feedback-network-based position error harmonic fluctuation elimination for sensorless IPMSM drives, IEEE Trans. Ind. Electron. 63 (2) (2016) 821–831.

[13] R.W. Hejny, R.D. Lorenz, Evaluating the practical low-speed limits for back-EMF tracking-based sensorless speed control using drive stiffness as a key metric, IEEE Trans. Ind. Appl. 47 (3) (2011) 1337–1343.

[14] K.W. Martin, Complex signal processing is not complex, IEEE Trans. Circuits Syst. 51 (9) (2004) 1823–1836.

Index

Note: Page numbers followed by *f* indicate figures, and *t* indicate tables.

D

Q

R

S

Z

FSC
www.fsc.org
MIX
Papier | Fördert
gute Waldnutzung
FSC® C083411

Kolibri 360 GmbH
Ferdinand-Jühlke-Straße 7
99095 Erfurt, Deutschland
produktsicherheit@kolibri360.de